ATOMIC PROCESSES IN PLASMAS

Previous Proceedings in the Series of Conferences on Atomic Processes in Plasmas

	Year	Held in	Publisher	ISBN
12th	2000	Reno, Nevada	AIP Conf. Proceedings vol. 547	1-56396-976-9
11th	1998	Auburn, Alabama	AIP Conf. Proceedings vol. 443	1-56396-802-9
10th	1996	San Francisco, California	AIP Conf. Proceedings vol. 381	1-56396-552-6
9th	1994	San Antonio, Texas	AIP Conf. Proceedings vol. 322	1-56396-411-2
8th	1991	Portland, Maine	AIP Conf. Proceedings vol. 257	0-88318-939-9
7th	1989	Gaithersburg, Maryland	AIP Conf. Proceedings vol. 206	0-88318-769-8

Other Related Titles from AIP Conference Proceedings

636   Atomic and Molecular Data and Their Applications: 3rd International Conference on Atomic and Molecular Data and Their Applications - ICAMDATA
Edited by David R. Schultz, Predrag S. Krstić, and Fay Ownby, October 2002, 0-7354-0091-1

611   Superstrong Fields in Plasmas: Second International Conference on Superstrong Fields in Plasmas
Edited by Maurizio Lontano, Gérard Mourou, Orazio Svelto, and Toshiki Tajima, April 2002, 0-7354-0057-1

606   Non-Neutral Plasma Physics IV: Workshop on Non-Neutral Plasmas
Edited by François Anderegg, Lutz Schweikhard, and Fred Driscoll, February 2002, 0-7354-0050-4

551   Atomic Physics 17: XVII International Conference on Atomic Physics; ICAP 2000
Edited by Ennio Arimondo, Paolo De Natale, and Massimo Inguscio, February 2001, 1-56396-982-3

543   Atomic and Molecular Data and Their Applications: ICAMDATA—Second International Conference
Edited by Keith A. Berrington and Kenneth L. Bell, November 2000, 1-56396-971-8

500   The Physics of Electronic and Atomic Collisions: XXI International Conference
Edited by Yukikazu Itikawa, Kazuhiko Okuno, Hiroshi Tanaka, Akira Yagishita, and Michio Matsuzawa, February 2000, 1-56396-777-4

477   Atomic Physics 16: Sixteenth International Conference on Atomic Physics
Edited by William E. Baylis and Gordon W. F. Drake, May 1999, 1-56396-752-9

To learn more about these titles, or the AIP Conference Proceedings Series, please visit the webpage **http://proceedings.aip.org/proceedings**

ATOMIC
PROCESSES
IN PLASMAS

13th APS Topical Conference on Atomic Processes in Plasmas

Gatlinburg, Tennessee   22–25 April 2002

EDITORS

David R. Schultz
Fred W. Meyer
Fay Ownby
Oak Ridge National Laboratory
Oak Ridge, Tennessee

SPONSORING ORGANIZATIONS
U.S. DOE Office of Fusion Energy Sciences
ORNL Controlled Fusion Atomic Data Center
ORNL Physics Division
Los Alamos National Laboratory
ORNL Fusion Energy Division
UT/Battelle
ORNL Physical Sciences Directorate

Melville, New York, 2002
AIP CONFERENCE PROCEEDINGS ■ VOLUME 635

Editors:

David R. Schultz
Fred W. Meyer
Fay Ownby

Oak Ridge National Laboratory
P.O. Box 2008
Oak Ridge, TN 37831-6372
USA

E-mail: schultzd@ornl.gov
          meyerfw@ornl.gov
          ownbyfm@ornl.gov

L.C. Catalog Card No. 2002112518
ISBN 0-7354-0090-3
ISSN 0094-243X
Printed in the United States of America

CONTENTS

LOW TEMPERATURE PLASMAS

INERTIAL CONFINEMENT FUSION

*Italicized names indicate authors who presented the papers.

*Italicized names indicate authors who presented the papers.

X-RAY SOURCES

MAGNETIC CONFINEMENT FUSION

*Italicized names indicate authors who presented the papers.

Preface

The 13th APS Topical Conference on Atomic Processes in Plasmas (APiP) was held April 22-25, 2002, in Gatlinburg, Tennessee, bringing together scientists working in the fields of atomic and plasma physics. The local committee was especially proud to welcome participants back to East Tennessee for what was the 25th anniversary of the first APiP held in 1977 in Knoxville. Over the years, the conference series has remained a vital forum for the communication between atomic physicists and researchers in plasma science, including notably those who are practicing members of both communities. Originally focused primarily on controlled fusion energy and astrophysics applications, APiP has evolved and grown to encompass a broad range of plasma physics, including laser and heavy ion inertial confinement fusion, Z pinches, x-ray sources, technical plasma processing, and lighting, for example. The conference is a unique meeting place at the cross roads of these fields united by common themes in plasma science and atomic physics.

The 13th APiP was attended by over one hundred participants and consisted of thirty-four invited oral presentations and forty-six poster presentations. The tradition of encouraging excellence in student research and participation in APiP was continued by the awarding of four cash prizes for student posters selected by the program committee. First prize was awarded to Alejandro Aguilar, second prize to Leslie Welser, and third prize shared by Abdur Rahman and Diana McCrorey.

The conference chair and co-chairs wish to express their thanks to the conference secretary Ms. Fay Ownby, who also served as chief editor of this volume, and to Ms. Carlene Stewart, the conference co-secretary. We also thank all of the attendees for their participation, the local conference committee for their invaluable assistance, the Oak Ridge Associated Universities for their fostering of university – national laboratory interchange through this meeting, and all the sponsors of the conference. We particularly acknowledge support of the publication of this volume by the US DOE Office of Fusion Energy Sciences. In anticipation of the lively and productive continuation of this unique and important conference series, we look forward to the next APiP to be held in Los Alamos in 2004.

David R. Schultz, Chair
Fred W. Meyer, Co-Chair
H. Kennon Carter, Co-Chair
Oak Ridge, Tennessee
July 2002

Dedication

This Proceedings is dedicated to the memory of a young and gifted researcher, Gwyneth Junkel-Vives, who tragically died as a result of complications following childbirth, December 21, 2001, in Los Alamos, New Mexico. Gwyn was known by many in our community and will be remembered not only for her creative work but also for her warm and enthusiastic character. She was a student of Chuck Hooper at the University of Florida and was working as a postdoc of Joe Abdallah at Los Alamos National Laboratory at the time of her death. Gwyn is survived by her husband Ted Vives and healthy son Peter Alexander Rheinhardt Vives.

Joe Abdallah
Jim Cohen
David Schultz

Program Committee

James Babb, Harvard-Smithsonian Center for Astrophysics

James Bailey, Sandia National Laboratory

James Cohen, Los Alamos National Laboratory

Gary Ferland, University of Kentucky

William Goldstein, Lawrence Livermore National Laboratory

Bruce Hammel, Lawrence Livermore National Laboratory

Ralph Isler, Oak Ridge National Laboratory

Yong-Ki Kim, National Institute of Standards and Technology

James Lawler, University of Wisconsin

Richard Lee, Lawrence Livermore National Laboratory

Roberto Mancini, University of Nevada, Reno

Earl Marmar, Massachusetts Institute of Technology

Ronald Phaneuf, University of Nevada, Reno

Mitch Pindzola, Auburn University

Jorge Rocca, Colorado State University

William Rowan, University of Texas

Phillip Stancil, University of Georgia

Jon Weisheit, Los Alamos National Laboratory

Local Committee

David Schultz, Chair
Fred Meyer, Co-Chair
Ken Carter, Co-Chair
Fay Ownby, Secretary
Carlene Stewart, Co-Secretary
Mark Bannister
Charles Havener

Herb Krause
Predrag Krstic
Joseph Macek
Serge Ovchinnikov
Carlos Reinhold
Lynda Saddiq
Randy Vane

Sponsors

US DOE Office of Fusion Energy Sciences

The ORNL Controlled Fusion Atomic Data Center

The ORNL Physics Division

Los Alamos National Laboratory

The ORNL Fusion Energy Division

UT/Battelle, managers of ORNL

The ORNL Physical Sciences Directorate

Meetings in This Series

1.	Knoxville, Tennessee	February 16-18, 1977
2.	Boulder, Colorado	January 17-19, 1979
3.	Baton Rouge, Louisiana	February 25-27, 1981
4.	Princeton, New Jersey	April 13-15, 1983
5.	Pacific Grove, California	February 25-28, 1981
6.	Santa Fe, New Mexico	September 28-October 2, 1987
7.	Gaithersburg, Maryland	October 2-5, 1989
8.	Portland, Maine	August 25-29, 1991
9.	San Antonio, Texas	September 19-23, 1993
10.	San Francisco, California	January 14-18, 1995
11.	Auburn, Alabama	March 23-26, 1998
12.	Reno, Nevada	March 19-23, 2000
13.	Gatlinburg, Tennessee	April 22-25, 2002

LOW TEMPERATURE PLASMAS

Fundamental Atomic Plasma Chemistry for Semiconductor Manufacturing Process Analysis

P.L.G.Ventzek, V. Kudrya[#], V. Astapenko[#], A. Eletskii[#], D. Zhang,
P.J. Stout, S. Rauf and M. Orlowski

Motorola Inc., Austin, Texas 78721 USA
[#]Soft-Tec, Moscow Russia

Abstract. An absence of fundamental atomic plasma chemistry data (e.g. electron impact cross-sections) hinders the application of plasma process models in semiconductor manufacturing. Of particular importance is excited state plasma chemistry data for metallization applications. This paper describes important plasma chemistry processes in the context of high density plasmas for metallization application and methods for the calculation of data for the study of these processes. Also discussed is the development of model data sets that address computational tractability issues. Examples of model electron impact cross-sections for Ni reduced from multiple collision processes are presented.

INTRODUCTION

The goal of simulations in the context of semiconductor manufacturing process development beyond adding to existing intuition is to quantify the relationship between process settings (e.g., power, pressure, gas mix) and process impact at the feature-level (e.g. trenches and vias). It is unfortunate that the ability to bring basic computational and theoretical plasma physics and chemistry to bear on engineering applications in semiconductor manufacturing has lagged empirical approaches. Only recently have approaches integrated plasma equipment simulations and feature scale simulations for quantitative engineering applications [1].

A drag on development has been the lack of fundamental plasma chemistry data (e.g., electron-atom interaction cross-sections) to incorporate into these models beyond those of the most basic of rare gas systems. The lack of data is usually associated with the fact that processing plasmas are more often than not medium or high-density plasmas for which excited state plasma chemistry is important but almost totally unknown. What is more is that the number of states relevant to atoms used in plasma processes can be large (i.e., greater than 100) making the potential size of a multidimensional kinetics problem intractable.

This paper addresses data needs in the context of low temperature plasma conditions, the means for generating electron-impact cross section data and how that data is managed.

CP635, *Atomic Processes in Plasmas: 13th APS Topical Conference*, edited by D. R. Schultz et al.
© 2002 American Institute of Physics 0-7354-0090-3/02/$19.00

IONIZED METAL PLASMA PHYSICAL VAPOR DEPOSITION

Ionized Physical Vapor Deposition (I-PVD) technology was described in an early paper in 1989 [2] and is now a standard technology for the metal film deposition into high-aspect-ratio features [3]. A typical configuration (referred to as Ionized Metal Plasma Physical Vapor Deposition or IMP-PVD) includes induction coils to generate a high-density plasma. I-PVD technology will be used to put the relative importance of various collision processes in perspective and to rationalize how electron impact cross-section data is managed. The typical inductively coupled plasma (ICP) used in I-PVD processes possesses the following plasma characteristics [4-10]:

- Inert buffer gas densities ranging from 10^{14} to 10^{16} cm^{-3},
- Metal atom densities ranging from 10^{11} to 10^{13} cm^{-3}
- The electron number density in ICP discharges ranges between 10^{10} and 10^{12} cm^{-3}
- Mean electron energies of 3-5 eV with nearly Maxwellian electron energy distribution functions.
- Low degrees of ionization for the inert gas; the degree of ionization of the metal vapor may be such that the fraction of metal species incident on a wafer is 10's of percent.
- Typical dimensions range from 10 to 30 cm in height to on the order of 30 cm in diameter.

GENERAL ANALYSIS OF INELASTIC COLLISION PROCESSES OCCURRING IN INDUCTIVELY COUPLED PLASMAS USED FOR METAL DEPOSITION

Overview of Excitation and Ionization Processes in ICP Plasmas

The overriding characteristic of high-density metal inert gas plasmas is that they are electropositive (except for impurities).   As such, attachment processes are unimportant. There are many possible charged particle generation mechanisms but loss is generally by transport to walls. In this section, the relative importance of charged particle generation processes is considered with the aim of pointing out where fundamental electron-particle interaction studies need be focused.

Electron Impact Ionization

The main ionization mechanism in ICP sources is electron impact ionization of atoms from the ground and long-lived states. Using the Thomson formula for estimation of the atom ionization cross section:

$$\sigma_{ion}^{I} = \frac{\pi e^4}{\varepsilon}\left(\frac{1}{I_I} - \frac{1}{\varepsilon}\right),$$

(1)

where $I_l = I - E_l$ is the ionization potential of the lth atomic state, one obtains the following expression for the ionization rate constant of the lth state calculated for the case in which electrons are described by a Boltzmann electron energy distribution function:

$$k_{ion}^l = \frac{\pi e^4}{I_l^2} \sqrt{\frac{8T_e}{m\pi}} e^{-I_l / k_B T_e} . \tag{2}$$

The total free electron production rate due to the electron impact mechanism is calculated through summing expression (2) over all long living excited states with a an atom population whose density of states is described by a Boltzmann distribution and the Maxwellian electron energy distribution function:

$$\left(\frac{\partial N_e}{\partial t} \right) = N_0 N_e \pi e^4 \sqrt{\frac{8\pi k_B T_e}{m}} \sum_l e^{-I_l / T_e} \frac{g_l}{I_l^2} . \tag{3}$$

The contribution of excited levels into the net ionization rate increases with decreasing level ionization potential I_l accompanying by the rise in the level degeneracy g_l. Using expression (3), the net charged particle production in ICP sources is approximately:

$$\left(\frac{\partial N_e}{\partial t} \right)_{el} \sim 10^{16} \, cm^{-3} s^{-1} . \tag{4}$$

This magnitude is typical of what is required to sustain plasmas at typical high-density plasma conditions when loss is only by transport

Excitation to states with forbidden transitions to ground or lower states implies production of metastable or long-lived states and excitation to states with allowed transitions implies short-lived state production.

Ionization Involving Heavy Particles

Additional ionization mechanisms derive from collisions between excited and ground state heavy particles. There are two kinds of collisions: *Penning ionization,*

$$A^{**} + B \rightarrow A + B^+ + e, \tag{R1}$$

and *associative ionization*

$$A + B^* \rightarrow AB^+ + e. \tag{R2}$$

Process (R1) can proceed if the excitation energy of the excited state A** exceeds the ionization potential of the atom B. In the case of Ar-metal plasmas this corresponds to collisions between the metastable $Ar(3p^5 4s)$ states and a ground state metal atom. An upper estimate of the cross section of this process relates to the capture of atoms in an attractive van der Waals potential,

$$k_c = 3[2C]^{1/3}\sqrt{\frac{2\pi}{\mu}}\Gamma\left(\frac{5}{3}\right)(k_BT)^{1/6}, \quad (5)$$

where C is the Van der Waals constant. Expression (5) results in the estimate $k_c \approx 10^{-9}$ cm^3/s for the collision between a metastable Ar atom and a ground state metal atom resulting in ionization of the metal atom. Using this estimate, the net charge particle production rate due to the Penning ionization mechanism is approximately,

$$\left(\frac{\partial N_e}{\partial t}\right)_{PI} \sim N_{Ar*}N_M k_c \sim 10^{15}\,\text{cm}^{-3}\text{s}^{-1}. \quad (6)$$

Here $N_M \approx 10^{12}$ cm^3/s is the number density of the metal vapor and N_{Ar*} is the number density of the metastable argon atoms which Boltzmann equilibrium suggests $N_{Ar*} \approx N_{Ar}\exp(-E*/k_BT_e) \approx 10^{12}$ cm^{-3}.

The contribution of associative ionization (R2) into the net charged particles production rate has an upper bound of $k_{AI} \approx 10^{-9}$ cm^3/s. From this estimate the charged particle production rate is approximately,

$$\left(\frac{\partial N_e}{\partial t}\right)_{AI} \sim N_{Ar*}N_{Ar} k_{AI} \sim 10^{15}\,cm^{-3}s^{-1}. \quad (7)$$

Here $N_{Ar*} \approx 10^{10}$ cm^{-3} is the number density of the highly excited Ar atoms with an electron binding energy less than 1 eV which are those permitted to participate in an associative ionization process. The contribution of metal atom associative ionization to the volumetric charged particle production rate is at least two orders of magnitude lower than that of Ar atoms. This is due to the rather low number density of metal vapor.

While ion-neutral processes are not negligible, comparing volumetric production rates indicates that the electron-impact processes are dominant in determining plasma behavior.

CROSS-SECTION CALCULATIONS

Kim and Rudd [11] describe the methodology for calculating the electron impact ionization cross-section calculations used in this work. Specifically, here the Binary-Encounter-Bethe approximation is used (cross-section: cm^2, energy: eV):

$$\sigma_{ionization} = 6.51\times10^{-14}\sum_{i=1}^{n}\frac{N_i}{B^2(1+t_i+u_i)}\left[\frac{1}{2}\left(1-\frac{1}{t_i^2}\right)\ln t_i + 1 + \frac{1}{t_i} - \frac{\ln t_i}{t_i+1}\right], \quad (8)$$

where n is the number of subshells, N_i is the occupation number, B_i is the electron orbital binding energy, $t_i=E/B_i$, $u_i=U/B_i$, and U_i is the average electron orbital kinetic energy.

Eletskii and Smirnov [12] describe the interpolating formula used here for calculating the electron impact excitation cross-section for optically allowed transitions:

$$\sigma_{allowed} = 1.3 \times 10^{-13} \frac{Nf}{(\Delta E)^2} \frac{\ln(1+0.5\sqrt{t-1})}{t+3},$$ (9)

where f is the transition oscillator strength, ΔE is the threshold of the excitation process and $t=E/\Delta E$.

The method for calculating electron impact excitation of optically forbidden transitions has been described by Astapenko $et\ al.$, [13] and the calculated cross-section is represented as,

$$\sigma_{forbidden} = c \frac{\sqrt{t-1}}{a+t^n},$$ (10)

where a and c are fitting parameters and n=3 except for n=5 for inter-combination transitions.

GENERAL APPROACH TO THE CROSS SECTION DATA COMPRESSION PROBLEM IN PLASMA METAL DEPOSITION SIMULATION

This section deals with the issue of reducing cross-section sets to manageable "model" cross-section sets containing reduced numbers of atomic states without loss if accuracy when they are used in plasma equipment simulations.

Determination of the Number of Relevant Atomic Energy Levels

In the case of non-isolated atoms in real plasmas, the number of levels is finite, an effect referred to as ionization potential reduction implying the existence of a maximum principal quantum number k_{max}. There are two main mechanisms for limiting the principal quantum number may be limited: level merging due to the Stark broadening effect and electrostatic interactions. In low temperature plasmas the first mechanism is dominant. The following relationship describes the maximum principle quantum number [14]:

$$k_{max} = \frac{1.04 \times 10^3}{n_e^{2/15}},$$ (11)

where n_e is the electron number density. For $n_e = 10^{10}$-10^{12} cm^{-3}, $k_{max} = 25$-50. Note that this formula was derived for hydrogen-like systems. In cases of other atoms, the fine splitting decreases k_{max} slightly more. Even so, the number of atomic energy levels within the predicted maximum for a high-density low temperature plasma is still rather large.

Effective Energy Levels

Groups of some levels can be compressed in single effective levels. Suppose there is a many-level atomic system and E_i and g_i are the energy and degeneracy degree of the ith level, respectively. Further assume that we can calculate the electron-atom inelastic impact cross-sections σ_{ij} for transitions between any two levels taking into account their electron configurations and multiplet characteristics.

First, the levels to be compressed should have similar radiative lifetimes. In this case, their radiative decay behaviors will be similar. Secondly, the levels are desired to be of the same parity. It provides a proper account of the important radiative decay channels. Finally, the levels should have close energies. For example, these can be fine-structure splitting energy levels that for light elements would be close enough. An energy proximity criterion is now presented.

Let us consider a system of levels. Define the compressed level's effective degeneracy degree:

$$g_{eff} = \sum_i g_i, \tag{12}$$

and its effective energy in terms of:

$$g_{eff} \exp\left(-\frac{E_{eff}}{k_B T_e}\right) = \sum_i g_i \exp\left(-\frac{E_i}{k_B T_e}\right), \tag{13}$$

with $g = 2J + 1$ in the case of the LS-bond. If the condition of energy level proximity

$$\left|E_{eff} - E_i\right| << k_B T_e \tag{14}$$

is satisfied, then the temperature-independent effective energy is

$$E_{eff} = \frac{\sum_i g_i E_i}{\sum_i g_i}. \tag{15}$$

This is a weighted average energy for which (14) is the combination criterion.

Effective Cross-Sections

Consider two systems of levels in close proximity: lower (n_i) and upper (n_j). Using the kinetic equation for the upper effective level population rate, define the effective rate constant:

$$\frac{d}{dt}\sum_j n_j = \sum_{ij} k_{ij} n_i - \dots = k_{eff} \sum_i n_i - \dots,$$

(16)

where k_{ij} is the rate constant for transition between two levels. Then, using the Boltzmann distribution and the level energy proximity condition (14), we obtain

$$k_{eff}(E) = \frac{\sum_{ij} g_i k_{ij}(E)}{\sum_i g_i}.$$

(17)

Accounting for the linear dependence of the rate constant on the cross-section, we find:

$$\sigma_{eff}(E) = \frac{\sum_{ij} g_i \sigma_{ij}(E)}{\sum_i g_i}.$$

(18)

This formula can also be used to calculate the ionization cross sections. In this case, the summation over j is not performed. To calculate the cross-section for optically allowed transitions, we can define the effective oscillator strength:

$$f_{eff} = \frac{\sum_{ij} g_i f_{ij}}{\sum_i g_i}.$$

(19)

Here f_{ij} is the oscillator strength for transition between two levels. For the reverse process cross-section, we use:

$$\sigma_{eff}^{rev}(E) = \frac{\sum_{ij} g_i \sigma_{ij}(E + \Delta E)}{\sum_j g_j}\left(1 + \frac{\Delta E}{E}\right),$$

(20)

where ΔE is the direct process threshold, E>0.

Lifetime Classification

The relative contribution of some excited state to the total kinetic picture depends on the relationship between the rates of radiative decay and loss through collisions. This relationship can be characterized by the dimensionless parameter,

$$M_n = \tau_n N_e \sum_m k_{nm} , \qquad (21)$$

where,

$$\tau_n^{-1} = \sum_{m<n} A_{nm} , \qquad (22)$$

is the radiative lifetime of the nth state, k_{mn} is the rate constant for the transition between the relevant atomic states under the action of electron impact, A_{nm} is the Einstein coefficient for the spontaneous radiative transition.

If $M_n \gg 1$, the stationary population of the nth state is close to the Boltzmann equilibrium magnitude,

$$N_n = N_0 \frac{g_n}{g_0} \exp\left(-\frac{E_n}{k_B T_e} \right), \qquad (23)$$

where N_0 is the ground state population, N_n , g_n, and E_n are the nth state population, statistical weight (degeneracy degree), and energy, respectively. This is the detailed balance principle. The excited states obeying this condition do not contribute to the electron energy but make a notable contribution to the total ionization rate because their population can be compared with that of the ground state.

In the case of $M_n \ll 1$, the main channel for state loss is spontaneous emission and the stationary population of this state is much less than the equilibrium population according to the Boltzmann distribution (23). Thus, the excitation of such short-lived states makes some contribution to the electron energy balance but does not make a notable contribution to the total ionization rate.

Taking into account these considerations, we can divide the total set of the excited states in three groups:

The first group includes short-living states characterized by a low magnitude of $M_n \ll 1$. The electron impact excitation of these states from the ground state should be taken into consideration in the electron energy balance equation. But they do not influence on the ionization rate because of their low population.

The second group of states (long-lived) involves mainly the metastable states characterized by a high value of M_n. As these states are in the equilibrium with the ground state, and their populations are described by the Boltzmann expression (23).

Some states fall in-between short and long and should be treated both in terms of kinetics and number continuity.

Consider the excitation cross section for an optically allowed transition [12]

$$\sigma_{nm}(E) = \frac{2\pi e^4 N_{eq} f_{nm}}{\Delta E_{nm}^2} \frac{\ln\left(1 + 0.5\sqrt{E/\Delta E_{nm} - 1}\right)}{E/\Delta E_{nm} + 3}, \tag{24}$$

where f_{nm} is the absorb oscillator strength of the transition, N_{eq} is the number of equivalent electrons, $\Delta E_{nm} = E_m - E_n$. The cross section of the reverse transition (quenching) is

$$\sigma_{mn}(E) = \frac{g_n}{g_m} \frac{E + \Delta E_{nm}}{E} \sigma_{nm}(E + \Delta E_{nm}). \tag{25}$$

For the energy range where $E/\Delta E_{nm} \ll 1$, the rate constant does not depend on the EEDF:

$$k_{mn} = \int_0^\infty dE \sigma_{mn}(E) \sqrt{\frac{2E}{m}} f(E) \approx \frac{1.2 \times 10^{-55} N_{eq}}{\tau_{mn} \Delta E_{nm}^{7/2}}, \tag{26}$$

where the radiative time of the transition

$$\frac{1}{\tau_{mn}} = \frac{g_n}{g_m} \frac{2e^2 f_{nm} \Delta E_{nm}^2}{mc^3 \hbar^2}. \tag{27}$$

Finally, we can obtain (ΔE in eV)

$$M_{nm} \approx 2.3 \times 10^{-14} \frac{n_e N_{eq}}{\Delta E_{nm}^{7/2}}. \tag{28}$$

Simple calculations show that all the resonant states of the atoms involved can be considered as short-lived in absence of significant radiation trapping.

The effective spontaneous radiation time of resonant states is prolonged due to the resonance radiation trapping and depends on resonance line (Doppler) broadening [15]. For Doppler broadening, the effective lifetime τ_{eff} of a resonance transition is expressed by the following equation:

$$\tau_{eff}/\tau_{mn} \approx 2.72 k_0 R_0 \sqrt{2\ln(k_0 R_0)}, \tag{29}$$

where R_0 is the characteristic size of ICP, k_0 is the absorption coefficient for the center-of-line resonant radiation,

$$k_0 = \frac{\lambda^3 N_a}{8\pi\tau_0 \sqrt{\frac{8kT}{M\pi}}}, \tag{30}$$

11

N_a is the atom number density, λ is the wavelength of the resonant radiation, M is the mass of the radiating atom.

Resonance radiation trapping is important if the parameter $k_0R_0 \gg 1$. For high density ICP sources if $k_0R_0 \approx 10^5$ for the strongest resonant transition of Ar ($\lambda = 104.8$ nm) and $k_0R_0 \approx 100$ for the strongest resonant lines of metal vapor at $N_a = 10^{12}$ cm^{-3}. In both the cases, the effective lifetime of a resonantly excited state τ_{eff} depends on the parameter k_0R_0 and for Ar $\tau_{eff} \approx 3\times10^{-4}$ s and strong metal absorbers $\tau_{eff} \approx 10^{-6}$ s. Therefore, resonantly excited states of Ar should be considered as long-lived whereas those of a metal atom are approximately short-lived states.

Consider that long-lived states are depleted mainly by the electron impact ionization. Using only ionization, the parameter M_n is,

$$M_n \approx \tau_n N_e k_n . \tag{31}$$

Letting $T_e \geq I_n$, where I_n is the ionization potential of the nth state one finds ionization rate constant:

$$k_n \approx \sigma_{n\max} <v_e> \approx \frac{e^4}{I_n^2}\sqrt{\frac{T_e}{m}} \approx \frac{10^{-6}\sqrt{T_e}}{I_n^2} . \tag{32}$$

where $\sigma_{n\ \max}$ is the maximal ionization cross section of the nth state, $<v_e>$ is the average electron velocity, T_e is the electron temperature (in eV). M_n is found as

$$M_n \approx \tau_n N_e \sqrt{\frac{T_e}{m}}\frac{e^4}{I_n^2} \approx \frac{10^{-6}\tau_n N_e\sqrt{T_e}}{I_n^2} . \tag{33}$$

Assigning a value for M_n (e.g., 5-10) now defines what long-lived means in the context of data management.

EXAMPLE

Let us consider a 3d-atom such as Ni comprised of more than 190 identified levels. Using the approach described in this paper, all the levels are be compressed to a reduced scheme which for the range of $T_e = 5$ eV contains 8 effective levels: ground, 3 long-lived, and 4 short-lived (see Fig. 1). It should be noted that every effective level has a rather high degree of degeneracy. For lower values of T_e the number of effective levels should be increased.

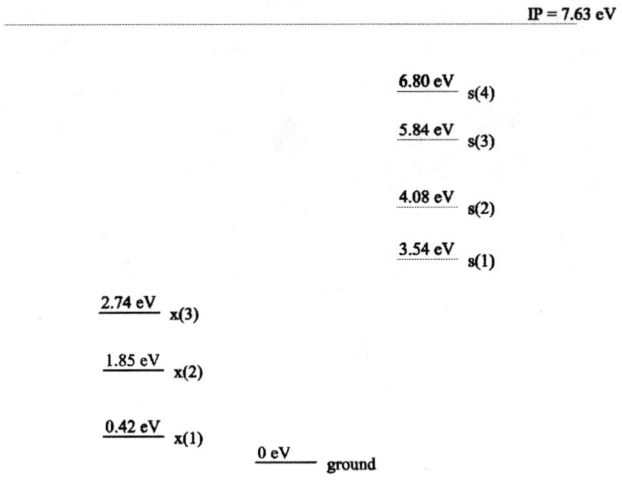

6.80 eV s(4)

5.84 eV s(3)

4.08 eV s(2)

3.54 eV s(1)

2.74 eV x(3)

1.85 eV x(2)

0.42 eV x(1)

0 eV ground

Fig. 1. The reduced level scheme of Ni with long-lived (denoted by x) and short-lived (denoted by s) effective levels.

Electron impact excitation and ionization cross sections can be calculated for all Ni multiplets using appropriate methods followed by "effective" cross sections computation. Examples are included in Fig. 2.

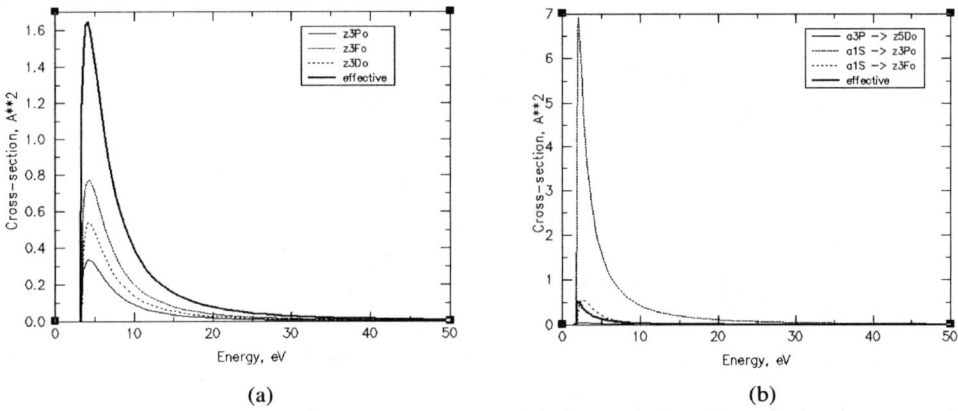

(a)   (b)

Fig. 2. Examples of actual and effective cross-sections for electron impact (a) excitation from ground and (b) excitation among different states of Ni.

Excitation from ground to an ensemble of states is represented in Fig. 2(a) while excitation among an ensemble of states is represented in Fig. 2(b).

Examples of simulations employing cross-sections sets calculated in the way described in this paper are presented in Fig. 3. Figure 3(a) contains the step coverage associated with different species and energy fluxes incident on a wafer. Figure 3(b) contains the density field of Tantalum in an IMP-PVD chamber calculated using the Hybrid Plasma Equipment Model (developed at the University of Illinois) [8]. The accuracy of the output of the equipment model has a direct correlation with the

13

accuracy of the electron impact cross-section data set used in the model and finally a direct impact of the accuracy of feature scale predictions contained in Fig. 3(a).

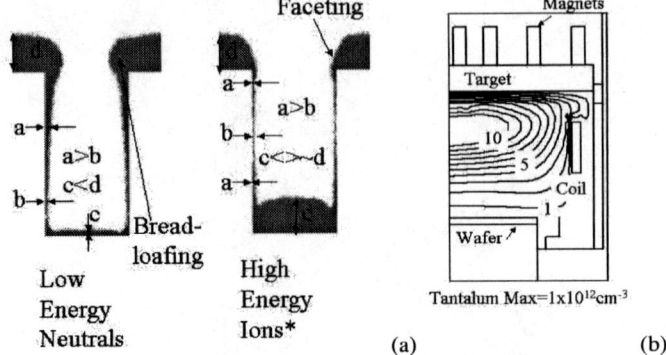

(a)   (b)

Fig. 3. (a) Examples of the impact of different species fluxes in a feature on the step-coverage. Low energy neutrals are typical of a case without plasma and high energy ions are typical of deposition from a plasma with very high plasma densities and very large powers coupled through a wafer. (b) Tantalum neutral species densities in an IMP-PVD chamber.

SUMMARY

Fundamental atomic plasma chemistry data plays a key role in the enabling of the use of integrated plasma process models for process engineering applications in semiconductor manufacturing. This paper dealt with filling in the gaps in the data for especially critical processes of relevance to excited state plasma chemistry. In addition to filling in the gaps, managing the wealth of data is a challenge to employing this data in modeling applications. A methodology for this has been developed.

REFERENCES

1. Arunachalam, V., Rauf, S., Coronell, D.G., Ventzek, P.L.G., J. Appl. Phys. **90**, 64 (2001).
2. Yamashita, M. J., Vac. Sci. Technol. A **7**, 151 (1989).
3. Rossnagel, S.M., IBM Journal of Research & Development **43**, 163 (1999).
4. Rossnagel, S.M., Hopwood, J., Appl. Phys. Lett. **63**, 3289 (1993).
5. Dickson, M., Qian, F., Hopwood, J., J. Vac. Sci. Technol. A **15**, 340 (1997).
6. Hopwood, J., Qian, F., J. Appl. Phys. **78**, 758 (1995).
7. Hopwood, J., Phys. Plasmas **5**, 1624 (1998).
8. Grapperhaus, M.J., Krivokapic, Z., Kushner M.J., J. Appl. Phys. **83**, 35 (1998).
9. Lu, J., Kushner, M.J., J. Appl. Phys. **87**, 7198 (2000).
10. Lu, J., Kushner, M.J., J. Appl. Phys. **89**, 878 (2001).
11. Kim, Y.K., Rudd, M.E., Phys. Rev. A **50**, 395 (1994).
12. Eletskii, A.V., Smirnov, B.M., Soviet Physics - JETF **57**, 955 (1983).
13. Astapenko, V.A., Eletskii, A.V., Kudrya, V.P., Ventzek, P., Laser Physics **10**, 1220 (2000).
14. Inglis, D., Teller, E., Astrophys. J. **90**, 439 (1939).
15. Sobel'man, I.I., Vainshtein, L.A., Yukov, E.A., "Excitation of atoms and broadening of spectral lines" 2nd edit. Springer Series on Atoms and Plasmas, v.15, 1995.

Ultracold Neutral Plasmas: Progress and Prospects

T. C. Killian

Department of Physics and Astronomy and Rice Quantum Institute, Rice University, Houston, TX 77005

Abstract. By photoionizing a sample of laser-cooled xenon atoms, we create ultracold neutral plasmas with initial temperatures of 1-1000 K and densities as high as $10^{10}\,cm^{-3}$. The plasma is formed by the trapping of electrons by the residual positive charge that is left after some electrons initially leave the sample. We excite plasma oscillations with applied radio frequency fields and use this to monitor the expansion of the unconfined plasma. We have observed significant recombination of the plasma into Rydberg atoms (up to 20 %). At these low temperatures, the only traditional form of recombination that could be significant is three-body recombination (TBR). Recent theoretical studies of ultracold plasmas, and prospects for future experiments, are discussed.

Recent experiments have opened a new regime of ultracold neutral plasmas with temperatures as low as a 1 K. Studies of the methods and conditions for forming the plasma[1], excitation and detection of plasma oscillations[2], dynamics of the plasma expansion[2], and collisional recombination into Rydberg atomic states[3] have demonstrated that ultracold neutral plasmas provide a powerful and flexible environment in which to test our fundamental understanding of plasmas physics.

The recipe for an ultracold neutral plasma starts with laser-cooled and trapped neutral atoms[4]. In a table-top apparatus, with a proper configuration of laser beams and magnetic fields, a few million atoms are laser cooled to approximately $10\,\mu$K. The peak density is about $2 \times 10^{10}\ cm^{-3}$ and the spatial distribution of the cloud is Gaussian with an rms radius $\sigma \approx 200\,\mu$m. Many different elements can be laser cooled, and ultracold plasma experiments conducted by the author and colleagues at the National Institute of Standards and Technology in Gaithersburg were performed with metastable xenon. More information on laser cooling and trapping of metastable xenon can be found in [5].

To produce the plasma, atoms are photoionized barely above threshold with a narrow-bandwidth pulsed laser. The kinetic energy of the original neutral atoms is negligible, so the energetics of the plasma is entirely determined by the photoionization process. Because of the small electron-ion mass ratio, the electrons have an initial kinetic energy (E_e) approximately equal to the difference between the photon energy and the ionization potential. E_e/k_B can be as low as the bandwidth of the ionizing laser, which is $\sim 100\,$mK with standard pulsed dye lasers, but most studies so far have dealt with E_e/k_B between 1 and 1000 K. The initial kinetic energy for the ions is in the mK range.

Immediately after photoionization, the charge distribution is everywhere neutral. Due to the kinetic energy of the electrons, the electron cloud expands, but on this time

CP635, *Atomic Processes in Plasmas: 13th APS Topical Conference*, edited by D. R. Schultz et al.
© 2002 American Institute of Physics 0-7354-0090-3/02/$19.00

scale the ions are essentially immobile. The resulting local charge imbalance creates an internal electric field that produces a Coulomb potential energy well for electrons. If the well never becomes deeper than E_e, all the electrons escape. If enough atoms are photoionized, however, only an outer shell of electrons escapes, and the well becomes deep enough to trap the rest. After the untrapped fraction has escaped, the cloud as a whole is no longer strictly neutral. Simulations show that electrons escape most easily from the edges of the spatial distribution, however, and the center of the cloud is well described as a neutral plasma.

The number of atoms ionized (N_i), and thus the density of the plasma (n), is controlled by varying the energy of the photoionizing laser pulse. Lower E_e and higher N_i lead to a greater fraction of the initially created electrons being trapped. For the coldest and densest conditions, over 90% of the electrons are confined.

For a given E_e there is a threshold number of positive ions required for trapping electrons. This threshold was demonstrated and studied in [1]. Theoretically and experimentally, it can be shown that trapping occurs when the Debye screening length, $\lambda_D = \sqrt{\varepsilon_0 k_B T_e / e^2 n}$, becomes less than the size of the sample σ. Here, ε_0 is the electric permittivity of vacuum, k_B is the Boltzmann constant, and e is the elementary charge. The electron temperature, T_e, is approximately equal to E_e/k_B [6]. An ionized gas is normally not considered a plasma unless the Debye length is smaller than the size of the system [7], so the threshold for electron trapping is also the threshold for the formation of a plasma. In our experiment, the Debye length can be as low as 500 nm, while the size of the sample is $\sigma \approx 200\,\mu m$. The condition $\lambda_D < \sigma$ for creating a plasma is thus easily fulfilled. Electron trapping can also be interpreted as a form of ambipolar diffusion.

Figure 1a shows electron signals from an ultracold neutral plasma created by photoionization at time $t = 0$. A small DC field (about 1 mV/cm) directs free electrons to a single channel electron multiplier for detection. The first peak at about 1 μs represents electrons that leave the sample and create the charge imbalance and Coulomb potential well. On a longer time scale, the plasma expands and the depth of the Coulomb well decreases, allowing the remaining electrons to leave the trap. This produces the broad peak at $\sim 25\,\mu s$. Colder and denser plasmas survive for as long as 300 μs before all the electrons escape.

It is possible to perform experiments on the system during the expansion. In [2], plasma oscillations were excited during this time interval by applying a radio frequency (rf) electric field to the plasma. The oscillations were used to map the plasma density distribution and reveal the particle dynamics and energy flow during the expansion of the ionized gas.

In the absence of a magnetic field, the frequency of plasma oscillations is given by $f_e = (1/2\pi)\sqrt{e^2 n_e / \varepsilon_0 m_e}$ [8], where n_e is the electron density and m_e is the electron mass. This expression is valid in a homogeneous medium and for excitations localized in regions of constant density in a spherically symmetric plasma. Discussions with various theorists have highlighted the need for a detailed analysis of the mode structure of plasma oscillations in this system, but we have so far assumed that we excite localized modes during our experiment.

Plasma oscillations with frequencies from 1 to 250 MHz have been observed, corresponding to resonant electron densities, n_r, between $1 \times 10^4\,cm^{-3}$ and $8 \times 10^8\,cm^{-3}$.

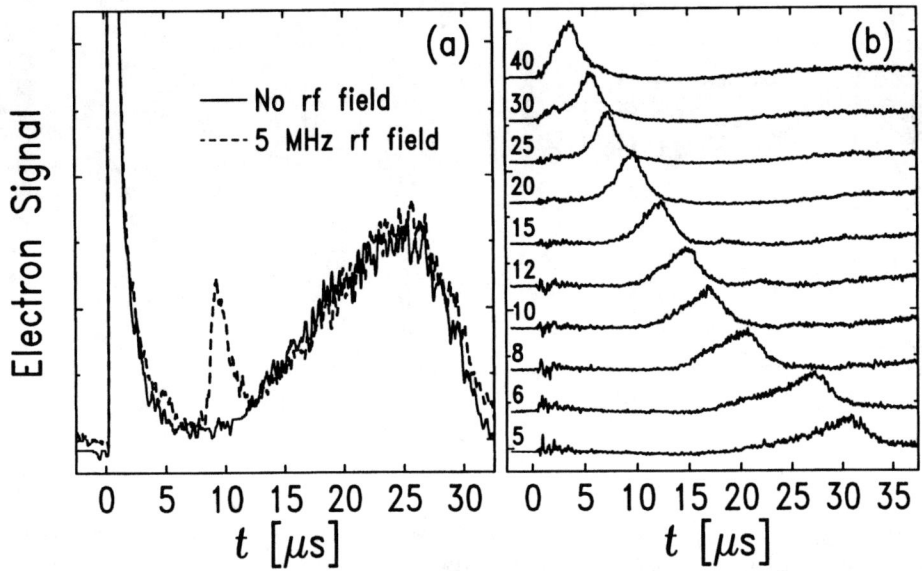

FIGURE 1. Electron signals from ultracold plasmas created by photoionization at $t = 0$. (a) 3×10^4 atoms are photoionized and $E_e/k_B = 540\,\text{K}$. If an rf field is applied during the expansion, resonant excitation of plasma oscillations produces an extra peak on the electron signal. (b) 8×10^4 atoms are photoionized and $E_e/k_B = 26\,\text{K}$. For each trace, the rf frequency in MHz is indicated, and the nonresonant response has been subtracted. The signals have been offset for clarity. The resonant response arrives later for lower frequency, reflecting expansion of the plasma. For 40 MHz, $n_r = 2.0 \times 10^7\,\text{cm}^{-3}$, and for 5 MHz, $n_r = 3.1 \times 10^5\,\text{cm}^{-3}$.

There are small corrections to the expression for f_e due to finite temperature [9] and strong coupling [10, 11]. These were neglected in the analysis of [2], but if these effects could be observed and measured, they would provide a great deal of information on the dynamics of ultracold neutral plasmas and perhaps on the physics of two-component strongly coupled systems.

The applied rf field efficiently excites plasma oscillations and pumps energy into the electron gas when the frequency is resonant with the average density in the plasma ($\bar{n}$). Collisions redistribute this energy and heat the electrons. This increases the evaporation rate of electrons out of the Coulomb well, which produces the plasma oscillation response on the electron signal (Fig. 1a). The resonant response arrives later for lower frequency (Fig. 1b) as expected because $\bar{n}$ decreases in time. With some analysis, such data implies that after a few μs the plasma expands with a constant velocity. A hydrodynamic model developed in [2] shows that the pressure of the electron gas drives the expansion, and the expansion velocity is a sensitive probe of the electron thermal energy at early times.

Recombination into Rydberg atoms in an ultracold neutral plasma was studied in [3]. At temperatures ranging from 1-1000 K, and densities from 10^5-$10^{10}\,\text{cm}^{-3}$, up to 20% of the initially free charges recombine on a timescale of $100\,\mu$s. Figure 2 is an electron signal from an ultracold neutral plasma that shows the formation of Rydberg atoms. The

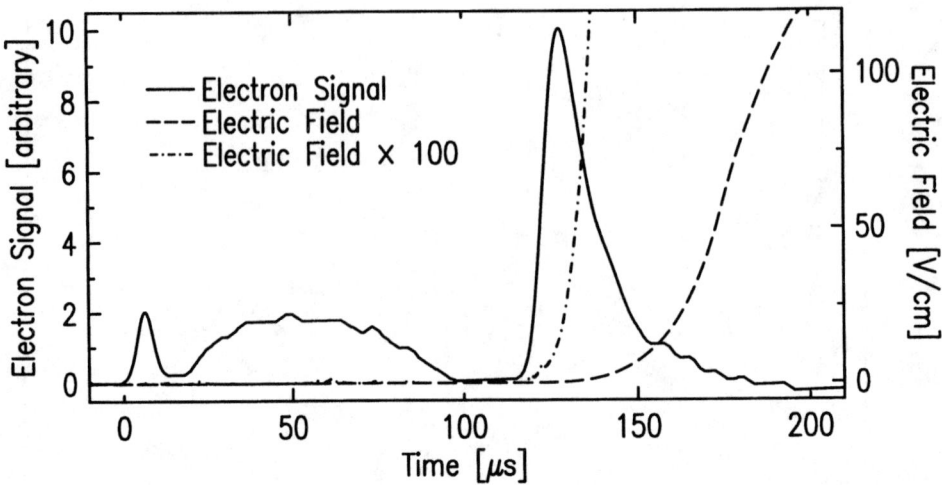

FIGURE 2. Electron signal from a plasma created by photoionizing 10^5 atoms at $t = 0$, with $E_e/k_B = 206\,K$. The first and second features represent free electrons escaping from the plasma. The third feature arises from ionization of Rydberg atoms. A 5 mV/cm field is present before the large field ramp commences at about 120 μs, and the collection and detection efficiency for the first and second features is approximately 10% of the efficiency for electrons from Rydberg atoms.

plasma is formed as described above. After the plasma has expanded so that the ions no longer form a Coulomb well and all free electrons have escaped, the electric field is increased to 120 V/cm in $\sim 100\,\mu s$. This field can ionize Rydberg atoms bound by as much as 70 K, corresponding to a principle quantum number of about $p = 47$. The number of Rydberg atoms formed is inferred from the number of electrons reaching the detector, and the distribution of Rydberg atoms as a function of p is constructed from the fields at which the atoms ionize.

Figure 3 shows typical Rydberg atom data. As N_i increases, or E_e decreases, a greater fraction of charges recombine and the Rydberg atom distribution shifts toward more deeply bound levels. The integral of each curve yields the total number of Rydberg atoms formed. The expected rates for radiative recombination or dielectronic recombination are many orders of magnitude too low to account for the observed Rydberg atom formation. We turn to three-body recombination (TBR), which is expected to dominate at ultracold temperatures.

Models of population distributions in equilibrium plasmas take into account TBR, collisional ionization, and collisional and radiative population redistribution. They predict a density-independent maximum in the Rydberg atom distribution at levels bound by a few $k_B T$ [12]. This contradicts the trend observed in this experiment toward more deeply bound levels as N_i increases or E_e decreases (Fig. 3). An ultracold neutral plasma is not formed in equilibrium, however, and a recent theoretical study of plasma dynamics [13] found good agreement with observed Rydberg atom distributions by including the effects of state changing collisions that drive population to lower levels.

Several theoretical studies have focused on the equilibration of the system after its

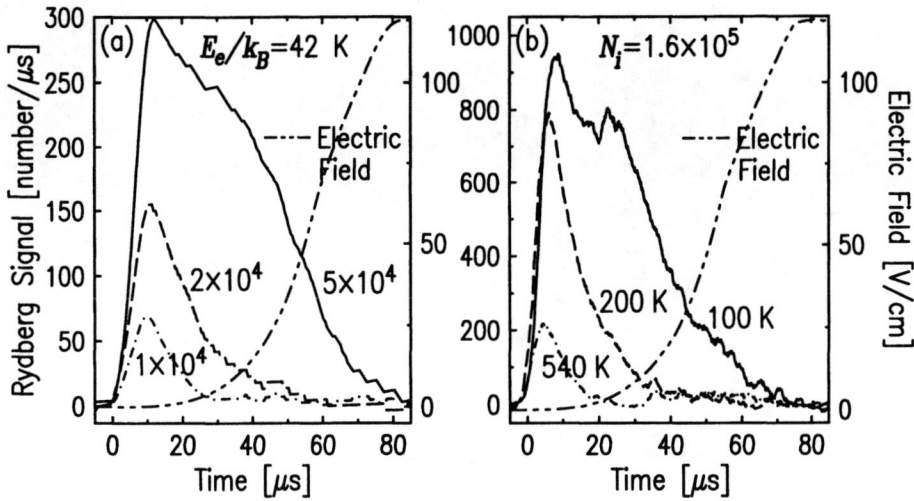

FIGURE 3. Rydberg ionization signals for various plasma conditions approximately 100 μs after photoionization. The time origin is the start of the electric field ramp. (a) Constant $E_e/k_B = 42$ K. N_i is indicated near each curve. (b) Constant $N_i = 1.6 \times 10^5$. E_e/k_B is indicated near each curve.

initial formation. Numerical simulations [13, 14, 15, 16] using several different methods find that electrons in ultracold neutral plasmas initially heat to $\Gamma_e \lesssim 1$ within a few plasma oscillation periods. This effect has been attributed to continuum lowering [15], recombination [13], and a combination of disorder-induced heating and recombination [16]. These explanations may all be different descriptions of the same process. The results seem to preclude the existence of electron-electron correlations in the experiment in its present form.

Murillo [17] found that the ions reach equilibrium in a liquid state ($\Gamma_i \sim 20$), although Kuzman and O'Neil [16] disagreed ($\Gamma_i \lesssim 1$.) Murillo pointed out that ion-ion correlations would be enhanced if the plasma were created by photoionizing atoms that already possessed spatial order - such as in a quantum degenerate Fermi gas. The Pauli exclusion principle prohibits two particles in such a system from occupying the same region in space, and such systems have been created experimentally [18]. Starting with atoms in three-dimensional lattice potentials [19] formed by standing waves of light may be another way to form plasmas with pre-existing correlation.

These studies have demonstrated the fascinating behavior of ultracold neutral plasmas and shown that much more needs to be done in order to understand this system. Improved diagnostics would be invaluable for studying the temperature and density of the plasma as it equilibrates. It would also facilitate attempts to create and observe spatial correlations arising from strong-coupling.

Besides the work described here that was conducted at NIST, studies of ultracold plasmas formed by spontaneous ionization of cold clouds of Rydberg atoms [20] have been conducted at the University of Virginia. Professor T. Gallagher has discussed these experiments at this conference. Two new experiments are coming on line. Professor S.

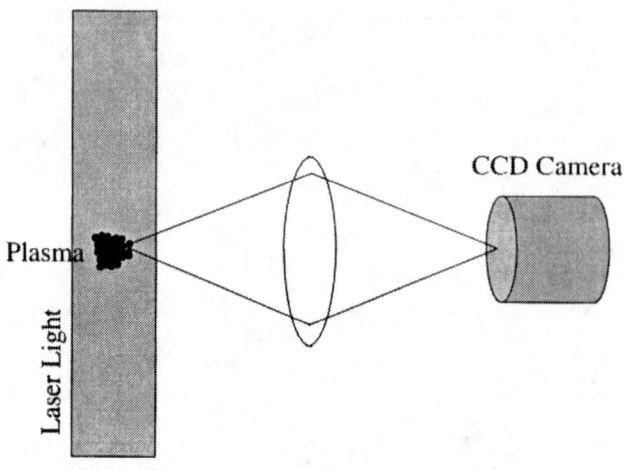

FIGURE 4. Imaging the plasma ions through resonant light scattering.

Bergeson is building an apparatus at Brigham Young University to work with calcium. The author is setting up an experiment at Rice University with strontium.

Strontium and calcium open up specific opportunities for ultracold plasma research. The ability to directly image the ion density profile through light scattering (Fig. 4) or absorption imaging would allow simple monitoring of the plasma density as a function of time. The plasma is optically thick [21] during much of its evolution, which means that optical imaging is straightforward in principle. Such imaging was impossible in previous studies because xenon ions, like most singly charged ions, have principle electronic transitions in the deep ultraviolet - outside the regime of simple lasers and imaging systems. Strontium and calcium ions possess strong electric dipole transitions at blue wavelengths. Lasers and standard CCD cameras are available for blue light.

The proposed imaging is similar to diagnostics used to study clouds of Be^+ ions in a Penning trap [22]. In these experiments, it was possible to resolve individual ions in a crystalline lattice with $12\,\mu m$ interparticle spacing. Such resolution would be more than sufficient for a host of scientific studies.

It is an added attraction of working with strontium that extremely high densities of laser-cooled atoms have been attained with this atom. It is possible to laser cool and trap 10^8 strontium atoms at densities of $10^{12}\,cm^{-3}$ [23]. This will translate into plasma densities about two orders of magnitude greater than attained so far with xenon, which will facilitate imaging the plasma and may help with experiments on strong coupling. The construction of the apparatus at Rice University is proceeding well. Currently, about 10^8 atoms are laser cooled and trapped. The density must still be measured. The pulsed laser for ionizing the atoms is being installed, and we should have ultracold strontium plasmas by the end of the summer!

REFERENCES

1. T. C. Killian, S. Kulin, S. D. Bergeson, L. A. Orozco, C. Orzel, and S. L. Rolston, Phys. Rev. Lett. **83**, 4776 (1999).
2. S. Kulin, T. C. Killian, S. D. Bergeson, and S. L. Rolston, Phys. Rev. Lett. **85**, 318 (2000).
3. T. C. Killian, M. Lim, S. Kulin, S. D. Bergeson, and S. L. Rolston, Phys. Rev. Lett. **86**, 3759 (2001).
4. H. J. Metcalf and P. van der Straten, *Laser Cooling and Trapping*, (Springer, New York, 1999).
5. M. Walhout, H. J. L. Megens, A. Witte, and S. L. Rolston, Phys. Rev. A **48**, R879 (1993).
6. As discussed later in the paper, theoretical results indicate that when the system's initial density and kinetic energy correspond to the region of strong coupling for a single component system in thermal equilibrium, the charged particles heat rapidly as potential energy associated with disorder is converted into kinetic energy.
7. F. F. Chen, *Introduction to Plasma Physics*, (Plenum Press, New York, 1974).
8. L. Tonks and I. Langmuir, Phys. Rev. **33**, 195 (1929).
9. D. Bohm and E. P. Gross, Phys. Rev. **75**, 1851 (1949).
10. Charged particles in a plasma are strongly coupled when their thermal energy is less than the Coulomb interaction energy between nearest neighbors. The situation is characterized quantitatively by the Coulomb coupling parameter, $\Gamma_x = (Z_x^2 e^2/4\pi\varepsilon_0 a_x)/k_B T_x$, where Z_x is the charge of species x, $a_x = (4\pi n_x/3)^{-1/3}$ is the Wigner-Seitz radius for density n_x, and T_x is the temperature. In a one-component strongly coupled plasma, $\Gamma > 1$ for only one species. In a two-component strongly coupled system, $\Gamma > 1$ for positive and negative charges.
11. G. Kalman, K. I. Golden, and M. Minella, in *Strongly Coupled Plasma Physics*, edited by H. M. Van Horn, and S. Ichimaru, (University of Rochester Press, Rochester, 1993), p. 323.
12. J. Stevefelt, J. Boulmer, and J-F. Delpech, Phys. Rev. A **12**, 1246 (1975).
13. F. Robicheaux and J. D. Hanson, Phys. Rev. Lett. **88**, 55002 (2002).
14. A. N. Tkachev and S. I. Yavlenko, Quantum Electronics **30**, 1077 (2000).
15. S. Mazvet, L. A. Collins, and J. D. Kress, Phys. Rev. Lett. **88**, 55001 (2002).
16. S. G. Kuzmin and T. M. O'Neil, Phys. Rev. Lett. **88**, 65003 (2002).
17. M. S. Murillo, Phys. Rev. Lett. **87**, 115003-1 (2001).
18. B. DeMarco and D. S. Jin, Science **285**, 1703 (1999).
19. M. T. Depue, C. McCormick, S. Lukman Winato, S. Oliver, and D. S. Weiss, Phys. Rev. Lett. **82**, 2262 (1999).
20. M. P. Robinson, B. Laburthe Tolra, M. W. Noel, T. F. Gallagher, and P. Pillet, Phys. Rev. Lett. **85** 4466 (2000).
21. For optically thick plasmas $n\sigma_\lambda d > 1$, where n is the plasma density, σ_λ is the ion's photon absorption cross section, and d is the plasma diameter.
22. J. J. Bollinger, T. B. Mitchell, X.-P. Huang, W. M. Itano, J. N. Tan, B. M. Jelenković, and D. J. Wineland, Phys. Plasmas **7**, 7 (2000).
23. H. Katori, T. Ido, Y. Isoya, and M. K-Gonokami, Phys. Rev. Lett. **82** 1116 (1999).

Evolution of Cold Rydberg Atoms into an Ultracold Plasma

T.F. Gallagher[ab], M.P. Robinson[a], B. Laburthe-Tolra[ab], M.W. Noel[a], and P. Pillet[b]

[a]*University of Virginia, Charlottesville, VA 22904, USA*
[b]*Laboratoire Aime Cotton, 91405 Orsay, France*

Abstract. Cold dense samples of Rydberg atoms spontaneously evolve into ultracold plasmas. Initially some cold ions are formed by black body photo ionization and collisions, and they trap electrons freed later. It appears that these trapped electrons initiate an avalanche of ionization of the remaining Rydberg atoms. How quickly the avalanche occurs is determined in part by how rapidly energy is supplied to the system.

INTRODUCTION

Usually a cold plasma is a plasma with a temperature of 10,000 K, or roughly 1 eV. It is, in general, not possible to have plasma much colder for the simple reason that it cannot sustain itself. Since the plasma is sustained by electron impact ionization, and the lowest ionization potentials are 5 eV, there must be some electrons with at least this much energy. Consequently, it is not a surprise that for conventional plasmas 10,000 K is cold [1].

Laser cooling and trapping has made it possible to make ultracold plasmas, and the first experiments were done with trapped Be^+ ions [2]. In these strongly coupled one component plasmas crystallization of the ions into regular lattices has been observed, and many of the properties of the ionic lattice studied. The first nearly neutral ultracold plasmas were made by Killian et. al., at NIST starting from cold (30 µK) trapped Xe atoms [3]. The atoms were photoionized slightly above threshold with a pulsed laser, yielding a cold ion and an electron with an energy equal to the difference between the photon energy and the ionization potential. This energy could be very small, ~ 3 cm^{-1}. When they produced only a small number of photoelectrons the electrons were detected promptly. However when more than a threshold number of photoelectrons were formed many of the photoelectrons did not leave the trap volume but remained for up to 100 µs, and they could only be ejected by applying an electric field. Due to the departure of the prompt photoelectrons there was a net positive charge (100-1000 ions), which held the majority of the photoelectrons in a slowly expanding ultracold plasma. The NIST group carried out an elegant series of experiments showing the expansion of and recombination in these plasmas [3-5]. In several laboratories,

CP635, *Atomic Processes in Plasmas: 13th APS Topical Conference*, edited by D. R. Schultz et al.

in Virginia, Laboratoire Aime Cotton, Michigan, and Connecticut, it has now been observed that a cold sample of Rydberg atoms [6,7] spontaneously evolves into an ultracold plasma. Although hot Rydberg atoms at much higher densities had been observed to evolve into a plasma, that the much lower densities of cold Rydberg atoms would evolve into a plasma was a surprise. In this report we suggest a mechanism for how this happens and identify the outstanding questions.

I. Experimental approach and initial observations

In our experiments we have cold Rb (Cs) atoms in a magneto optical trap (MOT), which is loaded from the 300 K vapor which has a pressure of 10^{-8} torr. There are 10^{10} trapped atoms in a sphere of diameter 1 mm. The atoms are continuously excited to the Rb $5p_{3/2}$ (Cs $6p_{3/2}$) state by the trap lasers, and, typically, one third of the atoms are in the $5p_{3/2}$ ($6p_{3/2}$) state. A 5 ns dye laser pulse from a laser running at a 20 (10) Hz repetition rate is used to excite the atoms to a Rydberg state. After a time delay T, from 0 –50 μs, we apply a field ramp, rising to 1000 V/cm in 2 μs, to the atoms to analyze the population in the Rydberg states. The time resolved signal implies the population distribution. For example the 40d state ionizes at a field of 125 V/cm. Higher lying states ionize earlier in the field ramp, and more deeply bound states ionize later.

Naively, 10 μs after exciting atoms to the 40d state, which has a 0 K lifetime of 70 μs [8], we would expect to see 5% of the population redistributed to nearby states by 300 K black body radiation along with a smaller amount of black body photoionization. What we actually see is shown in Fig. 1, a set of oscilloscope traces. With the delay $T = 2$ μs we see primarily the Rb 40 d state, at 280 ns, with a small amount of black body induced transfer to nearby states, and a small free ion signal, at 100 ns. This signal is just what is described above. However at $T = 5$ μs there are half ions, and at $T = 12$ μs almost the entire signal is due to ions, the Rydberg atoms are gone!

The signals shown in Fig. 1 are Rb^+ signals. If there were only ions present, their space charge should broaden the ion signals at $t = 100$ ns in Figs. 1b and 1c. However, the signals are not broadened relative to the signal in Fig. 1a, which implies that the ions must be in a macroscopically neutral sample and that there are electrons present as well. This notion is confirmed by detecting the electrons explicitly. Consistent with Fig. 1, an electron signal at the beginning of the field ramp appears after 5 μs and persists for tens of microseconds. These electrons are essentially free, so we apparently have a plasma.

II. Experiments to determine how the plasma is formed

Fig. 1 shows that the plasma forms after a time delay, and to understand how it forms we need to make more controlled measurements of the plasma formation as

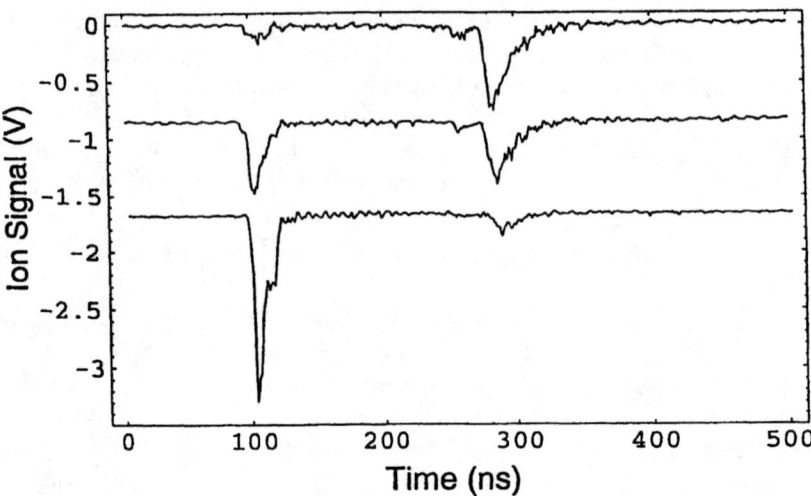

FIGURE 1. Oscilloscope traces for three different interaction times showing the evolution to a cold plasma. These are for the ^{85}Rb 36d state with a density of 1.5 x 10^9 cm^{-3}. From top to bottom the traces correspond to a time delay between excitation and field ionization of 2, 5, and 12μs. In the upper trace there is very little early ion signal and a large late Rydberg atom signal while the reverse is true for the lower trace indicating the formation of a plasma.

a function of delay time and number of atoms. The most straightforward approach would be to fix the number of atoms excited and vary the delay time T. This approach fails miserably because of the huge shot to shot variation in the number of atoms excited. What does work is to bin the data in the following way. In Fig. 1 it appears that the signal is either ions, at t = 100 ns, or 40d Rydberg atoms, at t = 280 ns. We assume that the sum of these two signals gives the number of atoms initially excited to the 40 d state. We fix T and detect both the ion and Rydberg signals on each shot of the laser, and let the laser run. Plotting the ion signal vs the sum of the ion and Rydberg signals, assumed to be equal to the initial Rydberg population N_0 gives the result shown in Fig. 2a for T = 2 μs and 12 μs. It is evident that for T = 2 μs something happens at N_0 = 2 x 10^5, and for T = 12 μs at N_0 = 0.8 x 10^5. By collecting data such as those of Fig. 2a for many values of T we can build up the time dependence of the signal for many initial populations, and representative examples are shown in Fig. 2b. It is apparent that for N_0 = 3.3 x 10^5 atoms that the number of ions increases slowly for 0.6 μs after which it rises rapidly until all the atoms are ionized. For N_0 = 2.4 x 10^5 the initial rise takes longer but appears to be to the same number of ions.

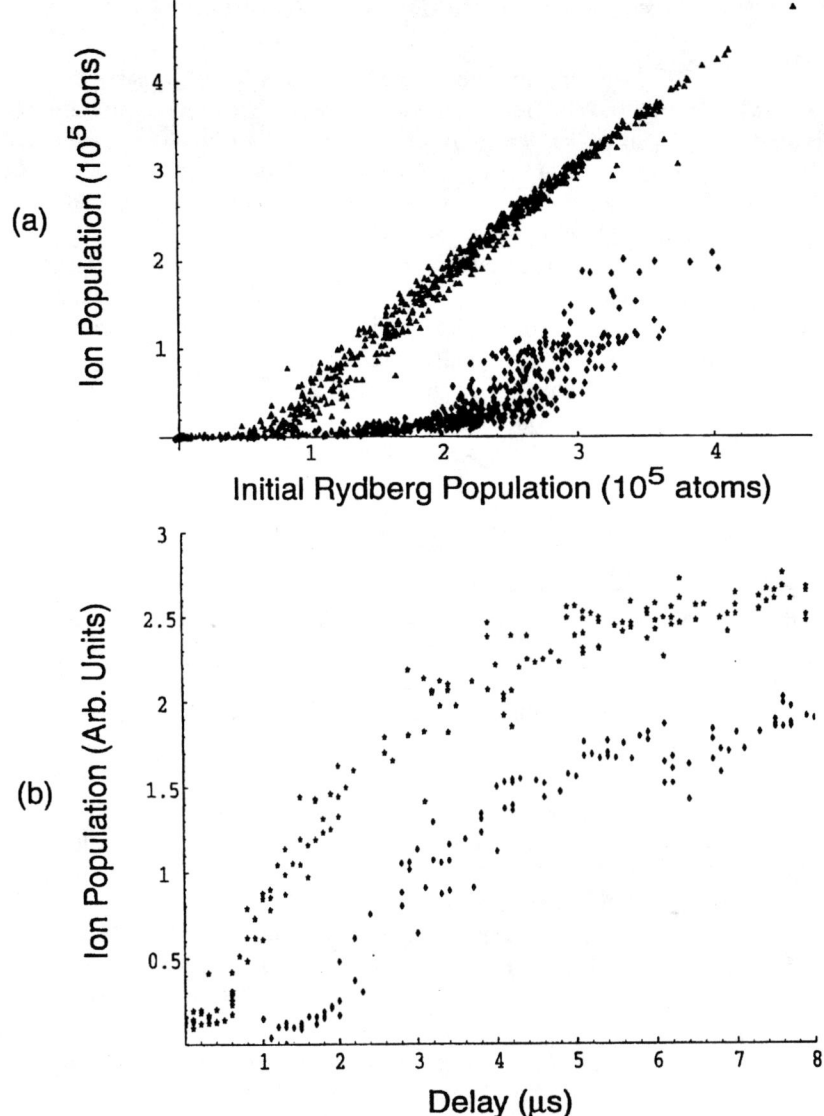

FIGURE 2. (a) Ion population as a function of initial population of the 36d Rydberg state for two time delays, (♦) 2 μs and (▲) 12 μs. (b) Ion population as a function of interaction time for two initial populations of the 40d Rydberg state, (★) 3.3 x 10^5 atoms or 10.5 x 10^9 cm^{-3} and (♦) 2.4 x 10^5 atoms or 7.6 x 10^9 cm^{-3} offset by 1μs.

III. Physical procession in the Rydberg to plasma evolution

As shown by Fig. 2b, initially the number of ions rises linearly, and we attribute it to black body photoionization and collisions between Rydberg atoms. Cold Rydberg atoms have velocities of 30 cm/s, and only close pairs could contribute to this ionization. On the other hands, hot Rydberg atoms excited from the 300 K background vapor have velocities in excess of 10^4 cm/s and can produce ionization at an appreciable rate. Once there are enough cold ions present, about 1000, their charge traps subsequent electrons, assuming that the electrons are liberated with energies of ~ 100 cm^{-1}. This is precisely the same process which led to the formation of the plasma in the NIST experiments. Once electrons are trapped they pass repeatedly through the cloud of Rydberg atoms, ionizing them rapidly. The onset of this rapid avalanche ionization is quite apparent in Fig. 2b. Note that once the macroscopic charge is sufficient to hold the first electron it is sufficient to keep all subsequent electrons freed since the net charge does not change.

IV. Verification

The initial 300 K black body photoionization rate can be calculated easily [9], and the calculated rates for n ~ 50 are 500 s^{-1}. Equally important the energies of the photoelectrons are less than the binding energy of the initial state, so a 40d atom leads to photoelectrons with energies below 70 cm^{-1}. Using the Rydberg-Rydberg cross sections of Olson [10], ten times the Rydberg geometric cross section σ_g, and the hot atom density 5 x 10^7 cm^{-1}, we can estimate the hot-cold Rydberg atom ionization rate for Γ Rb 40d atoms to be 10^3 s^{-1}. Of course we can also measure the initial ionization rates, and for the Cs 39d state the initial ionization rate is 2 x 10^3 s^{-1}, and if we remove the hot 300 K atoms the ionization rate is 1 x 10^3 s^{-1}. These rate are reasonably close to the calculated rates so these processes seem to be good candidates for the initial ionization, at least for n = 40. We further note that no plasma forms for n < 40 if we remove the hot atoms, whereas for n > 50 it does not seem to matter if hot atoms are present or not.

The effect of an electric field supports the notion that the macroscopic charge of ~ 1000 produces a potential well which traps low energy electrons. As shown by Fig. 3 a field of 1-2 V/cm completely inhibits the formation of the plasma. From this field and the 0.2 mm diameter of the cloud of Rydberg atoms we can estimate the depth of the ion potential produced by the ions to be 160 cm^{-1}, a number consistent with the binding energies of the expected photoelectrons from the 32d state. Curiously, it takes a slightly stronger field to suppress the formation of the plasma for a more highly excited state.

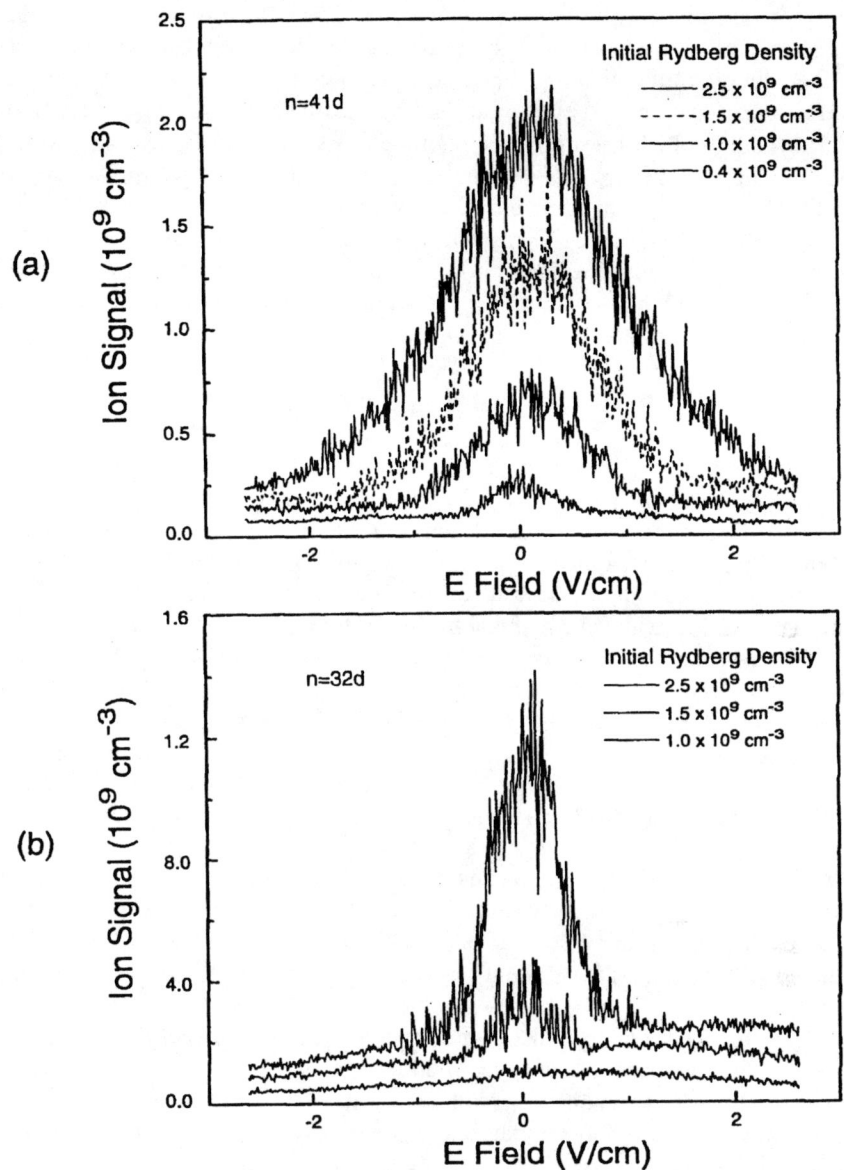

FIGURE 3. Electric field dependence of the ion signal for two different n states, (a) 41d and (b) 32d states. Each graph plots several different initial Rydberg population densities for a 10μs interaction time and is averaged thirty times.

Finally, it is interesting to compare our observations to a quantitative model. We assume that the atoms are initially ionized by black body photoionization and hot-cold Rydberg atom collisions, this process having a cross section ten times the geometric cross section. We assume that the avalanche ionization occurs by electron-Rydberg collisions, a process having a cross section $\gamma \sigma_g$, where γ is a fit parameter. In principle γ should be ~1. The results of this model are in reasonably good agreement with our experimental results, as shown by Fig. 4. However, in virtually all cases, i.e., from 32d to 48d, the value of γ is 0.2, not 1, indicating that the avalanche ionization is slower than expected. We do not account for the energy loss of the free electron in these ionizing collisions, and we interpret the value $\gamma \sim 0.2$ as reflecting this fact.

V. Outstanding problems

There are several outstanding problems. The primary one is an energy accounting problem. Inspecting Fig. 2b we can see that more than 95% of the atoms are ionized in the avalanche. Each 40d electron is initially bound by 70 cm^{-1}, but in the final plasma it is essentially free. It appears that we need an energy source. Presumably the rate of finding this energy is what limits the ionization rate in the avalanche and is why we find $\gamma = 0.2$, not 1, in our model.

To verify that adding energy makes the plasma form more rapidly we have added a radio frequency field, typically at a frequency of 100 MHz with an amplitude of 1 V/cm. To an electron initially at rest this field adds, on average, the ponderomotive energy $U_P = <1/2 \ mv^2> = 9$ cm^{-1}. The maximum kinetic energy is of course twice this value. If the electron is already moving, a much more likely case, the rf field can substantially increase the peak kinetic energy. For an electron with 20 cm^{-1} of motion in the field direction, the peak energy could be increased to 56 cm^{-1}, a reasonable fraction of that required for ionization for n = 30-40 atoms. In Fig. 5 we show the ion signals observed at delay T = 10 μs after excitation of the Rb 32d state with and without the rf field. The difference is quite apparent. Without the field the plasma does not form but with it it does.

It is clear that energy must come from somewhere to fuel the avalanche, but where? The most physically appealing idea is that it is superelastic collisions. The most likely of these would seem to be ones like e$^-$ + Rb 40d $\rightarrow$ e$^-$ + Rb n<40. These processes should have a large cross section. We have systematically looked for signals from lower lying Rydberg states, to no avail, but it may be that the final states are so spread out that they are hard to detect. Note that we have assumed that this process does not occur in our method of binning the data. It might be that superelastic collisions with the Rb 5p or Cs 6p are important, but the cross sections should be ~ 10^{-14} cm^2, too small to produce appreciable rates. An available source of energy is the 300 K black body radiation, but it is difficult to see how it is coupled to the atoms well enough.

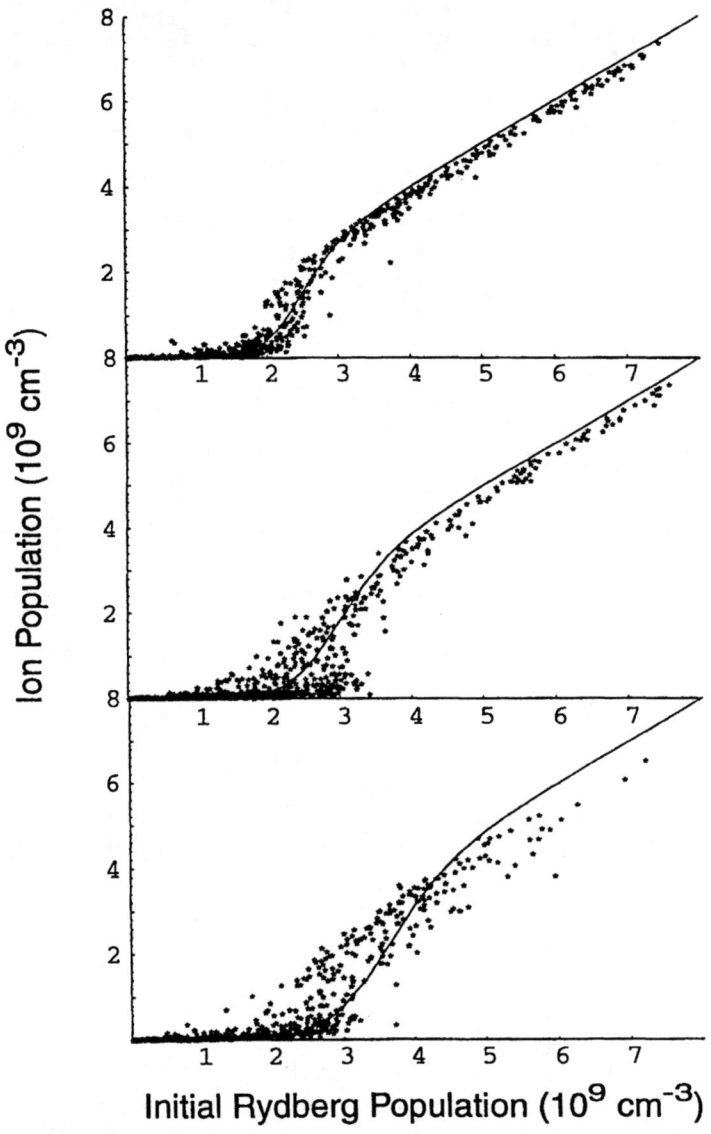

FIGURE 4. Comparing density scans to the rate equation for a time delay of 10 μs. From top to bottom are the 44d, 42d, and 40d states. The fit parameters are A = 10, and the number of hot atoms is 3.6% of the initial density for all the curves. B = 0.244, 0.230, and 0.217 from top to bottom respectively.

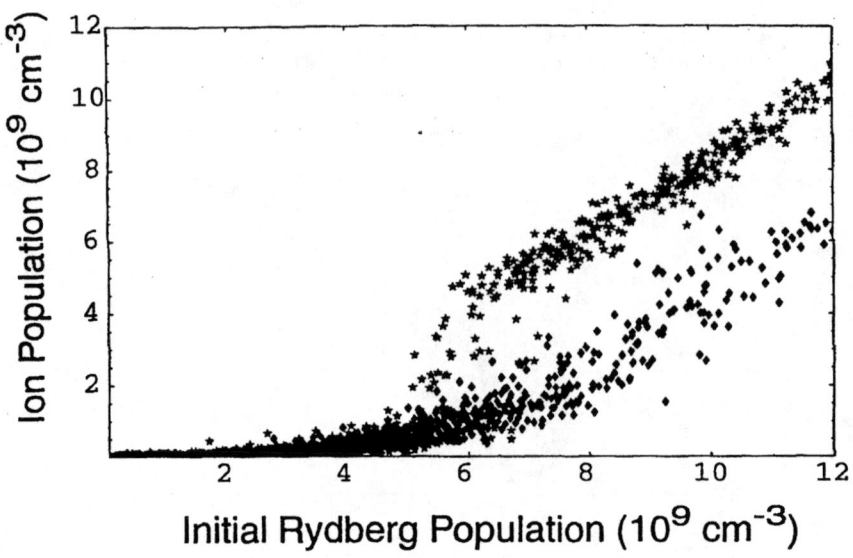

FIGURE 5. Comparing density scans with and without RF for the n = 32d state with a 10 μs delay. (★) 100 MHz, 0.5 V/cm. (♦) No. RF.

A second very puzzling observation is that the signals shown in Fig. 1 are, indeed, typical. As noted above, there is no signal at t > 280 ns from n < 40 states, suggesting the absence of superelastic collisions. Also, there is virtually no signal between the ion signal at t = 100 ns and the 40d signal at 280 ns. Since electron impact ionization is the extension of the excitation cross section over the limit one could reasonably expect to see a range of n > 40 states in the T = 5 μs and T = 12 μs traces. Occasionally we have observed signals which are filled in between the ion signal and the initial state, but far more often than not the signals look like those shown in Fig. 1. It may be that our notion of what is happening in the avalanche is wrong, and something more interesting is happening, such as an expanding cloud of plasma growing from a small seed to engulf the trapped Rydberg atom cloud.

VI. Conclusion

We have described the experimental observations which show that cold Rydberg atom samples spontaneously evolve into an ultracold plasma. While we can construct a reasonable scenario for how the evolution occurs, it is evident that we do not completely understand it, and it may be far more interesting than the simple picture we have presented suggests. To remove the present uncertainties requires more sophisticated experiments which are being prepared.

This work has been supported by the Air Force Office of Scientific Research and the Centre National de la Recherche Scientifique. We are pleased to acknowledge useful discussions with R.R. Jones, J.L. Cecchi, F. Robicheaux, E. Kolomeisky, T.C. Killian and S. Mazevet.

REFERENCES

1. E. Nasser, Fundamentals of gaseous ionization and plasma electronics. (Wiley, New York, 1971).
2. Huang, X.P., Bollinger, J.J., Mitchell, T.B., and Itano, W.M., *Phys. Rev. Letters* **80**, 73 (1998).
3. Killian, T.C., Kulin, S., Bergeson, S.D., Orozco, L.A., Orzel, C., and Rolston, S.L., *Phys. Rev. Letters* **83**, 4776 (1999).
4. Kulin, S., Killian, T.C., Bergeson, S.D., and Rolston, S.L., *Phys. Rev. Letters* **85**, 318 (2000).
5. Killian, T.C., Lim, M.J., Kulin, S., Dumke, R., Bergeson, S.D., and Rolston, S.L., *Phys. Rev. Letters* **86**, 3759 (2001).
6. Dutta, S.K., Feldbaum, D., Walz-Flannigan, A., Guest, J.R., and Raithel, G., *Phys. Rev. Letters* **86**, 3993 (2001).
7. Estrin, A., Tong, D., Ensher, J.R., Cheng, C.H., Eyler, E.E., and Gould, P.L., *Bulletin of the APS* **46**, H4 (2001).
8. T.F. Gallagher, *Rydberg Atoms*, Cambridge University Press (1994).
9. Spencer, W.P., Vaidyanathan, A.G., Kleppner, D., and Ducas, T.W., *Phys. Rev. A* **25**, 380 (1982).
10. Olson, R.E., *Phys. Rev. Letters* **43**, 126 (1979).

Spectroscopic Analysis of a Low Fluence Li-Ag Laser Driven Plasma Plume

M.E. Sherrill[1], R.C. Mancini[1], J.E. Bailey[2], A. Filuk[2], B. Clark[2], P. Lake[2], J. Abdallah[3]

[1]Department of Physics, University of Nevada, Reno, NV 89557
[2]Sandia National Laboratories, Albuquerque, NM 87185
[3]Los Alamos National Laboratory, Los Alamos, NM 87550

Abstract. Low fluence laser produced plasmas are used in many applications: from ion sources to material synthesis. Our work focuses on developing a quantitative description of these ablation plasmas through the interpretation and analysis of time- and spatially-resolved spectroscopic measurements with detailed spectral modeling. To this end, in a series of experiments performed at Sandia National Laboratories, laser generated Li-Ag plasma plumes were produced by irradiation of solid targets using a Nd pulsed laser. Time- and spatially-resolved optical spectra were recorded with a framing spectrograph. In order to limit the gradients along a direction perpendicular to the target's normal, targets with strips of Li-Ag coated on top of Pt were used. The Pt plume collisionally confines the Li-Ag, thus reducing the Li-Ag lateral expansion. The spectra display line transitions in Li and Ag atoms. A spectroscopic model based on time-dependent collisional-radiative atomic kinetics, detailed line shapes, and radiation transport was used to describe plasma parameters both spatially and temporally. In particular, this analysis has revealed that level populations in laser-ablated plumes may behave in a time-dependent manner, i.e. not in Local Thermodynamic Equilibrium (LTE). The time-scales associated with these phenomena and the interpretation of spectral data critically depends on the details of the atomic kinetic model and the quality of the rate coefficients. In order to generate accurate atomic data for atoms present in the plasma, a semi-empirical technique has been implemented in the Los Alamos suite of atomic structure and electron scattering codes. Details of the spectral model and analysis results will be discussed.

INTRODUCTION

Laser ablation refers to the process of removing material from a solid or liquid target, with a low intensity laser ranging from 1×10^7 W/cm^2 to 1×10^9 W/cm^2. Typically, a pulsed laser is used to irradiate the target. It deposits the bulk of its energy in the skin depth region of the target where this volume of material is heated and then undergoes melting, ablation and possibly plasma formation. The material in the gaseous state then forms a plume that expands away from the target with typical velocities of a few 10 µm/nsec normal to the target's surface. When a series of ablation events are performed in which the duration of the laser irradiation on the target is allowed to stay constant (i.e. full width at half maximum (FWHM) of the laser pulse is fixed) while the fluence is allowed to increase, a transition between evaporative plume to plasma plume formation can be observed.[1] Our work is concerned with plumes of the latter type.

CP635, *Atomic Processes in Plasmas: 13th APS Topical Conference*, edited by D. R. Schultz et al.

Though there has been a large body of experimental and theoretical work dedicated to describing the expansion of the ablation plume far from the target surface (~ 0.5 cm) for various target materials, little work has been done to describe the detailed state of the plasma during its early evolution near the surface[2]. Due to the non-intrusive nature of emission spectroscopy, spectroscopic modeling has been shown to be successful in quantitatively describing a wide variety of plasmas. A common concern in performing spectroscopic measurements in plasmas is the ability to produce homogeneous plasmas to limit gradients in temperature and density--at least along the line of sight of the spectrometer. This desire to reduce inhomogeneities in the plasma stems from the difficulty of interpreting spectra from non-uniform plasmas due to the convolution of contributions of characteristically different emission/absorbing regions. In this work we discuss a laser-ablation experiment, a spectroscopic technique to record time- and space-resolved spectra, and the development and application of a spectral model to extract information about the state of the plasma plume in the early stages of its formation and expansion away from the surface.

EXPERIMENT

In a series of experiments performed at Sandia National Laboratories, a binary alloy target composed of 47% Li and 53% Ag was irradiated using a Nd-YAG laser with a Gaussian pulse of 10 ns, at an angle of incidence of 30° to the target's normal and a fluence of 0.7 J/cm^2, which corresponds to an intensity of 1×10^8 W/cm^2. Time- and spatially-resolved (along the direction normal to the target's surface) optical spectra in the 4000 Å to 7000 Å range were recorded with a framing spectrograph. In order to limit the effects of plasma gradients on the spectra due to lateral expansion along a direction perpendicular to the target's normal, a novel target design was implemented. Several 1.5 mm wide, 0.5 μm thick thin film strips of Li-Ag were deposited on top of a 0.2 μm thick Pt-coated substrate. The 6x6 mm^2 focal spot of the laser illuminated both the Li-Ag strip and the surrounding Pt base as illustrated in Fig. 1. During the expansion of the metallic plume, the Pt plasma collisionally confined the Li-Ag core, thus reducing the Li-Ag lateral expansion as seen in Fig. 2. This technique represents a variation on the microdot spectroscopy method commonly employed in X-ray spectroscopy of high-temperature laser-produced plasmas[3], and it permitted a more accurate characterization of the Li-Ag plasma plume by reducing plasma gradients along the line of sight of the spectrometer. A package of laser-beam smoothing optics, based on diffuser and homogenizer optical elements, was used to improve the laser irradiation uniformity on the target. A gated spectrograph, equipped with a F5.6 Nikon lens and a 159 g/mm diffraction grating coupled to a charged coupled device (CCD), was set up to record a time-resolved 1D image along the target normal direction, i.e. a single time snapshot with a 1D spatial image. Each data image is integrated over a 2 ns time interval, and multiple shots with the spectrometer triggered at 10, 20, 30, 40, 50, 75, 100 ns after the end of the laser pulse were recorded to construct the spectral time history. Spatial resolution was approximately 60 μm, while the spectral resolution was 600. As illustrated in Fig. 3, the horizontal axis of each

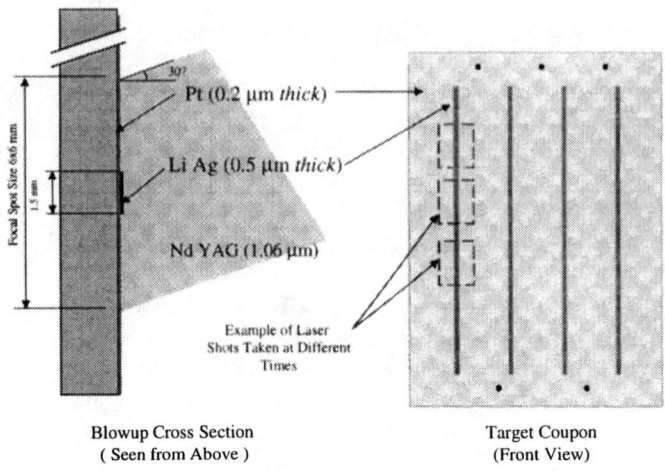

Blowup Cross Section
(Seen from Above)

Target Coupon
(Front View)

FIGURE 1. Schematic illustration of the Li-Ag strips on the Pt coated target. · Each laser shot irradiates a fresh portion of the target coupon.

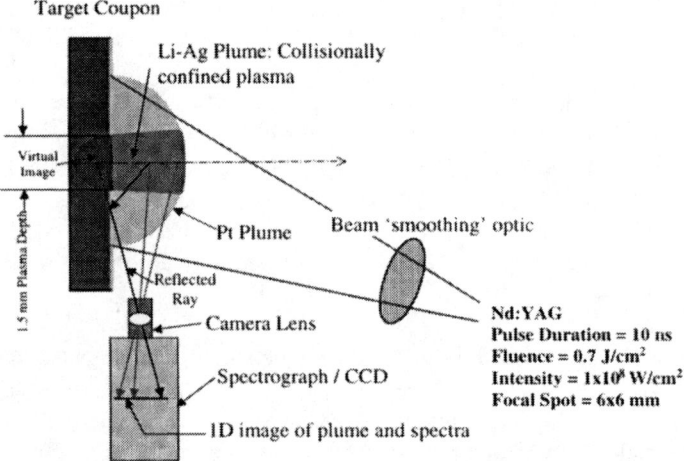

FIGURE 2. Schematic illustration of the experimental layout. The Li-Ag plume is collisionally confined by the Pt blowoff.

CCD image represents the wavelength from 4000 Å to 7000 Å, while the vertical axis represents the distance from and normal to the surface of the target from -1000 μm to +1000 μm, i.e. x = 0 characterizes the location of the target's surface. The instrument records the real as well as the virtual image due to the reflection of the plume emission from the mirror-like surface of the target as described in Fig. 2. Early in time, only continuum emission is observed. Later, the images display the characteristic line emissions from transitions in Li and Ag atoms. First, the line emissions are localized close to the surface and the lines are broad. This is indicative of the formation of a dense surface plasma. For later times, the spatial intensity distributions move away from the surface and the lines become narrower. This is an indication of the plasma plume expansion.

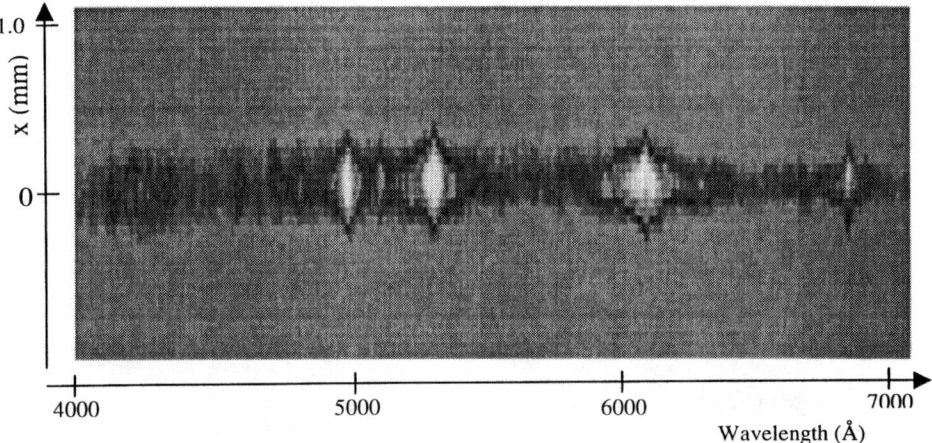

FIGURE 3. Experimental CCD images taken at 20 ns after the end of the laser pulse.

Spatially-resolved spectral lineouts can be generated from the data images. Early in time and close to the surface, as seen in Fig. 4., lineshapes are broadened by Stark and resonance broadening effects. A forbidden line transition (Li I: $1s^2$ 3p – $1s^2$ 3p) is apparent in the spectra. This is an indication that plasma ions microfields are field-mixing neighboring states and thus leading to the emission of forbidden line transitions. Furthermore, the self-reversal of the Li I: $1s^2$ 3d – $1s^2$ 2p transition, due to opacity effects, lends evidence to the existence of a slight gradient in the density after collisional confinement by the platinum. Later in time and away form the surface, the broadening of the lines does not change, and the line shapes are dominated by the characteristic instrumental function. From the data taken at 100 ns, the average FWHM of the Li I: $1s^2$ 3d -- $1s^2$ 2p was measured to be 0.00411 eV and this value was used as an estimate of instrumental broadening.

SPECTROSCOPIC MODEL

The interpretation of the data was done with the aid of detailed atomic kinetics, line profile and radiation transport calculations.

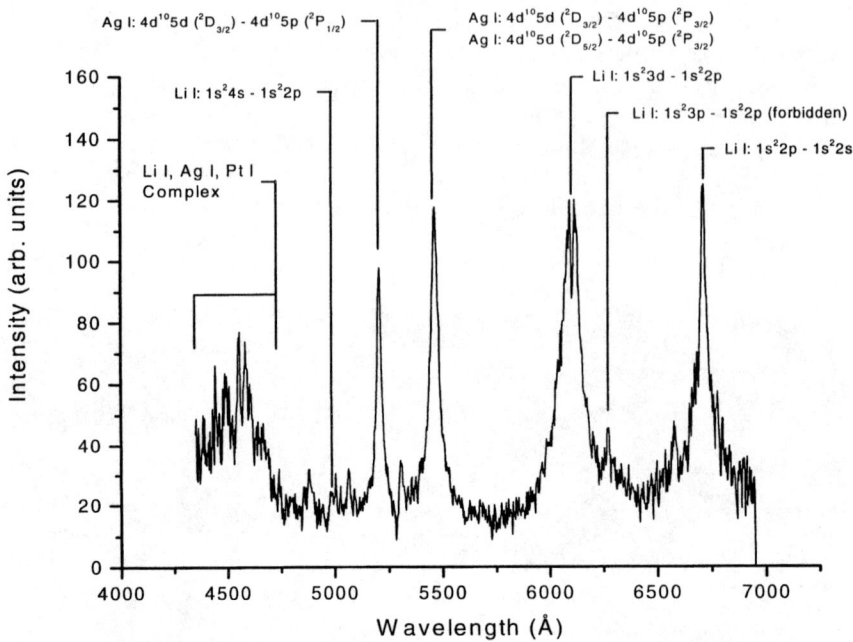

FIGURE 4. Spatially-resolved lineout taken from the +20 nsec CCD image at x = 30 μm from the target surface.

Due to the nearly equal abundances of Li and Ag in the target, the influence in the level populations and ionization balance of each species due to the other species in the plasma had to be taken into account. The Li and Ag population kinetics are coupled through the common pool of free electrons in the plasma. Therefore an atomic kinetic model was developed that would calculate, both in a time- dependent and time-independent manner, the ionization balance and level populations of all species simultaneously and self-consistently. Specifically, this model considers the Ag neutral atom and its first four ionization stages. The energy level structure of Ag I - Ag III is described up to and including states with principal quantum number n=10, while only the ground state of Ag IV is included. For Li, the neutral atom and its first three ionization stages have been considered with each having an energy level structure up to and including states with n=7.

To produce high-quality synthetic spectra for the analysis, fine-structure energy levels are used to describe excited states in Li and Ag neutral atoms. Configuration average energy levels are used for the first, second and third ionization stages of Ag, respectively. The use of a coarser grain atomic structure representation for non-spectroscopically-significant ionization stages in Ag arises from the compromise of having to include these ionization stages to obtain the appropriate ionization balance while overcoming the technical difficulty of representing the atomic structure of partially-filled d-shell ions. Ions such as Ag^{+2} would require 1100 levels to properly describe the energy level structure found below 90% of the ionization potential, not including autoionizing states. As an example, representing Ag^{+2} in the fine structure level of description would require the calculation of 605,000 electron-collisional cross sections while in the configuration average representation only 512 electron-collisional excitation cross sections are needed.

Several atomic processes linking the populations of ground and excited states in contiguous ionization stages (including the neutral atom) were considered in the model. These include electron impact excitation and deexcitation, electron impact ionization and recombination, photoionization and radiative recombination, spontaneous radiative decay and photoexcitation, dielectronic recombination and multiphoton ionization. Electron-collisional rates are calculated assuming a Maxwellian electron distribution function. However, this assumption can be easily removed and other electron distribution functions used instead.

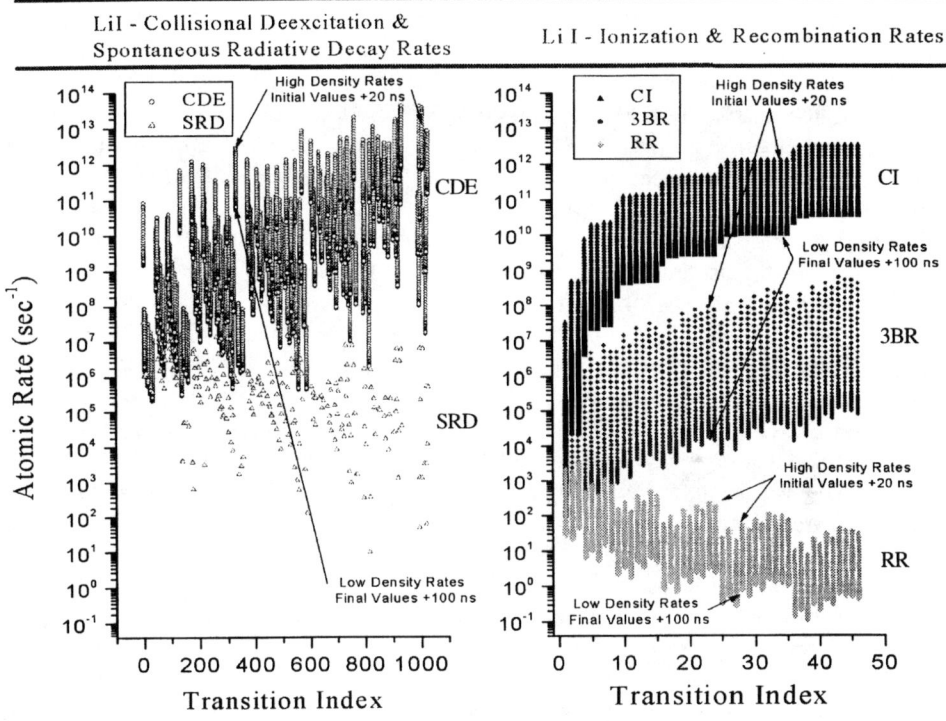

Figure 5. Evolution of Li atomic rates during the laser ablation of a Li-Ag plasma plume.

Modeling calculations performed using simple, scaled atomic rate coefficients describe a scenario for the laser ablation of a Li-Ag plume where the atomic kinetics is collisionally dominated but still time-dependent (i.e. non-equilibrium). To illustrate the relative magnitude of the rate coefficients, Fig. 5 displays the time histories of the electron collisional deexcitation (CDE), spontaneous radiative decay (SRD), electron collisional ionization (CI) and recombination (3BR, i.e. three-body recombination), and radiative recombination (RR) rates for several transitions during the ablation of a plasma fluid element. The data is for transitions in Li atoms and the time-histories are shown as vertical sequences with top values associated with early times and bottom values associated with late times. We note that collisional deexcitation dominates spontaneous radiative decay, and collisional recombination dominates radiative recombination.

Improving Atomic Structure and Collision Data for Neutrals

While several atomic structure and scattering codes are currently available to generate cross sections and rate coefficients for plasma atomic kinetics in cases where the ionization balance is dominated by highly-charged ions (i.e. high-temperature plasmas), computer codes to generate large amounts of atomic data for neutrals and low-charge ions typical of low-temperature plasmas are not readily available. Hence, special procedures or extensions have to be implemented in existing codes in order to make them useful for applications to atomic and low-charge ions population kinetics applications. To this end, we have worked with the suite of atomic structure and scattering codes developed by the atomic theory group (T4-group) at Los Alamos National Laboratory. These codes include a Hartree-Fock based atomic structure code (CATS), an electron impact excitation cross section code (ACE), and an electron impact ionization cross section, autoionization rate and photoionization cross section code (GIPPER). In addition, a general purpose code (TAPS) permits handling of output binary files and data extraction[4,5]. With the goal of making these codes suitable and enhancing the accuracy for applications in neutral atoms and low-charge ions, the atomic structure code CATS was modified to include a semi-empirical procedure that permits the calculation of self-consistent sets of eigenvalues and eigenvectors that match experimentally available data for energy level energies[6]. This procedure adjusts various radial energy parameters (configuration average energy E_{av}, spin-orbit parameter ζ, Slater integrals: R^k, F^k direct, and G^k exchange) in order to provide the best least-squares fit between calculated eigenvalues and the observed energy level values. The resulting radial energy parameters are then used to recalculate self-consistent wavefunctions. In this way, more accurate target-state wavefunctions can be computed by CATS and used as input for ACE and GIPPER for rates and cross-section calculations. The effect of the modified energy levels and wavefunctions is modest in Li atoms, but it is dramatic for the case of Ag atoms. Calculation of the Ag atomic structure without this correction leads to an energy level structure with incorrect energy level values and ordering. In turn, the improvement in target wavefunction characterization results in more accurate atomic rates and cross sections. To illustrate this point, Fig. 6 shows an electron-collisional excitation rate in the Ag atom as a function of electron temperature. Most of the relevant temperatures in the simulation fall in the range of about 1 eV (and below) where significant differences are found between rates calculated including and not-including the modified wavefunctions.

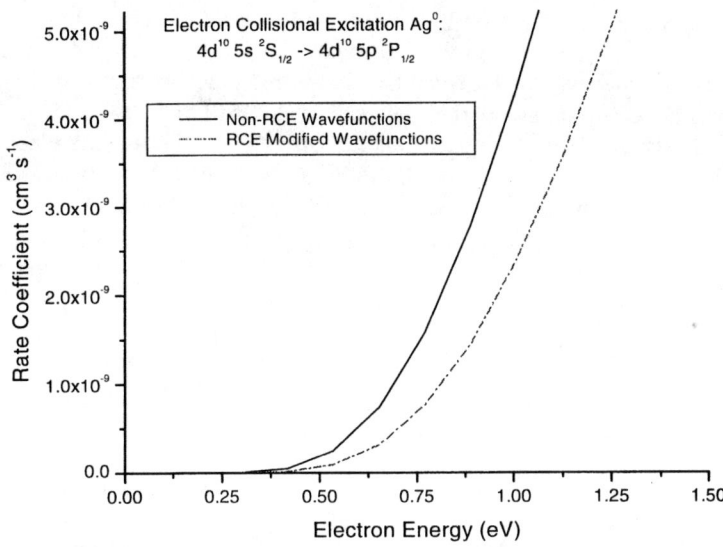

Figure 6. Comparison of electron collisional excitation rate coefficients computed with and without modified target wavefunctions.

DISCUSSION

The differences in energy level structure between Ag and Li ions drastically influence the level population distribution. At low temperatures (less than ~ 1 eV), Li is more ionized than Ag since the ionization potential of Li ($I_p = 5.392$ eV) is smaller than that for Ag ($I_p = 7.574$ eV). However, for temperatures in the 1 eV to 4 eV range, Ag is more ionized than Li since it is very hard to ionize Li^{+1} ($I_p = 75.368$ eV). As the temperature increases from 1 eV to 3 eV, Ag atoms ionize and effectively populate states in Ag^{+1} and Ag^{+2} while the population distribution of Li states remains relatively unchanged and mainly accumulates in the Li^{+1}-ion ground state. This fact and the simultaneous observation of strong Ag and Li atomic lines in the spectra can be used to estimate the temperature of the plasma close to the surface at about 1 eV. On the one hand, the temperature has to be high enough to effectively populate the atomic upper levels of the observed lines. On the other hand, the temperature cannot be too high or else the Ag atoms ionize and the Ag lines become relatively weak. This reasoning also relies on the assumption that steady-state atomic kinetics can be considered a good approximation under the high-density conditions near the target surface.

A density estimate near the surface can be obtained from the plasma broadening of Li and Ag lines. In particular, Stark broadening effects due to the plasma electron and ion electric microfields significantly perturb the energy levels of the Li $1s^23d-1s^22p$ line, and result in density-dependent line shapes. To this end, Stark broadened line profiles were calculated using a plasma line broadening code that simultaneously accounts for the perturbing effects from both plasma electrons and ions. Natural and

Doppler broadening were also included but their magnitude is negligible. Resonance broadening is much smaller than Stark broadening. In addition, the mixing between $1s^2\,3p$ and $1s^2\,3d$ energy levels, driven by the plasma electric microfields, gives rise to a forbidden line component on the low energy side of the Li $1s^23d-1s^22p$ line. Its location, intensity and broadening are also dependent on the plasma electron density. A non-negligible opacity effect in the Li $1s^23d-1s^22p$ line adds to its broadening while it also effectively enhances the relative intensity of the forbidden Li $1s^23p-1s^22p$ line by reducing the intensity of the Li $1s^23d-1s^22p$ line. Based on these considerations, an electron density in the order of 1×10^{17} cm^{-3} can be estimated for the early-in-time and close-to-the-surface plasma.

Later and further away from the surface, the spectra reveal other interesting properties of the plasma plume dynamics. A plasma fluid element has a plasma-plume expansion velocity of 10 μm/ns, which can be extracted from the experimental CCD images. This time scale is comparable to the time scales associated with the rates of atomic processes that control the level population distribution; particularly, at densities in the 1×10^{17} cm^{-3} to 1×10^{15} cm^{-3}. Hence, time-dependent effects can be expected to become more important as the plasma expands.

To test this idea, a time-dependent version of the atomic kinetics model was created. It was driven by temperature- and density- time histories that were motivated by the time-evolution of fluid elements observed in hydrodynamic simulations of this experiment. Fig. 7 shows the time evolution of neutral Li total fractional population (i.e. ground plus excited states) as produced by a temperature profile that drops linearly from 1 eV to 0.4 eV and a density profile that drops exponentially from 1×10^{17} cm^{-3} to 1×10^{15} cm^{-3}, over a time interval of 80 ns (t = 20 ns to t = 100 ns). The initial condition (i.e. 1 eV and 1×10^{17} cm^{-3}) is characteristic of the state of a plasma fluid element that has just been formed close to the surface.

For comparison, Fig. 7 also displays the results of the steady-state calculation driven by the same set of instantaneous temperature and density points. The results in Fig. 7 clearly show that time-dependent effects on the atomic kinetics are indeed important, and that the time-dependent result always lags behind that of the steady-state case. Initially, the sudden drop in density effectively drives ionization due to the weakening of three-body (collisional) recombination rates. Subsequently, the temperature drop dominates and the system recombines. In the time-dependent calculation this recombination proceeds at a much slower rate as compared to the steady-state case that instantaneously adjusts to the local temperature and density conditions.

These results are encouraging. They point to the fact that the combination of time- and spatially-resolved spectroscopic measurements and detailed spectral modeling and simulation constitute a powerful technique for the characterization of plasma plume formation and dynamics.

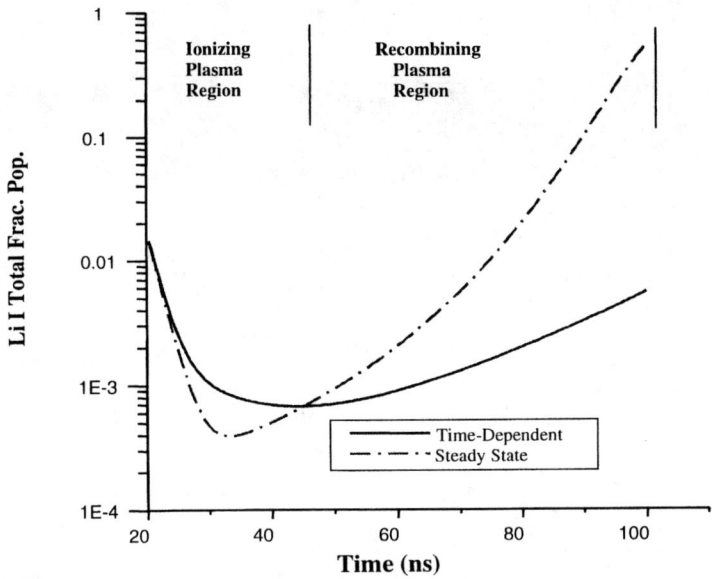

FIGURE 7. Comparison of total fractional population of neutral Li for steady-state and time-dependent simulations.

REFERENCES

1. Radziemski, L. J., and Creamers, D. A., editors. *Laser-Induced Plasmas and Applications*, New York, New York: Marcel Dekker, 1989, pp. 1-67.
2. Miller J. C., and Haglund, R. F., editors. *Laser Ablation and Desorption,* New York, New York: Academic Press, 1998, pp. 255-298.
3. Herbst, M. J., Burkhalter, P. G., Grun, J., Whitlock, R. R., and Fink, M., *Rev. Sci. Instrum.* **53**, 1418-1422 (1982).
4. Abdallah, J., Jr., Clark, R.E.H., and Cowan, R. D. *Los Alamos National Laboratory Report* LA-11436 M vol. I (1988).
5. Clark, R.E.H., Abdallah, J., Jr., Csanak, G., Mann, J.B., and Cowan, R.D., *Los Alamos National Laboratory Report* LA-11436 M vol. II (1988).
6. Cowan, R.D., *The Theory of Atomic Structure and Spectra,* Berkeley, California: University of California Press, 1981.

X-treme Diagnostics for HID Lamps?

Helmar G. Adler

OSRAM SYLVANIA
Lighting Research Center, 71 Cherry Hill Drive, Beverly MA 01915

Abstract. X-ray induced fluorescence has been used to measure elemental number densities of all constituents in a ceramic metal-halide discharge lamp. High-energy synchrotron radiation generated at the Sector 1 Insertion Device beam line at the Advanced Photon Source induced K-shell fluorescence in the high-pressure arc plasma. High signal-to-noise fluorescence spectra of Hg, Dy, Cs and I could be simultaneously obtained. A method for absolute calibration enabled us to map density distributions of all constituents throughout the whole lamp with a spatial resolution of 1 mm^3. Elemental ratios show clear evidence of radial de-mixing of metal additives. A possible explanation for the on-axis depletion is ambipolar transport of positive ions out of the center of the arc. The highly successful experiments demonstrate the utility of modern x-ray sources and methods for diagnostics as new tools for improved understanding of high-pressure arc discharges.

INTRODUCTION

High intensity discharge (HID) light sources, especially those based on the principle of the metal halide lamp, play an ever-increasing role in lighting applications. These lamps have reached a development level at which their high efficiency means a strong economic value with regard to energy consumption. Typically HID lamps are used in areas of General Lighting (indoors and outdoors), in photo-optics and more and more in automotive applications.

In the past most lamp developments were based on good and sometimes even ingenious engineering approaches. To allow faster tailoring of lamps to specific needs and guidance in lamp design, a better understanding of the underlying principles is still desired. A substantial number of diagnostics in lamp research to measure particle density and temperature distributions are based on optical techniques, utilizing emission or absorption characteristics of the arc discharges. These approaches can be very fruitful if the arc tube material is made of transparent quartz that allows good optical imaging with sufficient spatial resolution. Very often results are confined to the arc core region since probing of the outer mantle proves to be more difficult due to the closeness of the arc tube wall with its added optical distortions. More recently, the introduction of polycrystalline alumina (PCA, chemically Al$_2$O$_3$) as an arc tube material for many metal halide lamps has added another obstacle to optical diagnostics. PCA is - opposed to transparent quartz - only translucent, and optical measurements requiring spatial resolution are usually impossible due to its strong scattering characteristics.

CP635, *Atomic Processes in Plasmas: 13th APS Topical Conference*, edited by D. R. Schultz et al.
© 2002 American Institute of Physics 0-7354-0090-3/02/$19.00

To overcome these difficulties we have - in a collaborative effort between the University of Wisconsin and OSRAM SYLVANIA - explored alternative diagnostics utilizing x-rays. In particular we could demonstrate very successfully the feasibility of x-ray induced fluorescence and absorption imaging to probe species distributions in operating HID lamps. This contribution will describe some of our studies at the Advanced Photon Source (APS), Argonne National Laboratory.

HIGH PRESSURE ARC LAMP

A typical high-pressure arc lamp as we have used for our experiments is shown in Figure 1. The lamp consists of a cylindrical ceramic arc tube (polycrystalline alumina, PCA) with a total length of approximately 2 cm and a diameter of 0.9 cm. The arc length is defined by the spacing of the two tungsten electrodes, here 0.85 cm. The sealed arc tube is filled with 16 mg of Hg, 6.8 mg of DyI_3, 1.2 mg of CsI and 90 Torr of Ar. The arc tube is mounted inside a low expansion vacuum jacket to prevent oxidation of the electrical mechanical connections, to reduce heat loss by the arc tube, and to absorb ultraviolet radiation from the arc plasma. We operated the lamp with a 300 Hz square-wave ballast at a constant power of 150 W.

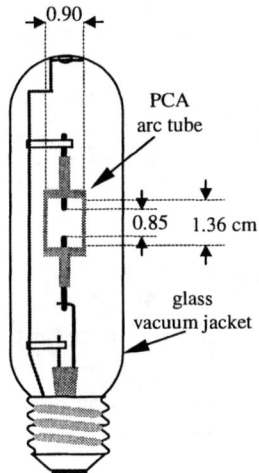

FIGURE 1. Basic construction of the 150 W ceramic metal-halide lamp used in the experiments.

X-RAY INDUCED FLUORESCENCE

Recently x-ray absorption imaging has been applied to measure the distribution of Hg in the gas-phase of an operating high-pressure arc [1]. Maps of Hg densities and gas temperature were derived. In addition to Hg, metal-halide (MH) discharge lamps contain other elements, such as Cs or rare-earth elements, which enhance the color properties and efficacy of these lamps. To get a better understanding of the complex transport mechanisms in MH discharges, it is desirable to determine the densities of the minority elements as well. As mentioned in the introduction, optical diagnostics are often not capable of delivering these measurements, in particular not in lamps with ceramic (PCA) arc tubes. Fohl et al. made attempts earlier to use x-ray induced fluorescence to detect additives in high-pressure arc lamps, but did not publish results [2].

In our experiments we were able to use an energy-dispersive fluorescence detector and higher incident photon energy. By these means we could separate element characteristic fluorescence radiation from Compton and elastic scattering [1].

Experimental Arrangement

All experiments were performed at the APS at the Sector 1 Beam Insertion Device (1-ID) beam line that is operated by the Synchrotron Radiation Instrumentation Collaborative Access Team (SRI-CAT). At this location a mono-energetic x-ray beam is available that is adjustable from 45 keV to 100keV with an energy spread of $dE/E=10^{-3}$ and a photon flux on the order of 10^{12} photons/s/mm^2 [3].

In analogy to laser-induced fluorescence experiments we set up a crossed-beam experiment, as shown in Figure 2. We mounted the discharge lamp on a computer-controlled x-y-z stage to allow positioning of the lamp with high precision. The cross-section of the x-ray beam was limited by a pair of precision tungsten-slits to a 1.0 mm by 1.0 mm square. We restricted the field-of-view of the cryogenic Ge-detector with two apertures to observe an approximately 1 mm long window along the beam. Thus the resulting spatial resolution of the measurement amounts to about 1 mm^3. The ionization chamber allows one to measure the incoming x-ray photon flux.

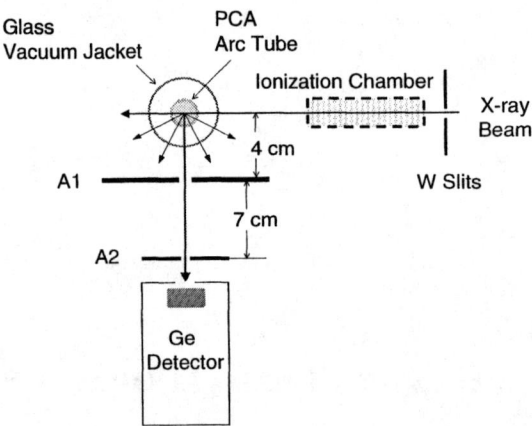

FIGURE 2: The experimental arrangement for x-ray induced fluorescence measurements in an operating HID-lamp. The x-ray beam is produced by the "Sector 1 Insertion Device" beam line at the Advanced Photon Source.

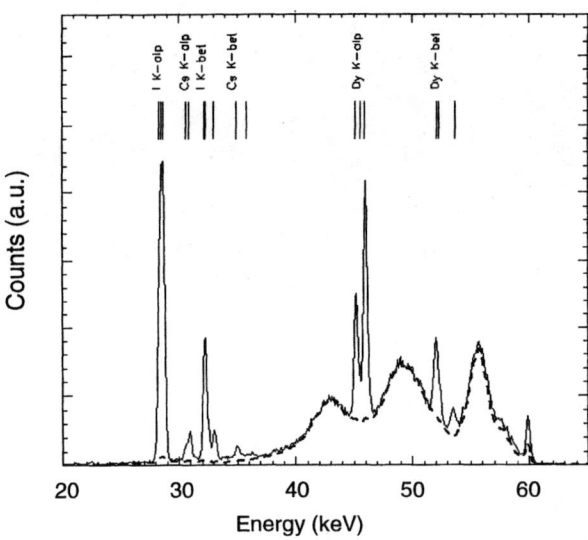

FIGURE 3. X-ray induced spectra of a 150 W ceramic metal-halide lamp; lamp off (dashed line), lamp on (solid line). Characteristic K-shell fluorescence lines from I, Cs and Dy, as well as Compton scattering from the arc tube material, and elastic scattering from the lamp envelope and Hg vapor can be seen.

A typical spectrum from an operating lamp, here at an excitation beam energy of 60 keV and an integration time of 200 s, is shown in Figure 3. A spectrum from a cold lamp is overlaid with a dashed line. All features in these spectra can be attributed to well-known processes.

Interpretation of Measured Spectra

The spectra in Figure 3 display a number of characteristic features. Element-specific fluorescence lines from Dy, Cs and I are easily identified. A small coherent scattering peak is noticeable at the incident photon energy of 60 keV. There appear to be some broadband features that are worth further explanation. Direct scattering from the solid arc tube can be excluded, since that could only be detected if the incident beam struck the arc tube within the narrow field of view of the detector. In general, the beam entered and exited the arc tube outside the field of view of the detector. However, there are additional paths possible for multiple Compton scattering from the arc tube wall, see Figure 4.

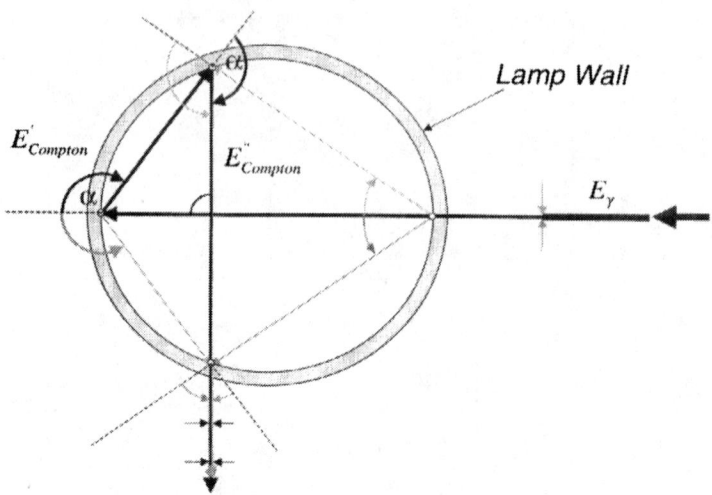

FIGURE 4. Enlarged cross-section of the arc tube and paths for double Compton scattering. The x-ray beam is entering from the right with energy E_γ, the detector would be at the bottom of the drawing, similar to Figure 2. Four paths are possible; application of Equation 1 twice for each path (double Compton-scattering) results in three different energy values, thus leading to three peaks in the measured x-ray spectrum.

In the Compton effect, an incident photon with an energy E_γ collides with a free or weakly bound electron. The scattered photon suffers a shift in energy that depends on the scattering angle α and the energy $E\gamma$. Applying principles of conservation of momentum and energy, the energy of the scattered photon can be calculated according to:

$$E_{Compton} = \frac{E_\gamma}{[1 + \frac{E_\gamma}{m_e c^2} \cdot (1 - \cos\alpha)]}. \tag{1}$$

To calculate the energy of photons reaching our detector after being Compton-scattered twice at the arc tube wall we applied Equation 1 twice. As can be seen from the geometry of the experimental set-up, see Figures 2 and 4, four paths are possible with two energies resulting in the same value. Thus three peaks appear in the measured spectra. We changed the energy of the incoming beam to four different values to test the validity of the double Compton scattering assumption. Figure 5 displays the resulting spectra. The shifts of the peak energies are clearly visible. Calculated results for the shifted scattering energies as a function of incoming beam energy show the good correspondence to the measured values in Figure 6. The peaks are broadened due to the thickness of the arc tube wall (about 1 mm).

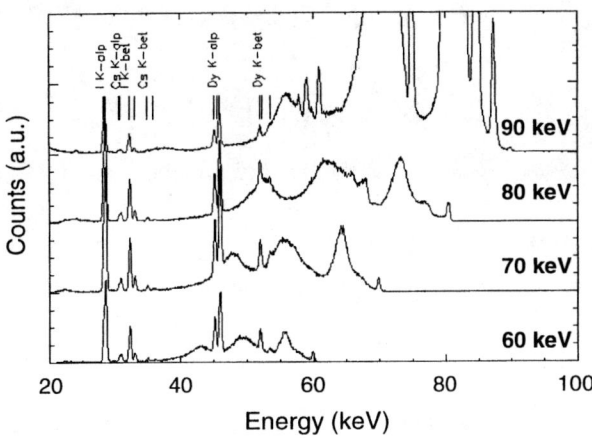

FIGURE 5. Measured x-ray spectra at four different incoming beam energies: 60, 70, 80 and 90 keV. Very strong Hg fluorescence lines appear at beam energies of 90 keV since K-shell fluorescence lines from Hg are not excited below 83.1 keV. Indicated are the positions of K-shell fluorescence lines for I, Cs, Dy and Hg.

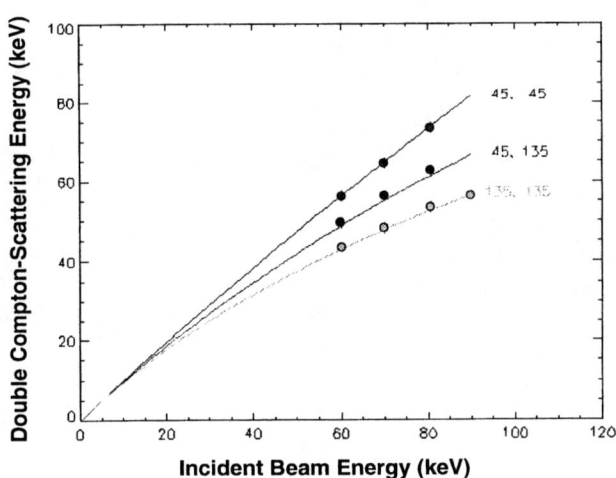

FIGURE 6. Energies of double-Compton-scattering peaks, incoming x-ray beam centered in lamp: calculations (lines), peak positions from measurements in Figure 5 (circles). The numbers close to the curves indicate the values for first and second scattering angle for this particular geometry.

MAPPING OF ELEMENTAL DENSITIES

Absolute Calibration

One of our goals was to determine the density distributions of various species in the gas phase, requiring an absolute calibration of the measurement system. In principle, the K_α fluorescence signals counted by the detector are related to the density N_a of an element a by

$$C(E_f) = \Phi_i \sigma_a^K(E_i) Y_a^K B_a^{K_\alpha} \frac{\Omega}{4\pi} VT(E_f) N_a , \qquad (2)$$

as shown in our first paper [1]. Here $C(E_f)$ is the number of counts per second integrated across an entire K_α fluorescence line feature at energy E_f; Φ_i is the incident photon flux density at the energy E_i; $\sigma_a^K(E_i)$ is the cross-section for K-shell absorption of species a at energy E_i; Y_a^K and $B_a^{K\alpha}$ are K-shell fluorescence yield and the fraction of K-shell fluorescence emitted on the α-lines, respectively, for each element a; V is the volume of the incident beam viewed by the detector; $\Omega/4\pi$ is the angular collection efficiency of the detector; $T(E_f)$ is the transmission of the fluorescence photons through the arc envelope and outer jacket. Absorption cross-sections, fluorescence yields and branching fractions are readily available from literature and, in our days, from the Internet. The angular collection efficiency of the detection system and the viewing volume can, in principle, be estimated from the geometry of the experimental set-up. However, since these estimates are only possible with relatively large uncertainties, we chose a different method for absolute calibrations. We fabricated a number of "fluorescence cells" from the same arc tubes as used for the lamps. The cells were filled with aqueous dilutions from standard solutions of various constituents, such as Dy, providing elemental densities with accurately-known values (for example: 1×10^{17} atoms/cm^{-3}, a typical value found in arc lamps). For the absolute calibrations we replaced the lamp arc tube with a "fluorescence cell". Fluorescence counts from these cells can be directly combined with Equation 2 to determine the angular collection efficiency and the collection volume.

Mapping

Probing of different volume elements in the lamp was achieved by moving the lamp in respect to the stationary incoming x-ray beam and the detector system. Due to the cylindrical symmetry of the arc discharge investigated, only axial and (one) radial movement was necessary. With only limited beam time available experimental procedures have to be planned carefully. We chose a distribution of probing locations according to Figure 7, which provided us with fairly complete maps of the whole interior of the lamp.

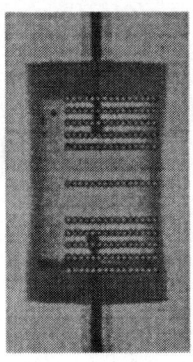

FIGURE 7. Distribution of probing locations for x-ray induced fluorescence measurements.

Two sets of data were collected, one for incident photon energies of 70 keV and another for 85 keV. 70 keV photons do not excite the Hg fluorescence and can be used to determine the additive densities, Dy, Cs and I, with increased sensitivity. Due to the much higher density of Hg compared to the additives, the Hg fluorescence became so strong at 85 keV that we had to reduce the incident photon flux by a factor of 10 to avoid detector saturation.

A side-by-side comparison of an x-ray image of a cold lamp, a false-color infra-red image of an operating lamp and a false-color image of the Hg distribution, determined by the x-ray induced fluorescence methods explained above, is quite instructive and is displayed in Figure 8. The infrared image gives an indication of the temperature distribution of the arc vessel. The hottest temperature appears to be shifted upward from the center of the arc due to convection effects. Similar convective behavior is found in the distribution of Hg where the blue color (dark gray in b/w) indicates the lowest density. The highest Hg concentrations (white) are found in the lower corners of the cylindrical lamp. Similarly, density distributions of Dy, Cs and I have been obtained.

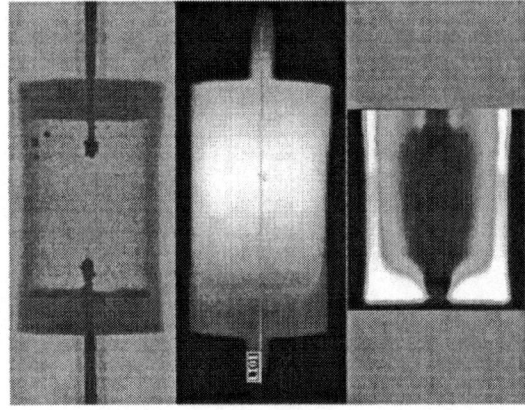

FIGURE 8. A side-by-side comparison of an x-ray image of a cold lamp, a false-color infrared image of an operating lamp and a false-color image of the Hg distribution.

ELEMENTAL RATIOS AND DE-MIXING

Many arc models are based on the assumption of thermodynamic equilibrium in which the vapor is fully mixed and characterized by spatially homogeneous elemental ratios. When we examined radial density profiles and elemental ratios in the midplane of the arc we found a little surprise [4]. By dividing the additive densities by the Hg density we essentially eliminate the 1/T variations (T is the gas temperature) of the additive densities. Here we assume that the Hg density is varying with the local gas temperature according to the Ideal Gas Law, an assumption that is probably valid for these high-pressure arc discharges.

Fully mixed vapors would produce constant ratios across the radius, as can be seen for the ratio of I:Hg. However, in the case of Dy:Hg, and in particular for Cs:Hg, the ratios show depletion in the arc core relative to mantle and wall values. Figure 9 displays the results for I:Hg and Cs:Hg. The depletion of Cs amounts to more than 60 %. A possible explanation for these de-mixing and depletion effects is ambipolar diffusion of the metal ions from the core of the arc. The ionization potentials of Dy and Cs are 5.94 and 3.89 eV, respectively, thus Cs might show higher depletion due to the lower value of its ionization potential.

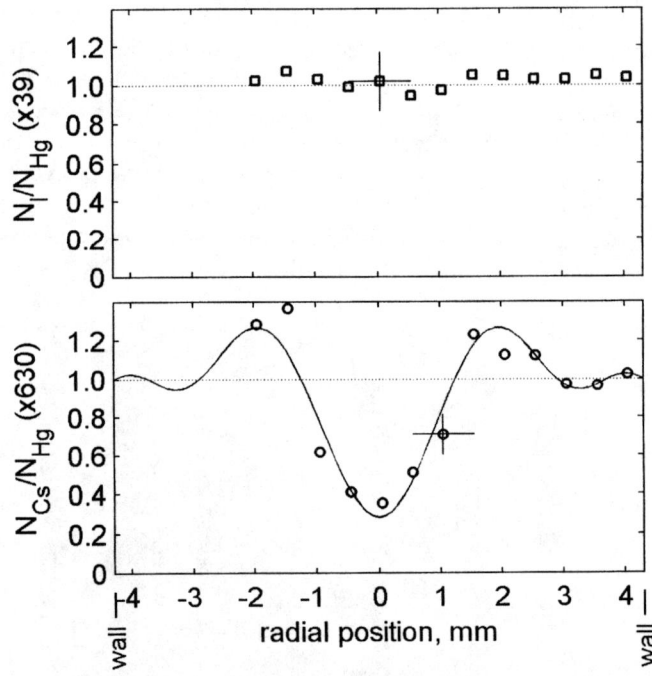

FIGURE 9. Elemental ratios at the midplane of the arc: I:Hg (top) and Cs:Hg (bottom) showing de-mixing effects and strong depletion of Cs from the arc core.

SUMMARY AND OUTLOOK

X-ray induced fluorescence measurements at the Advanced Photon Source have been used to determine "complete" elemental density distributions in operating high-pressure arc discharges. Radial de-mixing effects that are difficult to examine by other means were found and have the potential to trigger enhancements in lamp arc modeling. Future experiments could be geared to improve spatial, time and energy resolution to probe molecular species.

Even if the current experiments might appear to be on the extreme side for a lamp producing company they have opened the door for more, important research opportunities, in particular, but not only, for ceramic lamps.

ACKNOWLEDGMENTS

The success of these experiments was only possible due to the close collaboration of James E. Lawler's group at the University of Wisconsin and OSRAM SYLVANIA, with the very efficient support of the SRI-CAT team at the Advanced Photon Source. In particular I would like to mention John J. Curry (UW), who now has a position at NIST, Gaithersburg, and S. D. Shastri and W.-K. Lee (both APS). Additionally, discussions with OSI colleagues T. Dreeben, W. Lapatovich and R. Snellgrove and the support of J. Greene for preparation of the standard solutions are gratefully acknowledged.

Use of the Advanced Photon Source was supported by the U.S. DOE, Basic Energy Sciences, under contract No. W-31-109-Eng-38.

REFERENCES

1. Curry, J. J., Adler, H. G., Shastri, S. D., and Lawler, J.E., *Appl. Phys. Letter.* **79**, 1974-1976 (2001).
2. Fohl, T., Kramer, J. M., and Lester, J. E., *J. Appl. Phys.* **73**, 46 (1993).
3. Shastri, S.D., Mashayekhi, A., Fezzaa, K., Lee. W.-K., Fernandez, P.B., Tajiri, C., Ferguson, D. A., and Lee, P. L., APS Activity Report No. 1999, Argonne National Laboratory Report No. ANL-00/5 (2001) 258-259.
4. Publication in preparation.

Spectroscopic Characterization of Post-Cluster Argon Plasmas During the Blast Wave Expansion

H.-K. Chung*, K. B. Fournier*, M. J. Edwards*, H. A. Scott*, R. Cattolica[†], T. Ditmire** and R. W. Lee*

*Lawrence Livermore National Laboratory, Livermore, CA 94550
[†]University of California-San Diego CA 92093
**University of Texas, Austin, TX 78731

Abstract. In this work we present temperature diagnostics of an expanding laser-produced argon plasma. A short-pulse (35fs) laser with an intensity of $I = 10^{17}$ W/cm deposits ~ 100 mJ of energy into argon clusters. This generates a hot plasma filament that develops into a cylindrically expanding shock. We develop spectral diagnostics for the temperatures of the argon plasma in the shock region and the preionized region ahead of the shock.

A collisional-radiative model is applied to explore line intensity ratios derived from Ar II-Ar IV spectra that are sensitive to temperatures in a few eV range. The results of hydrodynamic simulations are employed to derive a time dependent radiative transport calculation that generates the theoretical emission spectra from the expanding plasma.

INTRODUCTION

Laser interaction with gas clusters of atoms provides a unique opportunity for studies of high temperature plasmas. This is due to the high absorption efficiency of laser light, in the range of 50%~100% [1] and the efficient transfer of absorbed laser energy into the kinetic energies of ions as well as electrons [2]. The strong absorption of 10-100 fs laser light by clusters shows great promise in the development of soft X-ray sources [3] and in the production of fast ions and electrons [4]. Moreover, an interesting application of the cluster-laser interaction arises in the study of laser-driven blast waves in rare gases that have parameters relevant to the evolution of astrophysical shocks [5, 6].

In the present laser-generated blast wave experiments, clusters are produced by discharging argon from a high-pressure gas jet with a backing pressure of 1000 psi and letting it expand adiabatically. A 35 fs laser pulse of 100 mJ is focused to irradiate a column 4mm (FWHM) in length with a 25 μm radius, and leads to ~ 50% absorption, i.e., ~ 12.5 mJ/mm energy deposition. The cylindrical shock formed in the gas by laser absorption is observed from the radial and temporal distributions of electron densities measured by interferometry [5, 6]. The peak electron densities in the compressed region are found to decrease from ~ 10^{19} cm^{-3} at the time of shock formation, 6-7 ns after the laser shot, to 10^{18} cm^{-3} at the end of blast wave expansion, ~ 200 ns later. Electron temperatures of the expanding plasmas also evolve spatially and temporally and are expected to decrease sharply from an initial value of $\geq$ 100 eV to ~ 2 eV as the shock dissipates. In the present work, temporal and spatial spectra from the expanding plas-

CP635, *Atomic Processes in Plasmas: 13th APS Topical Conference*, edited by D. R. Schultz et al.
© 2002 American Institute of Physics 0-7354-0090-3/02/$19.00

mas are recorded in the ultraviolet spectral range between 230 nm - 360 nm with 2 ns resolution.

We develop temperature diagnostics of the argon plasma from the UV spectra. As the plasma evolves in space and time atoms and ions experience distinct phases with respect to kinetic processes: 1) during the shock compression level populations are predominantly governed by collisions and 2) during the preheating, ahead of the shock, radiative processes are also important. A collisional-radiative model is applied to explore line intensity ratios in the Ar II - Ar IV spectra that are sensitive to temperatures in a few eV range. We present theoretical predictions of population ratios as a function of temperature relevant to the blast-wave plasmas employing the results of hydrodynamic simulations.

COLLISIONAL-RADIATIVE MODEL

We use a collisional-radiative model (CRM) to develop an electron temperature diagnostic [7]. Time-dependent level population distributions are obtained by solving rate equations incorporating collisional and radiative rates. The rate equations and radiative transport equation are solved self-consistently with the radiative transport code CRETIN [8]. For collisional processes, we include thermal electron excitation and de-excitation, ionization and three-body recombination processes, while for radiative processes we included spontaneous emission, as well as stimulated emission and absorption, photoionization, radiative recombination and dielectronic recombination processes. Since the plasma expands cylindrically, level population distributions are solved as a function of radius as well as time. As an input parameter, electron temperatures as a function of time and radius are taken from hydrodynamic simulations (HYADES) [9] and corresponding charge states and electron densities are computed from CRETIN's population densities. We use data from an atomic model that was generated at the University of Wisconsin at Madison for ion beam transport experiments [10, 11].

TEMPERATURE DIAGNOSTICS OF EXPANDING ARGON PLASMAS

Laser-produced argon plasmas during blast wave expansion experience distinct time histories of temperature and density. This is depicted in Figure 1 by plotting electron temperature and density as a function of radius at a given time obtained from a hydrodynamic simulation with HYADES [9]. Before the cylindrical shock wave arrives, atoms are preionized by radiation emitted from the shock, this is the preionized region. Once compressed, the region is heated rapidly, this is the shock region. As the shock wave passes by, the ion mass density decreases rapidly and the temperature decreases slowly to form the cavity region.

Robust temperature diagnostics using UV spectra can be developed by exploiting level population distributions as a function of temperature. We plot emission spectra from the expanding argon plasma at 6 ns and 20 ns in Figure 2. It shows that there are many Ar II

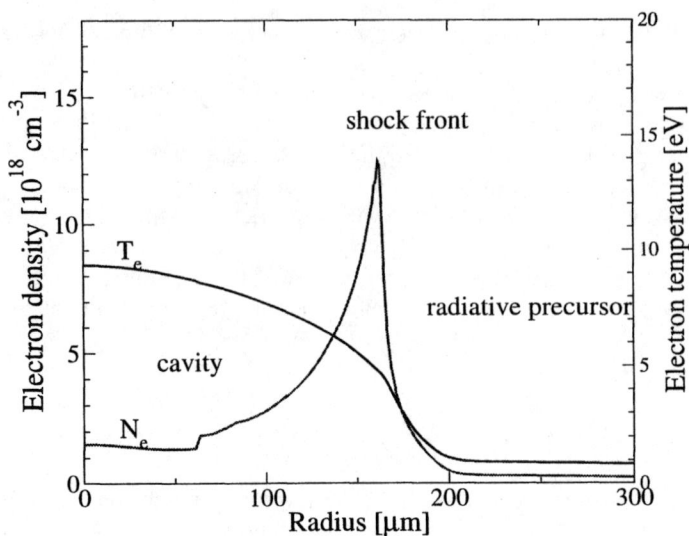

FIGURE 1. Spatial electron density and temperature distributions of plasma experiencing preionization, shock, cavity phases at a given time, 10 ns after the pulse in this case

and Ar III lines present and potentially they can be used for temperature diagnostics. Emission spectra can be categorized based on relative temperatures of the radiating regions: hotter shock and post-shock regions and the colder preionized region. The level population distribution is dominated by collisional processes in the hot regions, where ion and electron densities are of the order of 10^{19} cm^{-3} and the electron temperature is ≥ 2 eV. On the other hand, in the cold preionized region, radiation from the existing hot region plays an important role in the charge state distribution and detailed level populations. Here we present T_e diagnostics for these two different regions.

Shock and post-shock region

As temperatures are greater than 2 eV in the shock and post-shock regions, emission from Ar II to Ar IV lines is observed in the spectral range of 230 nm - 360 nm. In this case, the mean charge state moves to more than singly ionized and the electron density is sufficiently high for collision processes to be predominant. We find that Ar II and III spectra from the shock and post-shock region arise predominantly from line radiation of transitions of $3p^44p$–$3p^43d$, $3p^44p$–$3p^44s$, $3p^44d$–$3p^44p$ and $3p^45d$–$3p^45p$ levels. The energy spread of upper levels in strong transitions is 7.7 eV for Ar II and 8.1 eV for Ar III. Therefore, using Ar II and Ar III line intensities one can expect to have a good T_e diagnostic up to $T_e \sim 8$ eV. An example of a population ratio of two upper levels is shown as a function of temperature in Figure 3. Combining several populations ratios with different energy spacings, one can have a robust T_e diagnostic of plasmas in the shock and post-shock region.

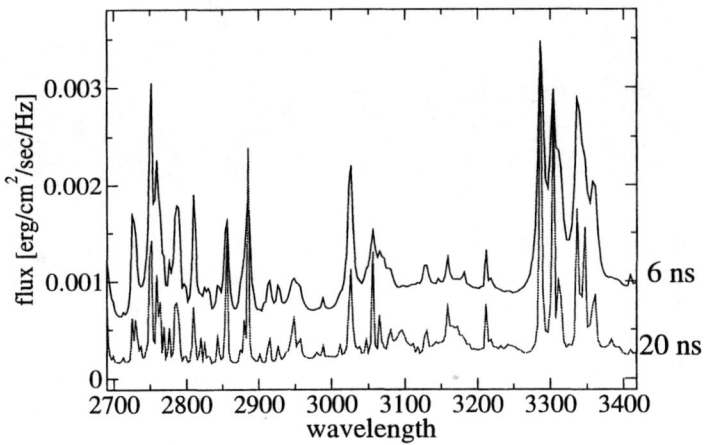

FIGURE 2. Emission flux calculations at 6 ns and 20 ns calculated at 500 μm away from the center

Preionized region

A layer ahead of the shock front is photoionized by radiation from the hot inner region before the shock wave arrives. The average charge state in the preionized region closely follows that of the shock due to photoionization by radiation from the shock region even though the electron temperature is much lower. Emission from this region will be affected by non-local plasma conditions, that is, T_e of the hot region. We tested time-dependent and steady-state models and found that the two results are in agreement after the first nanosecond. During the first nanosecond, the cold gas at the ambient temperature is being ionized and its level population is yet to evolve to steady-state.

Since the strong photoexcitation dominates the level population distributions within the same spin system, relative population densities of levels with the same spin represent non-local properties of radiation flux. Relative population distributions of those levels with different spins, however, represent collisional properties as each spin system couples to the thermal bath of local temperature.

A sample of a population ratio of Ar II upper levels, one belonging to quartet system and the other to doublet system is shown as a function of temperature in Figure 4. In the figure, we plot time-dependent (symbols) and steady-state (lines) population density ratios from different radial positions in the plasma. The few below the dominant curve represent the ratios in the first nanosecond in the time-dependent case before level populations reach a steady-state. Since temperature information of the radiating hot region is readily available from line diagnostics discussed in the previous section, local T_e diagnostics of preionized region will be also possible from steady-state results incorporating non-local radiative processes.

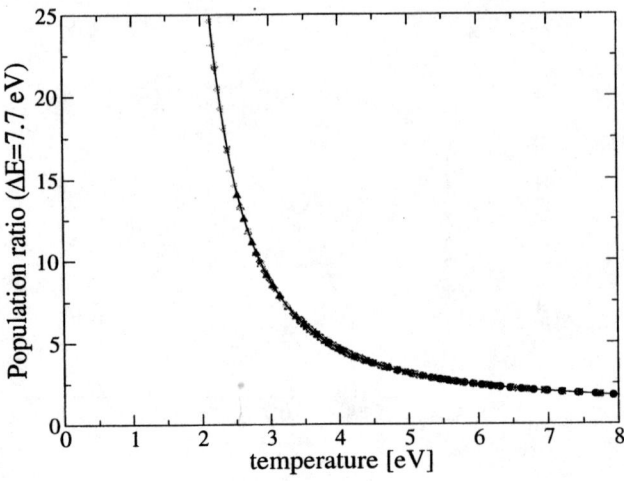

FIGURE 3. Ratio of population densities as a function of temperature between Ar II $3p^4 4p$ ^2P and $3p^4 5d$ ^2D levels.

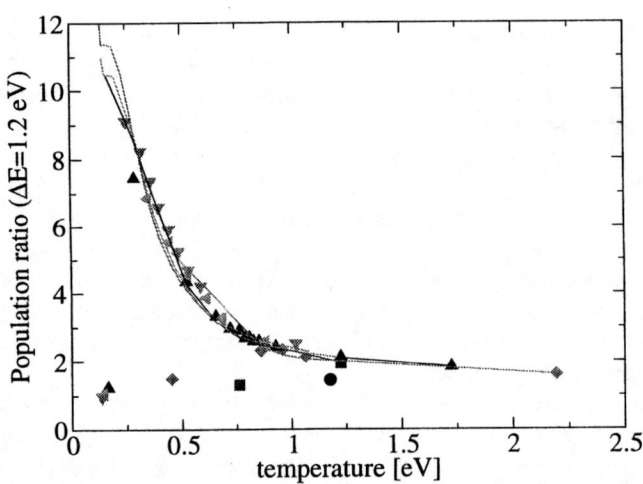

FIGURE 4. Ratio of population densities in Ar II $3p^4 4d$ ^4D and $3p^4 4d$ ^2D levels as a function temperature in comparison between time-dependent(symbols) and steady-state (lines) results.

SUMMARY AND FUTURE WORK

We have shown that line intensity ratios can be used as a local temperature diagnostic for this expanding argon plasma. While collisional processes are dominant in shock and post-shock region, radiative processes play a critical role in preionized region. Accounting for the difference in dominant atomic processes, a collisional-radiative model can be applied for T_e diagnotics of both regions. We will apply the developed T_e diagnostics for analysis of measured argon spectra in the spectral range of 230 -360 nm.

ACKNOWLEDGMENTS

This work was performed under the auspices of the U.S. Department of Energy by University of California Lawrence Livermore National Laboratory under contract No. W-7405-Eng-48.

REFERENCES

1. Ditmire, T., Smith, R., Tisch, J., and Hutchinson, M., *Physical Review Letters*, **78**, 3121 (1997).
2. Ditmire, T., Tisch, J., Springate, E., Mason, M., Hay, N., Marangos, J., and Hutchinson, M., *Physical Review Letters*, **78**, 2732 (1997).
3. Ditmire, T., Donnelly, T., Falcone, R., and Peny, M., *Physical Review Letters*, **75**, 3122 (1995).
4. Zweiback, J., Smith, R., Cowan, T., Hays, G., Wharton, K., Yanovsky, V., and Ditmire, T., *Physical Review Letters*, **84**, 2634 (2000).
5. Shigermori, K., Ditmire, T., Remington, B. A., Yanovsky, V., Ryutov, D., Estabrook, K. G., Edwards, M. J., MacKinnon, A. J., Rubenchik, A. M., Keilty, K. A., and Liang, E., *The astrophysical journal*, **533**, L159–L162 (2000).
6. Edwards, M., MacKinnon, A., Zweiback, J., Shigemori, K., Ryutov, D., Rubenchik, A., Keilty, K., Liang, E., Remington, B., and Ditmire, T., *Physical Review Letters*, **87**, 085004 (2001).
7. Mihalas, D., *Stellar Atmospheres*, W. H. Freeman and Company, San Francisco, 1978.
8. Scott, H., *Journal of Quantitative Spectroscopy and Radiative Transfer*, **71**, 689 (2001).
9. Larsen, J. T., and Lane, S. M., *Journal of Quantitative Spectroscopy and Radiative Transfer*, **51**, 179 (1994).
10. Wang, P., *Computation and application of Atomic Data for Inertial Confinement Fusion Plasmas*, Ph.D. thesis, Nuclear Engineering and Engineering Physics, University of Wisconsin-Madison, Madison, WI (1991).
11. Chung, H., MacFarlane, J., Wang, P., Moses, G., Bailey, J., Olson, C., and Welch, D., *Review of Scientific Instruments*, **68**, 350 (1996).

INERTIAL CONFINEMENT FUSION

Spectroscopic Determination of Core Gradients in Inertial Confinement Fusion Implosions

R.C. Mancini[a], L.A. Welser[a], I.E. Golovkin[b], Y. Ochi[c], K. Fujita[c],
H. Nishimura[c], R. Butzbach[d], I. Uschmann[d], E. Förster[d], F.J. Marshall[e],
J.A. Delettrez[e], J.A. Koch[f], H.E. Dalhed[f], R.W. Lee[f] and L. Klein[g]

[a]*Department of Physics, University of Nevada, Reno, NV 89557, USA*
[b]*Prism Computational Sciences, Madison, WI 53703, USA*
[c]*Institute of Laser Engineering, Osaka University, Osaka, JAPAN*
[d]*Institute of Optics and Quantum Electronics, Jena University, Jena, GERMANY*
[e]*Laboratory for Laser Energetics, University of Rochester, Rochester, NY 14623, USA*
[f]*Lawrence Livermore National Laboratory, Livermore, CA 94550, USA*
[g]*Department of Physics and Astronomy, Howard University, Washington, D.C. 20059, USA*

Abstract. We report on a collaborative effort that has led to the development of a spectroscopic method for the determination of the gradient structure in ICF implosion cores based on the self-consistent analysis of simultaneous X-ray monochromatic images and X-ray line spectra. This technique is applied to a series of stable and spherically symmetric implosion experiments where Ar-doped D_2-filled plastic shells were driven with the GEKKO and OMEGA laser systems. Argon K-shell X-ray line spectra were measured with streak crystal spectrometers, while X-ray monochromatic imagers recorded core images based on the Ar Heβ line. The analysis self-consistently determines the temperature and density gradients that yield the best fits to both the spatial distribution of monochromatic emissivity and spectral line shapes. A multi-objective genetic algorithm is used to efficiently perform the analysis. This measurement is critical for understanding the spectra formation and plasma dynamics associated with the implosion process. In addition, since the results are independent of hydrodynamic simulations they are important for the verification and benchmarking of detailed fluid dynamic models of high energy density plasmas.

INTRODUCTION

The determination of plasma core parameters during the implosion of a D_2-filled microsphere can provide critical information about the plasma state in the final stages of the implosion. A wide variety of particle- and radiation-based techniques are currently used to diagnose implosion dynamics and, in particular, to determine core conditions. In this connection, X-ray spectroscopic measurements have proved to be a very powerful diagnostic for these high-energy-density plasmas [1]. Over the years and with the aid of tracer elements, several types of emission and absorption X-ray spectral features have been observed and modeled in detail with the goal of understanding their plasma density and temperature sensitivity, and investigating their

CP635, *Atomic Processes in Plasmas: 13th APS Topical Conference*, edited by D. R. Schultz et al.
© 2002 American Institute of Physics 0-7354-0090-3/02/$19.00

potential for plasma spectroscopy diagnostics. Mostly, these X-ray spectroscopic diagnostics have relied only on the analysis of space-integrated spectra and have been applied to deduce emissivity-averaged or effective temperature and density [2-5]. However, when the line emission extends over the whole plasma source (i.e. implosion core) the usefulness of these measurements is questionable.

Since most of the diagnostic data from imploded cores has been used to estimate average or effective properties, the possibility of developing experimental and modeling techniques for the determination of core gradients has broad implications. On the one hand, it will provide a critical insight into the current capability to simulate the peak neutron production and peak compression stages of an Inertial Confinement Fusion (ICF) implosion using large-scale hydrodynamic codes. On the other hand, having a better characterization of the core plasma will help to address several issues of dense plasma spectroscopy with greater certainty, e.g. line formation, line shapes, level shifts and continuum lowering.

The formation and hydrodynamic evolution of the core plasma in ICF implosions depends on a combination of physical phenomena including laser-target coupling, thermal and radiation transport, compressible fluid dynamics, shock wave phenomena, and equation of state properties [6]. From the time when the main shock wave breaks into the core to the deceleration and stagnation stages, large-scale hydrodynamic simulations predict gradients in the core. The possibility of measuring these gradients offers a unique opportunity for a more thorough characterization of plasma dynamics, as well as an excellent source of data for more rigorous tests of hydrodynamic simulation codes.

CORE GRADIENT DETERMINATION

The determination of plasma temperature and density gradients in the core of an ICF implosion requires the self-consistent modeling and analysis of simultaneous X-ray line spectra and X-ray monochromatic images. This can be accomplished by doping the D_2-gas fill with small amounts of a suitable tracer element (e.g. Ar) that can provide adequate line radiation for the analysis without affecting the hydrodynamics. The Ar concentration should also be low enough to ensure that the Ar line emission is as optically thin as possible. Analysis of spatially-integrated line spectra results in spatially-averaged values of plasma density and temperature, and can provide no information concerning the gradients. As an illustration of this problem, Fig. 1 displays three combinations of one-dimensional (1-D) temperature and density gradients that result in almost identical space-integrated spectra. However, space distributions of line emissivity are different. Note that, although for illustrative purposes these calculations were performed for linear gradients, more complicated gradients may in actuality occur. Hence, to determine and characterize core gradients unambiguously, additional information has to be taken into account in the analysis.

Additional information for the characterization of core gradients can be obtained from the analysis of line-based X-ray monochromatic images. The two-dimensional monochromatic images provide a spatial map of the emissivity that is dependent on the gradients in temperature and density. In the direct- and indirect-drive implosions considered here, the monochromatic X-ray images are dominated by Ar Heβ ($1s^2$ ^1S-$1s3p$ ^1P) line emission and have a negligible contribution from continuum emission. Although the emissivity map provides important spatially-resolved information about the plasma source, it does not impose a sufficient constraint to provide spatial information on both temperature and density gradients. On the other hand, using the constraints imposed by self-consistently fitting the spectra and the spatially-resolved monochromatic emissivity at a given time provides, for that time, both the temperature and density as a function of spatial coordinate [7]. Thus, we can extract the electron temperature $T_e(r,t)$ and density $N_e(r,t)$ gradients from this analysis.

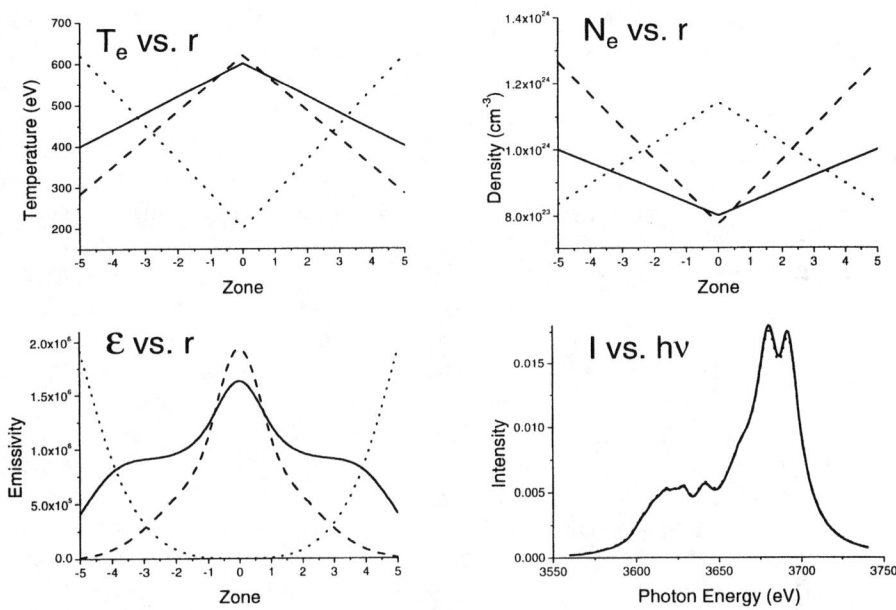

FIGURE 1. Ar Heβ line space-integrated spectra (in arb. units) and spatial emissivity profiles (in arb. units) for three combinations of temperature and density gradients. Zone in the plot refers to spatial zone. It is important to note that the space-integrated line emission spectra in the lower right would be observably the same for all cases. However, the space distributions of line emissivity are different.

Space-integrated time-resolved Ar Heβ, Heγ ($1s^2$ ^1S-$1s4p$ ^1P) and Lyβ ($1s$ ^2S-$3p$ ^2P) line spectra, and their associated Li- and He-like satellite lines, were recorded and used for the analysis. Opacity effects on these lines are smaller than for the Heα ($1s^2$ ^1S-$1s2p$ ^1P) and Lyα ($1s$ ^2S-$2p$ ^2P) lines; in addition, using a small concentration of dopant (Ar) further reduces possible opacity effects. To ensure that this condition is satisfied, the spectra were analyzed with a detailed Ar K-shell spectral model and code that can consider uniform and non-uniform plasmas, in the optically thin and optically thick approximations. Opacity effects in the model are taken into account by self-consistently solving the radiation transport equation and a set of collisional-radiative population kinetics equations [8]. First, for a given set of data, we perform the analysis of the spectra assuming that the plasma is uniform. This allows us to extract the emissivity-averaged electron temperature and density of the core, under the assumption that the lines are optically thin [8]. To check this assumption, the same spectrum is also analyzed considering uniform plasma conditions but in the optically thick approximation. The temperature and density extracted in this way are compared with those from the optically thin analysis. The difference between optically thin and thick analysis results can be used as a measure of the importance of opacity effects in the spectra. This idea was also successfully employed to systematically study opacity effects in the spectra from a series of Ar-doped implosions performed at the NOVA laser facility [9].

Next, we analyze the spectra using temperature and density gradients that are subject to the constraint of reproducing the value for the emissivity-averaged temperature and density obtained with the uniform model analysis. Further, the same gradients are also used to fit the 1-D spatially-dependent emissivity extracted from the analysis of the Ar Heβ X-ray monochromatic image. In this way, a set of self-consistent 1-D electron temperature and density gradients are extracted from the data that simultaneously fit the spectrum and the monochromatic emissivity spatial profile subject to the emissivity averaged constraints. The search in parameter space for suitable temperature and density gradient functions is performed with the aid of a genetic algorithm technique [10-12]. Due to the small concentration of Ar employed in these implosions, the Ar Heβ line remains optically thin ($\tau_o \leq 0.2$), and the emissivity profiles can be obtained from the monochromatic images using the Abel inversion procedure [13]. Although usually discussed for cases of cylindrical geometry, the Abel inversion method can also be considered for the case of spherical geometry [14]. We note that in Ar-doped deuterium plasmas electron and ion number densities can be considered equal since most of the electrons come from the ionization of deuterium. Hence, the electron density gradients extracted with this analysis are also characteristic of the ion number density (i.e. mass density) gradients.

EXPERIMENTS

We illustrate the application of the core gradient determination method with data from stable, low-convergence direct- and indirect-drive ICF implosions. The direct-drive implosion experiments were performed at the Osaka University GEKKO XII laser system [15]. The array of diagnostics instrumentation included a monochromatic X-ray framing camera, essential for the spatial resolution of the plasma [16]. For the first time in an implosion experiment, X-ray monochromatic images were used to obtain spatially and temporally resolved data on the collapsing core. In addition, time-resolved (but spatially-averaged) streak spectrograph spectral data for the usual spatially averaged analysis were also recorded. The drive consisted of a 12 beam, 2.55 kJ Nd glass laser operating at 526 nm. Random phase plates were used to smooth individual beams. The laser pulse was composed of a 0.2 ns pre-pulse followed by a 1.6 ns square pulse with rise time of 0.05 ns. The pre-pulse is important for the stabilization of the implosion [17]. Targets were plastic shells, 500 μm in diameter, with 8 μm wall thickness, filled with 30 atm of D_2 and doped with 0.075 atm of Ar. The implosion is diagnosed by recording both the compressed core image and the Ar K-shell line spectrum. In particular, time-resolved X-ray monochromatic images and simultaneous spatially integrated X-ray spectra of the Ar Heβ line were recorded. For the spatial information, a two-dimensional X-ray monochromatic framing camera imaging the central 19 eV of the Heβ line emitted by the Ar in the core was employed. This X-ray imager monitors the implosion symmetry and provides up to 5 frames with Δt=40 ps duration, 50 ps interframe time, and 10 μm spatial resolution [16]. Fig. 2 shows an example of a core image recorded by the X-ray imager in the time-interval 342-382 ps (i.e. before the peak of emission of the Heβ line). The image displays a central maximum in the intensity, as expected in a spherical implosion. The X-ray spectrometer consisted of a flat RbAP (100) crystal coupled to an X-ray streak camera with 10 ps time resolution and resolving power, $\lambda/\Delta\lambda$, of 600. It is important that the entire core of the implosion is in the field of view of the spectrograph so that a spatial average over the imploded core radius is recorded for the determination of both $<T_e(t)>$ and $<N_e(t)>$.

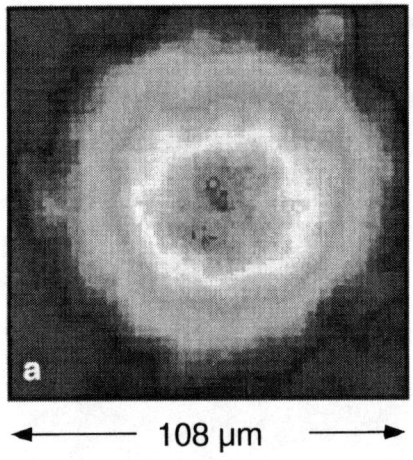

← —— 108 μm ——→

FIGURE 2. Implosion core Ar Heβ monochromatic image from direct-drive GEKKO shot 22091.

The indirect-drive implosion experiments were performed at the OMEGA laser facility of the Laboratory for Laser Energetics at the University of Rochester. The targets consisted of a Au-hohlraum with a plastic capsule inside. The Au-hohlraum was 2550 μm long, 1600 μm in diameter, and had 1200 μm Laser Entrance Holes (LEH). The plastic capsule had an external diameter of 510 μm with a wall thickness of 35 μm (i.e. initial core diameter is 440 μm), and it was placed at the center of the hohlraum. The capsules were designed to avoid burn through of the hohlraum X-ray radiation. Numerical simulation indicates that at the collapse of the implosion approximately 37% of the shell mass remained unablated and in some state of compression [18]. The core was filled with 50 atm of D_2 and 0.1 atm of Ar. The small (tracer) amount of Ar results in a typical optical depth of the Heβ line of 0.2, and even less for the Heγ and Lyβ lines. These hohlraum targets were irradiated with 30 UV OMEGA beams, split in 15 beams per LEH that were arranged in two cones of 5 and 10 beams each. The beam cones were pointed in such a way that they produced two rings of beams on each end of the hohlraum. The laser energy per beam was 500 J, for a total UV laser energy of 15 kJ. The hohlraum radiation temperature was 210 eV. Three diagnostic holes placed on the side walls of the holhraum provided lines-of-sight for a Gated-X-ray-Monochromatic–Imager (GMXI) [19], a new pinhole-array X-ray Multi-Monochromatic-Imager (MMI) [20], and a streaked X-ray crystal spectrometer. GMXI can record gated (Δt=80 ps) Ar Heβ images with a spatial resolution of 10 μm and a spectral resolution of 22 eV. MMI is a pinhole-array instrument that records numerous narrow-band (75 eV) X-ray images in the photon energy range from 3000 eV to 5000 eV with 10 μm spatial resolution. Data from MMI can be used to construct monochromatic images from several lines as well as continuum-based images. The streaked spectrometer uses a flat RbAP crystal, and has time and spectral resolutions of 30 ps and 500 (res. power), respectively. As an illustration of the monochromatic images recorded in the indirect-drive implosions, Fig. 3 shows a Heβ monochromatic image of the core recorded with GMXI in OMEGA shot 23686. In this case, the image is time-integrated over the period of emission of the Heβ line.

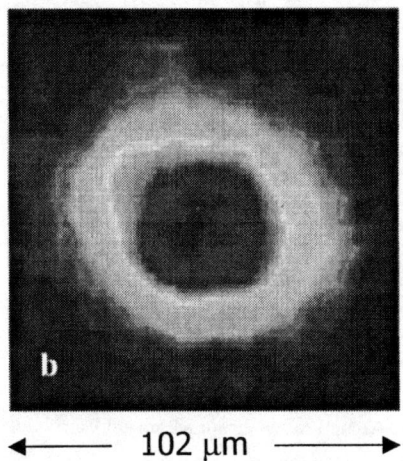

← —— 102 μm —— →

FIGURE 3. Implosion core Ar Heβ monochromatic image from indirect-drive OMEGA shot 23686.

ANALYSIS RESULTS

The problem of searching in parameter space for the gradients that produce the best self-consistent fits to the spectrum and the spatially-dependent emissivity, subject to the additional constraint of producing the correct emissivity-averaged values, is in general non-trivial. Our strategy is to use a "search engine" in parameter space driven

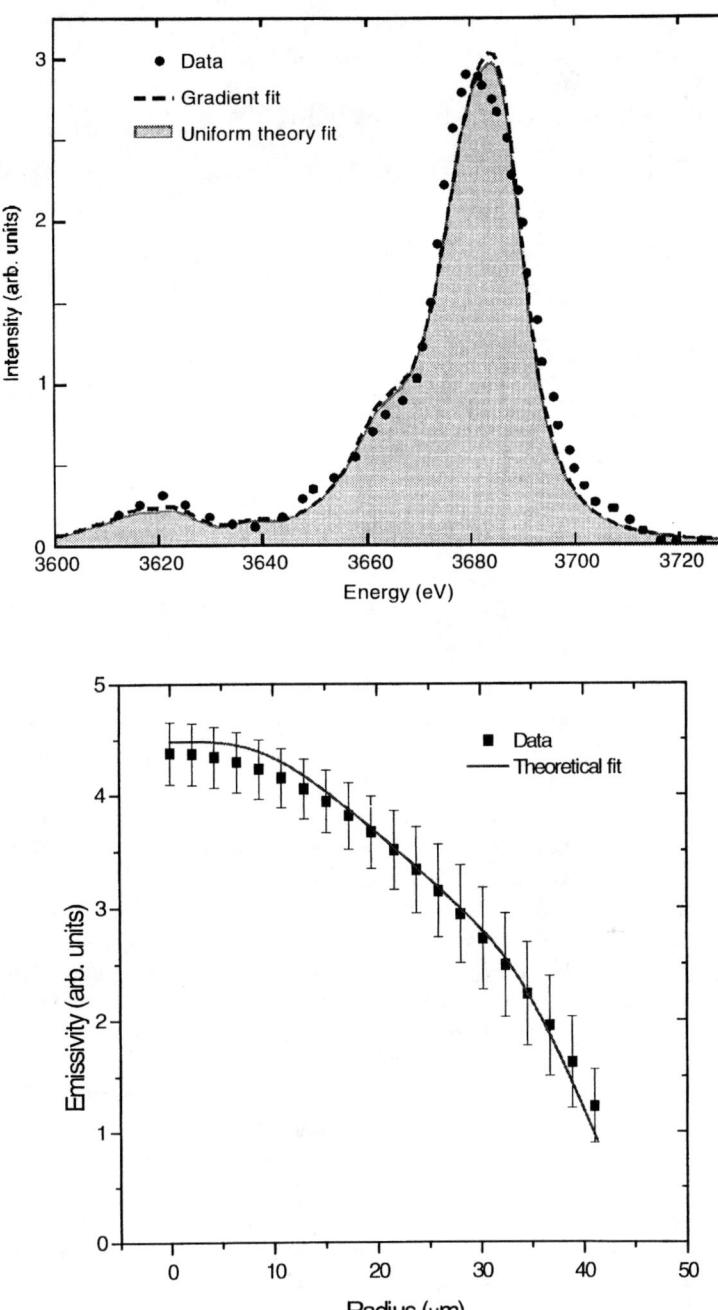

FIGURE 4. Self-consistent fits to the Ar Heβ line spectrum (top) and monochromatic emissivity (bottom) extracted from the image data of Fig. 2, and line spectra for direct-drive GEKKO shot 22091.

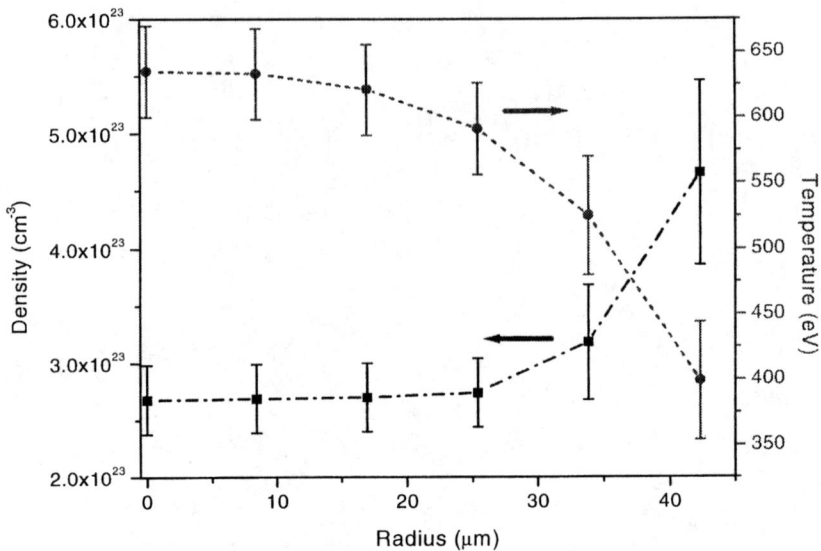

FIGURE 5. Self-consistent core gradient determination for direct-drive GEKKO data displayed in Figs. 2 and 4.

by a niched Pareto genetic algorithm [11,21]. Genetic algorithms (GA) provide robust search and optimization algorithms for complex problems, and can be adapted to situations involving multi-objective criteria and complex constraints in a straight-forward manner. They start from a randomly selected initial population and are not likely to get caught in local minima. The use of a computer driven, automated search procedure in parameter space allows one to conduct objective and systematic searches. This is important to address the issue of uniqueness of the solution [11].

The result of the gradient analysis for the GEKKO implosion core displayed in Fig. 2 is illustrated in Figs. 4 and 5 [7]. Because of the 10 μm resolution of the X-ray imager, the spatial dependence of the gradients is characterized by values given in six spatial zones. The uncertainty in the gradient values is related to the alternative gradients found in the vicinity of the optimal solution that produce comparable self-consistent fits using the range of input values produced by the deviation from spherical symmetry (see Fig. 4, bottom). In order to illustrate the accuracy of the fits, we also display in Fig. 4 the comparison between the data and the fits calculated with the self-consistent $T_e(r,t)$ and $N_e(r,t)$. Note that, for comparison with the image data, the streak spectrograph data must be integrated about the central time of the image over a time interval corresponding to the 40 ps frame time duration. As a check, the emissivity-weighted average values of the gradients shown in Fig. 5 were compared with the values of the electron temperature and density obtained from a fit to the Heβ line-out, presented in Fig. 4, using the same spectral model but assuming a uniform plasma core. These values, $<N_e>=3\times10^{23}$ cm^{-3} and $<T_e>=600$ eV, are consistent to within 4%

with those obtained with a uniform plasma model analysis. This level of consistency is typical of all the cases analyzed in this work.

As an illustration of 1-D core gradient determination in indirect-drive OMEGA implosions, Fig. 6 displays the results obtained for the analysis of the implosion core displayed in Fig. 3. In this case, Ar Heβ X-ray monochromatic image and X-ray line spectrum are both time-integrated over the interval of emission of the Ar Heβ, which for this implosion core is approximately 200 ps. The error bars in the gradient are again determined taking into account solutions in the vicinity of the optimal solution that yield fits to the data of comparable quality, and the deviation from (perfect) spherical symmetry. The emissivity-weighted averages of the gradients shown in Fig. 6 are 860 eV and $7.8 \times 10^{23} cm^{-3}$ which compare well with results from the uniform model analysis (i.e. 875 eV and $8 \times 10^{23} cm^{-3}$).

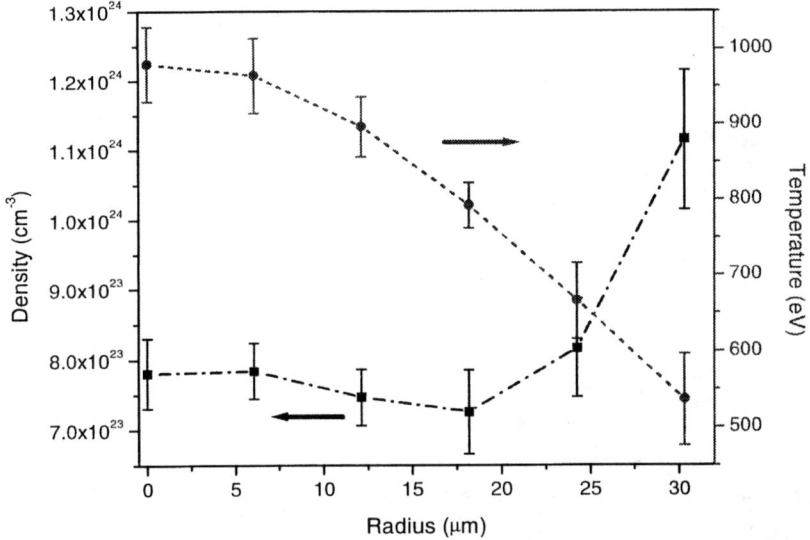

FIGURE 6. Self-consistent core gradient determination for the implosion core of Fig. 3.

CONCLUSIONS

We have discussed a spectroscopic method for core gradient determination during the collapse phase of direct- and indirect-drive ICF implosions. The method is based on the self-consistent reconstruction and analysis of simultaneous X-ray monochromatic images and X-ray line spectra. We note that temperature and density gradients obtained from this analysis are countercorrelated. This is consistent with an isobaric core at the collapse of the implosion, and it is also in qualitative agreement with results from 1D hydrodynamic code simulations.

ACKNOWLEDGMENTS

This work was supported by DOE NLUF Grant DE-FG03-01SF22225, DOE HEDS Grant DE-FG03-98DP00213, the LLNL under the auspices of DOE contract W-7405-ENG-48, the ILS, the Japan-Germany and Japan-US Collab. Prog. of JSPS, and the Deutsche Forschungsgemeinschaft contract FO186/3-1.

REFERENCES

1. Griem, H.R., *Phys. Fluids B* **4**, 2346 (1992).
2. Griem, H.R., *Principles of Plasma Spectroscopy,* Cambridge University Press, 1996.
3. Haynes, D.A., Jr., Garber, D.T., Hooper, C.F., Jr., Mancini, R.C., Lee, Y.T., Bradley, D.K., Delettrez, J., Epstein, R., and Jaanimagi, P.A., *Phys. Rev. E* **53**, 1042 (1996).
4. Woolsey, N.C., Hammel, B.A., Keane, C.J., Asfaw, A., Back, C.A., Moreno, J.C., Nash, J.K., Calisti, A, Mosse, C., Stamm, R., Talin, B., Klein, L., and Lee, R.W., *Phys. Rev. E* **57**, 4650 (1998).
5. Regan, S.P., Delettrez, J.A., Epstein, R., Jaanimagi, P.A., Yaakobi, B., Smalyuk, V.A., Marshall, F.J., Meyerhofer, D.D., Seka, W., Haynes, D.A., Jr., Golovkin, I.E., and Hooper, C.F., Jr., *Phys. Plasmas* **9**, 1357 (2002).
6. Lindl, J., *Phys. Plasmas* **2**, 3933 (1995).
7. Golovkin, I., Mancini, R., Louis, S., Ochi, Y., Fujita, K., Nishimura, H., Shirga, H., Miyanaga, N., Azechi, H., Butzbach, R., Uschmann, I., Förster, E., Delettrez, J., Koch, J., Lee, R.W., and Klein, L., *Phys. Rev. Lett.* **88**, 045002-1 (2002).
8. Golovkin, I.E., and Mancini, R.C., *J. Quant. Spectrosc. Radiat. Transfer* **65**, 273 (2000).
9. Golovkin, I.E., Mancini, R.C., Woolsey, N.C., Back, C.A., Lee, R.W., and Klein, L., in *Intertial Fusion Science and Applications*, edited by C. Labaune, W.J. Hogan and K.A. Tanaka, Elsevier, Paris, 2000, p. 1123.
10. Goldberg, D.E., *Genetic Algorithms in Search, Optimization & Machine Learning*, Addison Wesley, Reading, Massachusetts, 1989.
11. Golovkin, I.E., PhD Dissertation, University of Nevada, Reno, 2000.
12. Golovkin, I.E., Mancini, R.C., Louis, S., Lee, R.W., and Klein, L., *J. Quant. Spectrosc. Radiat. Transfer* **75**, 625 (2002).
13. Bockasten, K., *J. Opt. Soc. Am.* **51**, 943 (1961).
14. Yaakobi, B., Marshall, F.J., and Delettrez, J., *Opt. Commun.* **133**, 43 (1997).
15. Ochi, Y., Kawamura, T., Nishimura, H., Fukao, M., Fujita, K., Shiraga, H., Azechi, H., and Yamanaka, T., *J. Quant. Spectrosc. Radiat. Transfer* **71**, 531 (2001).
16. Uschmann, I., Fujita, K., Niki, I., Butzbach, R., Nishimura, H., Funakura, J., Nakai, M., Förster, E., and Mima, K., *Appl. Opt.* **39**, 5865 (2000).
17. Mima, K., Kato, Y., Azechi, H., Shigemori, K., Takabe, H., Miyanaga, N., Kanabe, T., Nishimura, H., Shiraga, H., Nakai, M., Kodama, R., Tanaka, K., Takagi, M., Nakatsuba, M., Nishihara, K., Yamanaka, T., and Nakai, S., *Phys. Fluids* **3**, 2077 (1996).
18. Dalhed, H.E., private communication.
19. Marshall, F.J., and Oertel, J.A., *Rev. Sci. Instrum.* **68**, 735 (1997).
20. Koch, J.A., private communication.
21. Coello-Coello, C.A., Van Veldhuizen, D.A., and Lamont, G.B., *Evolutionary Algorithms for Solving Multi-Objective Problems*, Kluwer Academic Pub., New York, New York, 2002.

The Production of Exotic Satellite Structures in Short Pulse Laser Heated Foils

R. Shepherd*, P. Audebert†, O. Peyrusse**, K. B. Fournier‡, H-K. Chung‡,
D. Price‡, J. C. Gauthier†, R. W. Lee‡, K. Widmann‡, P. Springer‡
and L. Klein§

*Physics and Advanced Technologies, Lawrence Livermore National Laboratory, M.S. L-43,
P.O. Box 808, Livermore, CA USA 94550, E-mail: shepherd1@llnl.gov
†Laboratoire pour l'Utilisation des Lasers Intenses, UMR7605, CNRS - CEA - Université
Paris VI - École polytechnique, 91128 Palaiseau, France
**°CEA-DIF, Bruyère-le-Châtel, France
‡Physics and Advanced Technologies, Lawrence Livermore National Laboratory,
Livermore, CA 94550
§Physics and Astronomy Department, Howard University, Washington D.C.

Abstract. The success in building ultrashort pulse lasers capable of delivering higher and higher intensities have pushed laser-solid interaction experiments into a new regime where v_{os}/c approaches 1. We have performed experiments to study the K-shell aluminum emission spectra from thin foils heated with an ultrashort pulse laser. The foils are illuminated at an intensity of 1×10^{19} W/cm². After heating, minimal radiation cooling and longitudinal heat conduction occur due to the targets' low atomic number and thickness, allowing hydrodynamic expansion to dominate the cooling process and simplifying the analysis. The time resolved Helium like $1s^2$-$1s2p(^1P)$ (He$_\alpha$), $1s^2$-$1s3p(^1P)$ (He$_\beta$), and $1s^2$-$1s4p(^1P)$ (He$_\gamma$) spectrum is collected with a 500 fs x-ray steak camera interfaced to a two crystal von Hamös spectrograph. The spectra from these plasmas have shown interesting and unusual features. In particular, exotic satellites have been observed when high intensity ultrashort pulse lasers interact with solids. The satellite emission brightness relative to the resonance lines suggest an expectedly large fraction of Li-like, Be-like, and B-like ions. An electron beam generated in the field of the laser is offered as an explanation of the observed satellite emission.

INTRODUCTION

Currently short pulse lasers are being used to generate high energy-density plasmas in laboratories around the world[1, 2, 3]. Because the time required to deposit the laser is significantly less than the time required for bulk motion of the inertial mass, solid matter can be heated to high temperatures while maintaining a density close to solid. As the power available with ultrashort pulse lasers has pushed focused laser intensities higher and higher, the interaction of these lasers with solids has moved experiments into a regime where the electrons oscillate with a velocity $v_{os}/c \approx 1$. The result is an extremely dynamic plasma far removed from ionization equilibrium. Further, temperature and density gradients rapidly evolve due to conductive cooling into the bulk solid and expansion of the heated solid layer, making it difficult to interpret spectral data.

We have performed systematic experiments on 500 Å aluminum foils to study the

CP635, *Atomic Processes in Plasmas: 13th APS Topical Conference*, edited by D. R. Schultz et al.
© 2002 American Institute of Physics 0-7354-0090-3/02/$19.00

K-shell emission from high energy density matter. The target thickness minimizes the temperature gradient produced by conduction cooling along the laser propagation direction. The dynamic evolution of the spectral emission is resolved with a 500 fs x-ray steak camera interfaced to a two crystal von Hamös spectrograph. The Helium like $1s^2$-$1s2p(^1P)$ (He$_\alpha$), $1s^2$-$1s3p(^1P)$ (He$_\beta$), and $1s^2$-$1s4p(^1P)$ (He$_\gamma$) were observed. The data show intense satellite emission distributed over a broad spectral range, suggesting the presence of doubly excited charge-states from various ionization states. Although these satellites have been observed in time integrated x-ray measurements on bulk laser heated solids[4, 5], this data represents the first time resolved observation of these satellites in a minimal longitudinal gradient environment where satellite production cannot be attributed to emission from colder material deep within the target.

EXPERIMENTAL LAYOUT

We performed our experiment at the Lawrence Livermore National Laboratory Ultra-Short Pulse laser facility. The laser has been described in detail elsewhere so only a brief description is given here [6]. The laser utilizes chirped pulse amplification to generate high energy, 150 fs pulses. The pulses from an 82 MHz Ti:Sapphire oscillator are temporally stretched and then amplified in three stages: first in a regenerative amplifier, followed by a five pass power amplifier, and finally in a large aperture two pass power amplifier. The pulses are compressed in a vacuum compressor and frequency doubled with a 1.5 mm thick KD*P crystal before focusing on the target with an f/3.5 off axis parabola. The measured energy on target is 150 to 200 mJ at 400 nm and is focusable to a FWHM spot size of 2 μm.

Several key diagnostics were instituted to insure high contrast, high intensity pulses. The pulse duration out of the oscillator was checked by computer using a portion of the zero order reflection off the grating and imaging it into a spectrometer. Pulses deviating more than 10% from the initial 100 fs Gaussian pulse are shuttered and prevented from further amplification, minimizing the chance of a pulse with temporal "wings" being used in the experiment. A Frequency Resolved Optical Gate (FROG)[7] was used to monitor the phase of the pulse on every shot so data could be discarded from a poorly compressed pulses. To suppress ASE, the laser pulse was frequency doubled with a 1.5 mm thick KD*P crystal. Frequency doubling, however, can introduce distortions in the pulse time and space profiles. Distortions in the pulse profile were controlled by measuring the spectrum and energy of the frequency doubled laser light, then the energy was adjusted until an undistorted frequency spectrum was produced. To check the effects of the ASE, the targets were observed after heating with the unseeded pulse and no damage was observed, suggesting no measurable effect from the focused ASE. A high dynamic range, third order, scanning autocorrelation measurement performed on the 800 nm pulse suggest the pulse has an intensity contrast of $I_{background}/I_{peak} \approx$ 10^{-5} at 1.5 ps before the peak of the laser pulse. The contrast is further enhanced (approximately squared) by the frequency doubling. A low dynamic range (10^{-4}), single shot autocorrelation was performed on every shot and used to monitor the pulse width while an energy calorimeter was used to measure the energy on target. An energy

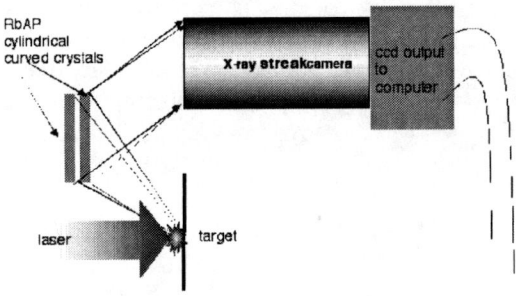

FIGURE 1. Experimental setup showing target, streak camera, and duel crystal von Hamös spectrograph.

monitor placed behind the target detected no transmitted energy through the foils, suggesting the target remained above the critical density during illumination.

The experiment was conducted with $\hat{S}$ polarized, 150 fs light with a peak intensity of $\approx 2 \times 10^{19}$ W/cm^2 ($I\lambda^2 \approx 5x10^{18} W\mu m^2/cm^2$). The spectroscopic measurements were made with a 500 fs X-ray streak camera interfaced to a dual crystal von Hamös spectrograph (see figure 1). The streak camera was sufficiently small that it fit inside the vacuum chamber, allowing the slit to lie on the focal axis of the crystals. The duel crystal design allowed simultaneous collection of time resolved data from multiple He-like spectral lines on a single shot. Two RbAP crystals were bent to 3.6 and 3.0 cm to collect time resolved data from the He$_\alpha$ transition and the He$_\beta$, He$_\gamma$ transitions, respectively. The reflecting surface of the crystals were oriented at approximately 83° to the target normal, increasing the target thickness t along the streak camera viewing cord by approximately $t/sin(17°)$. The spectral data were focused into the 25.4 mm long, 150 μm wide entrance slit to the streak camera. The high camera sweep speed (2.6 ps /mm) allowed the observation of the transit time difference between the x-rays reflecting off two crystals, producing a spatial gap between the streaked data from the respective crystals. Additionally, the transit time difference of the x-rays of different wavelengths reflected off a single crystal was used to calibrate and check linearity of the sweep for each shot. The streaked data was intensified and collected with a 1024 x 1024 CCD camera.

The X-rays emitted between 10 keV and 1 MeV were collected using a scintillator-diode array. The X-rays with energy above 30 keV were detected using 12 filtered photomultipliers while six filtered x-ray diodes were used for energies below 30 keV. These data were electrically relayed to analog to digital converters and stored on a computer. The targets consist of 250 Å, 500 Å, and 750 Å thick free-standing aluminum. Each foil was suspended over an array of 1.7 mm diameter holes in a ceramic substrate. The 1.7 mm diameter foil area was destroyed and replaced after each shot.

DATA AND ANALYSIS

Spectral data was collected in a single shot. Uncorrected, time resolved data from a 250 Å, 500 Å, and 750 Å thick foils is shown in figure 2. These data were collected with a laser intensity $I \approx 2 \times 10^{19}$ W/cm^2. Several spectral and temporal features are worth noting. The data shows satellite emission on the long wavelength side of He$_\alpha$ (1s2lnl' manifold) and between the He$_\beta$ and He$_\gamma$ transitions (Li-like and Be-like satellite manifolds). The satellite emission increases in number and intensity with target thickness while the resonance line emission tapers off, making it difficult to separate the satellite and resonance line emission with the 750 Å thick foil. This is consistent with a constant energy heating a larger mass, resulting in a lower peak temperature as the foil thickness increases. In general, the time history of the satellites is fairly short ($\leq 3\,ps$ FWHM) where as He$_\alpha$ tends to be relatively long ($\geq 7\,ps$ FWHM). The higher n-state resonance lines (He$_\beta$ and He$_\gamma$) also tend to be longer than the satellite emission, although not nearly so long.

Finally, the so-called Li-like "blue" satellite appears in the data. Although it is not used in this analysis, it is interesting to note several characteristics of this feature. This satellite has been attributed to the $1s2p[^1P]3d^2D_{5/2} - 1s^23d^2D_{5/2}$ and $1s2p[^1P]3d^2D_{3/2} - 1s^23d^2D_{3/2}$ transitions[8]. Although it seems to appear in all three foil thicknesses, it is particularly interesting in 250 Å foil. In this case, there is little ambiguity between the potentially broadened He$_\alpha$ (due to trapping and density) and the blue satellite. To our knowledge, this is the first time-resolved observation of this satellite in an optically thin case. The time history is noted as distinctly short lived ($\approx 1\ ps$), suggesting conditions suitable for its production being extremely transient in nature.

To analyze the data, corrections were made for the x-ray transit time difference, the photocathode response, the wavelength dependent solid angle, and the crystal reflectivity.

MODELING

Due to concerns of the He$_\alpha$ opacity, the focus of the data reduction was on the spectral region between the He$_\beta$ and He$_\gamma$ transitions. The spectral data was modeled by calculating the atomic structure and cross-sections for a set of plasma conditions (density and temperature). These data are input into a non-LTE, collisional-radiative model to generate simulated spectra, then a comparison is made to a temporal slice from the experimental data record. The process is repeated until a good agreement between experiment

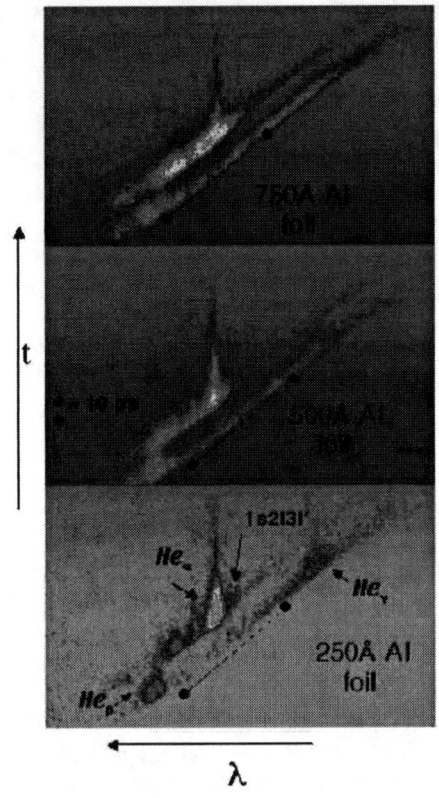

FIGURE 2. Data from 250 Å, 500 Å, and 750 Å thick aluminum foils

and simulation is found.

   As can be seen in the data, a near continuum of satellites features are produce between the He_β and He_γ transitions. Including all the necessary ions and levels quickly adds and makes a calculation of the level populations time consuming. Thus, the level populations were calculated using two methods; detailed configuration modeling and configuration averaged modeling. The detailed calculation was used to benchmark the configuration averaged model. When the benchmark was shown to be accurate, the configuration averaged model was used to generate synthetic spectra. The detailed spectral modeling was performed using the Hebrew University Lawrence Livermore Atomic physics Code

(HULLAC) to calculate the atomic structure and cross-sections. A detailed description of the HULLAC code is given in reference[9]. HULLAC includes the physical processes of collisional excitation and de-excitation, collisional ionization, radiative decay, autoionization, dielectronic, three-body, and radiative recombination. The H-like, He-like, Li-like, Be-like, and B-like ions were included in the simulations. A typical HULLAC simulation for $N_e = 3 \times 10^{23}$ e/cm^3 and $T_e = 200$ eV is shown in figure 3. For theses simulations, levels $1snln'l'(n, n' \leq 4)$ are included. The simulation shows large Li-like and Be-like satellite contributions between the He$_\beta$ and He$_\gamma$ transitions. With the addition of the n=5 level, further contributions appear from B-like satellites.

The configuration averaged calculation was performed using the code AVERROÈS. A detailed explanation of AVERROÈS is in reference [10, 11, 12]. In general, AVERROÈS forms "super levels" or superconfigurations from groups of configurations. The super-configurations are made of subshells that are grouped in "supershells" with an interger number of electrons. The level spacing within the individual groups are close enough to assume LTE statistics. AVERROÈS includes the same physical processes as HULLAC (collisional excitation and de-excitation, collisional ionization, etc.). With the data from AVERROÈS as input, the code TRANSPEC[13, 14] was used to generate the NLTE level populations and synthetic spectra. TRANSPEC is a 1-D, time dependent, collisional-radiative model that self-consistently couples atomic population data with radiation. The hydrodynamics code LASNEX[16] was also used and showed similar results. A comparison of HULLAC and AVERROÈS is shown in figure 3. The AVERROÈS approximation to the detailed HULLAC model is shown as an overlay in the figure. The 1-D conditions were obtained from the hydrodynamics code FILM[15]. The plasma conditions of the central cell are shown in figure 4. Using AVERROÈS and TRANSPEC, synthetic spectra were produced for each foil thickness. The data and synthetic spectra are displayed side by side in see figures 5. In general, for all three cases the agreement is fairly good. However, it is noted that the agreement between the AVERROÈS/TRANSPEC synthetic spectra and the data gets better as the foil thickness decreases. This could result from the increase in the longitudinal gradient with foil thickness, producing a greater contribution from lower charge states in the plasma.

DISCUSSION

The appearence of the large number of satellites between He$_\beta$ and He$_\gamma$ was not anticipated. The thin targets should eliminate the longitudinal temperature gradient that act as a source of cold plasma. However, the large number and brightness of the satellites suggest significant populations of lower charge states that should not be present at such a high irradiance. What is described below is one explanation for the strong satellite emission. For this explanation, a heuristic model has been formed that is consistent with these data.

At the high laser intensity used in this experiment ($\approx 2 \times 10^{19}$ W/cm^2), initially the electrons oscillate at the quiver velocity of the laser v_{os}. The laser intensity is reduced as the laser light is attenuated in the aluminum. However for foil thicknesses ≤ 500 Å, the energy remains high throughout the entire thickness of the foil. On average,

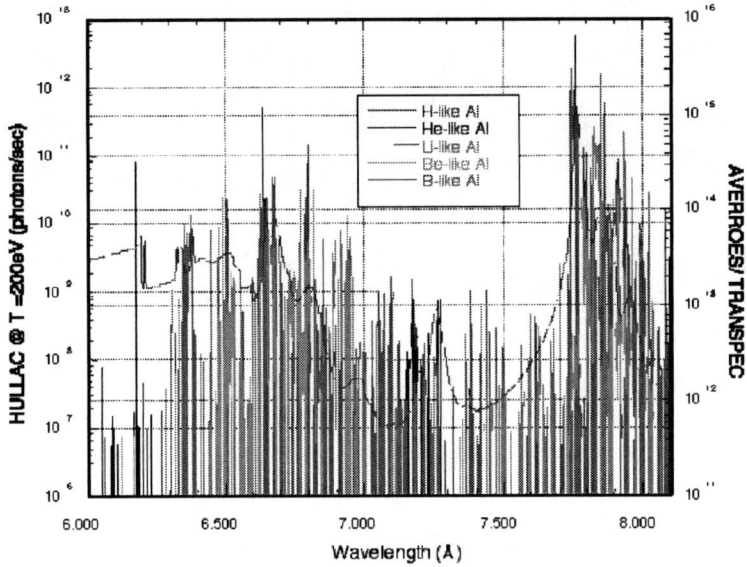

FIGURE 3. Comparison between AVERROÈS/TRANSPEC (solid line) and HULLAC (vertical lines) for $N_e = 3 \times 10^{23}$ e/cm^3 and $T_e = 200$ eV.

measurements with the scintillator/diode array suggest high energy electron temperature $E_{os} \approx 150$ keV. For $\hat{S}$ polarized laser light, these electrons oscillate in the plane of the target. In vacuum, the distance a 150 keV electron travels in the field of the laser is ≈ 500 Å which is small compared to the laser spot size ($\approx 2\,\mu$m). Thus the effects of the "oscillating electron beam" is initially only felt within the spot diameter. Although these electrons have a rather small collision cross-section, they dominate the ionization process. This is seen by considering two factors: 1) the low number of electrons heated by inverse bremsstrahlung and 2) the relative ionization rate of a thermal plasma versus that of a 150 keV electron beam. Inverse bremsstrahlung requires a non-ionizing, inelastic collision between an electron and an ion. However, this cross section falls rapidly with increasing electron velocity ($1/v^2$). As a result, the inverse bremsstrahlung absorption coefficient falls as $1/I^{3/2}$[17], significantly reducing the number of electrons heated by this process at high laser irradiance. Conversely, many of the free electrons in the focal spot are accelerated in the field of the laser to energy E_{os}. A comparison of the rate coefficient for ionization from a Maxwellian distribution of electrons (at various temperatures) to the ionization rate coefficient for a 150 keV electron beam illustrates the relative importance each plays in the ionization process. In figure 6, we compare the ionization rate coefficient for a 150 keV electron beam and the ionization rate coefficient

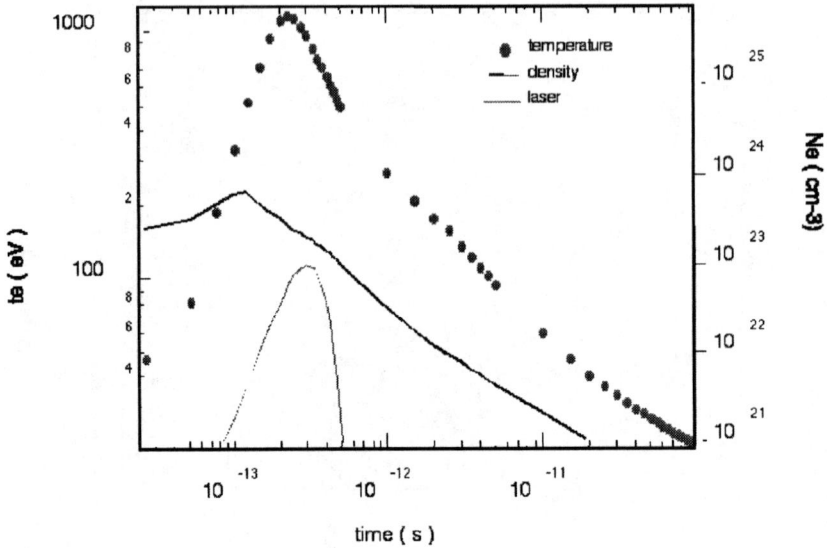

FIGURE 4. Hydrodynamics simulation of a 500 Å foil (100 fs pulse, $I\lambda^2 = 5 \times 10^{18}$ W/cm$^2\mu$m)

for a Maxwellian distribution ($T_e \leq 600\,eV$). For inner-shell ionization, the ionization rate coefficient for a 150 keV electron beam is always greater than the rate coefficient for ionization from a Maxwellian distribution. This suggest the dominance of the 150 keV electron beam in the ionization process. When laser heating subsides, energetic electrons damp their energy by ionization, electron-electron collisions, and electrostatic acceleation of ions. If we simply focus on the first two energy loss mechanisms, we find the range of these electrons is substantial. Using Bethe's stopping power theory[18], an estimate can be made of the range of the energetic electrons parallel to the target surface (once the laser is off) as $\approx 134\,\mu$m[19] . The mean-free path of the electrons is substantially larger than the diameter of the focal spot, suggesting a large emission outside the laser focus. Because much of the electron energy is lost at the end of the electron range (where the electron energy is relatively low), one would expect a larger fraction of low charge-state ions surrounding a small volume of high charge-state ions. This structure could explain the unexpected brightness of the satellite emission relative to the resonance lines from the plasma. Further discussion of the radial gradient in short pulse laser heated targets is described in reference[20]. Currently experiments are underway to further investigate this phenomena.

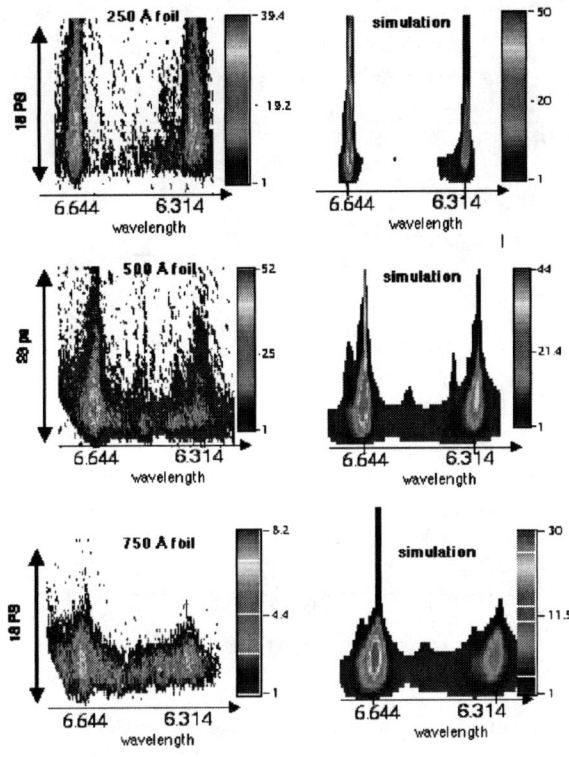

FIGURE 5. Comparison between data and AVERROÈS/TRANSPEC simulation

CONCLUSIONS

In conclusion, we have measured the time-resolved, K-shell emission spectra from 250 Å, 500 Å, and 750 Å thick aluminum foils. Detailed analysis was performed on the 500 Å foil. The data shows relatively large Li-like, Be-like, and B-like satellite emission. By bench-marking the collisional-radiative model AVERROÈS/TRANSPEC with HULLAC, the spectra was successfully modeled. The model suggests conditions of $T_e \approx 500$ eV and $N_e \approx 3 \times 10^{23}$ W/cm^2 after the laser pulse. A heuristic model has been suggested to help interpret the physics of the process, acknowledging that an alternative

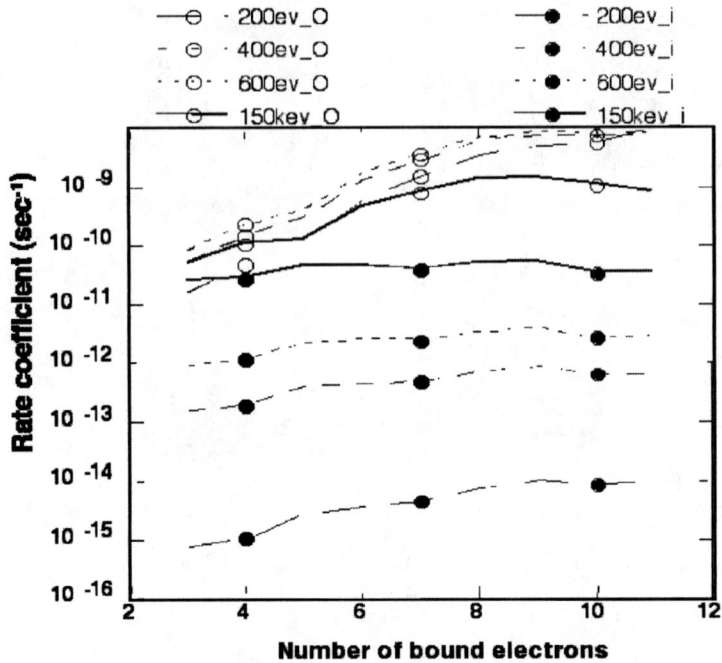

FIGURE 6. Comparison of the ionization rate coefficient for a 150 keV electron beam and the ionization rate coefficient for a Maxwellian distribution for the 1s inner bound electrons (labeled "i") and 2p outer bound electrons (labeled "O")

model is also being considered.

Future experiments are expected to help decipher the explanations and further the understanding the absorption and partition of energy at relativistic laser-matter interaction intensities.

ACKNOWLEDGMENTS

This work was performed under the auspices of the U.S. Department of Energy by University of California Lawrence Livermore National Laboratory under contract No. W-7405-Eng-48. The authors would like to thank the UltraShort Pulse laser staff at LLNL. L. K was supported in part by Howard University DoE HEDS Grant No. DE-FG03-98DP00213.

REFERENCES

1. Bastiani, S., Audebert, P., Geindre, J., Schlegel, T., Gauthier, J., Quoix, C., Hamoniaux, G., Grillon, G., and Antonetti, A., *Phys. Rev. E*, **60**, 3439 (1999).
2. Andiel, U., Eidmann, K., Witte, K., , and Forster, E., *App. Phys. Lett*, **80**, 198 (2002).
3. Dorchies, F., Forget, P., Gallant, P., , Jiang, Z., Kieffer, J. C., Pépin H., and Peyrusse, O., *Phy. of Plasmas*, **8**, 3862 (2001).
4. Rosmej, F., Faenov, A., Pikuz, T. A., Flora, F., and P, D. L., *J. Phys. B: At. Mol. Opt. Phys*, **31**, 1921 (1998).
5. Urnov, A. M., Faenov, A. Y., Pikuz, T. A., Skobelev, I. Y., Abdallah, J., Clark, R. E. H., Cohen, J., Johnson, R. P., Kyrala, G. A., and Osterheld, A. L., *JETP Letters*, **67**, 489 (1998).
6. Sullivan, A., Bonlie, J., Price, D. F., and White, W. E., *Optics lett.*, **21**, 603 (1996).
7. Kane, D. J., and Trebino, R., *IEEE Journal of Quantum Electronics*, **29**, 571 (1993).
8. Rosmej, F. B., and Abdallah, J., *Phys. Letters A*, **245**, 548 (1998).
9. Bar-Shalom, A., Klapisch, M., and Oreg, J., *J. of Quant. SPect. & Rad. Trans.*, **71**, 169 (2001).
10. Peyrusse, O., *J. Phys. B*, **32**, 169 (1999).
11. Peyrusse, O., *J. of Quant. SPect. & Rad. Trans.*, **71**, 571 (2001).
12. Peyrusse, O., *J. Phys. B*, **33**, 4303 (2000).
13. Peyrusse, O., *Phys. of Fluids B*, **4**, 2007 (1992).
14. Peyrusse, O., *J. of Quant. SPect. & Rad. Trans.*, **51**, 281 (1994).
15. Teubner, U., *Phys. Plasmas*, **3**, 2679 (1996).
16. Zimmerman G., Kershaw, D., Bailey, D., Harte, J. "The LASNEX code for inertial confinement fusion," in *Topical Meeting on Inertial Confinement Fusion, San Diego, CA, USA, 7-9 Feb. 1978.*, Topical Meeting on Inertial Confinement Fusion, Optical Society of America, 2010 Massachusetts Ave., NW, Washington, DC 20036, USA, 1977, vol.1-4., p.700.
17. Hughes, T. P., *Plasmas and Laser Light*, Adam Hilger Ltd., 29 King street, London WC2E 8JH, Winterstoke Rd, Bristol BS3 2NT, England, 1975.
18. Bethe, H., *Ann. der Physik*, **5**, 325 (1930).
19. Berger, M., and Seltzer, S., "Tables of energy-losses and ranges of electrons and positrons," in *Studies in penetration of charged particles in matter*, National Academy of Sciences-National Research Council, National Academy of Sciences, 2101 Constitutin Ave., N.W., Washington, D.C. 20418, USA, 1964, vol. 1133.
20. Audebert P., et. al., "Time-resolved plasma spectroscopy of thin foils heated by a relativistic-intensity short pulse laser". Submitted to Phys. Rev. E., 2002.

Spectral Studies of Short Pulse Laser Irradiated Argon Clusters

G.C. Junkel-Vives[1], J. Abdallah, Jr.[1], F. Blasco[2], F. Dorchies[2], C. Stenz[2], F. Salin[2], A.Ya. Faenov[3], A.I. Magunov[3], T.A. Pikuz[3], I.Yu. Skobelev[3], T. Auguste[4], S. Dobosz[4], P. D'Oliveira[4], S. Hulin[4], and P. Monot[4]

[1] Los Alamos National Laboratory, P. O. Box 1663, Los Alamos, New Mexico 87545, USA
[2] CELIA, Universite Bordeaux 1, 33405 Talence, France
[3] Multicharged Ions Spectra Data Center of VNIIFTRI, Mendeleevo, Moscow region, 141570, Russia
[4] Centre D'Etudes de Saclay, DSM/DRECAM, CEA, 91191 Cif-sur-Yvette, France

Abstract. The systematic experimental studies of plasma produced upon the interaction of ultra-short laser pulses with cluster targets are carried out using the methods of x-ray spectroscopy. The dependence of the plasma parameters on the initial properties of a cluster target such as the design of a supersonic nozzle, the average size of clusters, and their spatial distribution as well as the laser-pulse properties of duration and contrast are studied. The plasma diagnostics is performed using the model of spectra formation that provides a good agreement with the experimental data.

INTRODUCTION

Studies of high-power ultras-short laser pulse interaction with clusters have been of considerable interest during the last few years [1–8]. Cluster targets combine the main mechanisms of plasma formation in gaseous and solid media irradiated by lasers (optical field ionization and resonance absorption). It was found that cluster targets absorb the laser pulse energy more effectively. This is especially important for plasma x-rays sources, which can be used in micro-lithography, medicine, and biology. Unfortunately, the experimental results been obtained so far are not systematic because they depend on many parameters of a laser pulse and irradiated targets. This complicates an analysis of the specific features of plasma formation by ultra-short pulses.

A numerical model of cluster formation in a supersonic gas jet was considered in [8–13]. It was shown that two-phase gas dynamic processes in a jet forming by a nozzle led to inhomogeneous spatial distribution for parameters of the clusters, which are important for the correct calculation of the cluster heating by ultra-short laser pulses. A simple model of the cluster target heating by femtosecond laser pulses with a picosecond pre-pulse was proposed in these papers, which allowed us to describe the x-ray emission spectra observed in the plasma. The model is valid if the pre-pulse duration τ_{prep} does not exceed the decay lifetime of clusters in expanding plasma τ_{exp}. In this case the main femtosecond pulse interacts with a dense pre-plasma. The electron temperature in these regions jumps up to several keV for the laser peak

CP635, Atomic Processes in Plasmas: 13th APS Topical Conference, edited by D. R. Schultz et al.
© 2002 American Institute of Physics 0-7354-0090-3/02/$19.00

intensity of $\sim10^{18}$ W/cm^2. After the end of the main pulse the hot electrons partly penetrate to the inter-cluster plasma region where the electron temperature is between 100–200 eV. Therefore, the integral x-ray emission spectrum of the plasma should be contributed by less dense (subcritical electron density N_{e1}) and more dense (N_{e2} close to the critical density N_{cr}) plasma regions heated to different temperatures $T_{e1}<T_{e2}$. In both regions a fraction of hot electrons f_i has the mean energy $E_0>>T_{e,i}$. It is clear that $f_2>f_1$. The estimation of the collision ionization probability shows that for multiply charged ions with $Z=10$-20 and typical parameters of fs-laser pulses ($\tau_p=30$–60 fs, $q_p=10^{17}$–10^{18} W/cm^2 and contrast $I_p/I_{prep}\sim10^5$), the ionization state of plasmas in regions 1 and 2 will approximately correspond to its electron temperature, whereas the influence of hot electrons can be neglected in the first approximation. Thus the time-integrated emission spectrum of the plasma calculated within the framework of this model should depend on eight free parameters N_{e1}, N_{e2}, T_{e1}, T_{e2}, f_1, f_2, E_0, and α (the ratio of contributions from regions 1 and 2). Their values can be found from the best fit of the experimental spectrum. From general considerations one should expect that these parameters satisfy the conditions $N_{e1}<N_{cr}\leq N_{e2}$, $T_{e1}<T_{e2}<<E_0$, and $f_1<f_2<<1$. Moreover, because the temperature T_{e1} is attained due to the prepulse with typical intensity $I_{prep}=10^{12}$–10^{13} W/cm^2, its value should be between 100–200 eV, while the energy E_0 of hot electrons produced during the action of the main pulse should be of the order of several keV. Note that the correspondence of plasma ionization state to the temperature T_e is modified by the value of electron density N_e. At low plasma density the ionization state could not achieve the stationary value during the action of a prepulse, thus the value of T_{e1} used in the calculation of the emission spectrum will correspond rather to the ionization temperature than to the electron temperature.

In this paper we used this model (see [9,10] for more details) to interpret the x-ray spectra in the experiments performed with supersonic nozzles of different types and laser pulses with different duration and contrast. In most experiments described below we used argon target. Other gases (CO_2, Kr, and Xe) were also used.

EXPERIMENTAL SETUP AND METHOD OF CALCULATION

The set of experiments were performed using two fs-laser installations at Centre D'Etudes de Saclay and Universite Bordeaux (France).

In the first one (UHI10), a 0.8-µm, 10 TW Ti:sapphire laser was used [9] with the pulse duration $\tau_p=60$ fs and energy $E_p=600$ mJ. The pre-pulse duration was about 1 ps and contrast was $\sim10^5$. The laser beam was focused in a cluster target with an off-axis parabolic mirror. The focal beam spot diameter was about 25 µm, which provided the laser intensity on the target up to 10^{18} W/cm^2.

The second setup also used a Ti:sapphire laser, but with a lower power, with $E_p=15$ mJ and τ_p varying in a the range from 20 fs to several picoseconds [14]. The pulse contrast could be varied from 10 to 10^6. An off-axis parabolic mirror focused laser radiation into a spot of diameter 12 µm, providing the power density up to $4\cdot 10^{17}$ W/cm^2.

In both cases, two types of a supersonic nozzles were used: the Laval nozzle and a conical nozzle. The gas pressure in a valve could be varied up to 100 bar. Using different nozzles with different pressure, we could vary both the average size of clusters in a broad range and their spatial distribution in the interaction region.

The plasma x-ray emission was detected simultaneously with several spectrographs with spherical mica crystals (radius of curvature 100 and 150 mm). The spectrographs were arranged in the FSSR-2D scheme (see [15–17]) and were tuned to the spectral ranges containing resonance transitions in the H- and He-like Ar XVIII and Ar XVII ions. The spectral resolution was $\lambda/\Delta\lambda\approx4000$ and the spatial resolution was better than 80 µm. The spectra were recorded either on a photographic film or using a CCD camera.

The plasma emission spectra were calculated using the level population kinetic code developed at the Los Alamos National Laboratory (USA). In these calculations, the H-like through Be-like argon ions were taken into account with electronic configurations of the principal quantum numbers $n\leq5$, including autoionizing states. The kinetic matrix included 1500 levels coupled by radiative, collision, and autoionization transitions. The collision rates were calculated using the model electron distribution function [18,19], which included the hot component with the average energy $E_0=5$ keV. The shape of spectral lines was assumed instrumental with account of the Doppler and Stark broadening when it was necessary.

EFFECT OF CLUSTER PROPERTIES ON THE PLASMA EMISSION SPECTRA

The calculations of formation of clusters in a supersonic jet performed in [10-12] show an important role of the type of the nozzle and gas pressure value in the valve in the formation clusters. A change in the nozzle design leads to less obvious results. It was found that for the conical and Laval nozzles, with all other factors being the same, the first one provides considerably larger clusters with more uniform distribution across the gas jet. The properties of the cluster target should affect the radiation emission characteristics of the laser-produced plasma.

Distribution of Clusters in Gas Jet

The results of calculations of cluster parameters in a gas jet emerging from a conical nozzle and Laval nozzles [10-12] provide a helpful guide for understanding the x-ray emission characteristics of plasma in the laser–cluster experiments. In the case of the Laval nozzle a target is not uniform along the laser beam with a local minima of cluster concentration at the gas-jet axis. Thus the laser pulse propagating perpendicular to the gas-jet axis passes through two regions with the maximal cluster abundance. The observed spatial dependence of the plasma emission should correspond to the initial distribution of clusters, which determines the efficiency of the laser pulse energy absorption. On the other hand, the emission depends not only on plasma density but also on its temperature defined by the laser intensity. The latter decreases at distances further than the Rayleigh length from the focal plane. For this reason, the plasma emission spatial distribution will be pressed to the jet axis more strongly than the cluster concentration.

These arguments are illustrated by the experimental results that were obtained with Laval nozzles at Bordeaux and Saclay [14,19]. Jet targets of Kr, Xe, and CO_2 along with Ar were used in these experiments. The plasma x-ray emission was detected with a spatial resolution in different spectral lines using crystal spectrometers and pinhole cameras. The results are presented in figure 1. One can see that the spatial distribution of the plasma emission is indeed non-monotonic and the distance between the peaks is approximately 0.7 mm. Note, that the coincidence of the results obtained for spectral lines of different ions means that non-monotonic emission is related to the general parameters of the plasma such as its temperature and density and does not reflect the specific features of the level kinetics.

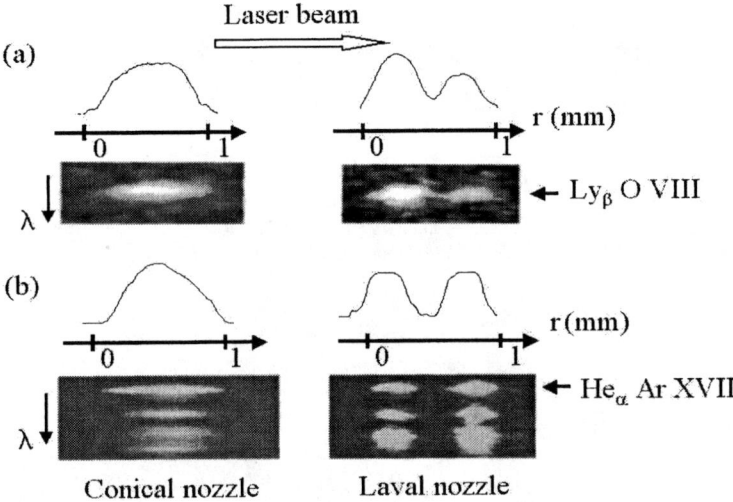

FIGURE 1. Spatial distributions of plasma emission measured with a spectrograph with a one-dimensional spatial resolution for the Laval and conic nozzles in experiments with (a) CO2 and (b) argon clusters.

It follows from the theoretical and experimental studies that the use of a conical nozzle for the cluster target formation is preferable. First, in this case a more homogeneous target can be produced and second, clusters of considerably larger size can be obtained. Using long-focus lenses, we can easily obtain a very long (~4–6 mm) plasma object. Figure 2 shows, for example, the results obtained for CO_2 and Xe clusters using interferometer imaging and an x-ray pinhole camera. One can see that the length of plasma stripe is about 4 mm with a width less than 0.7 mm.

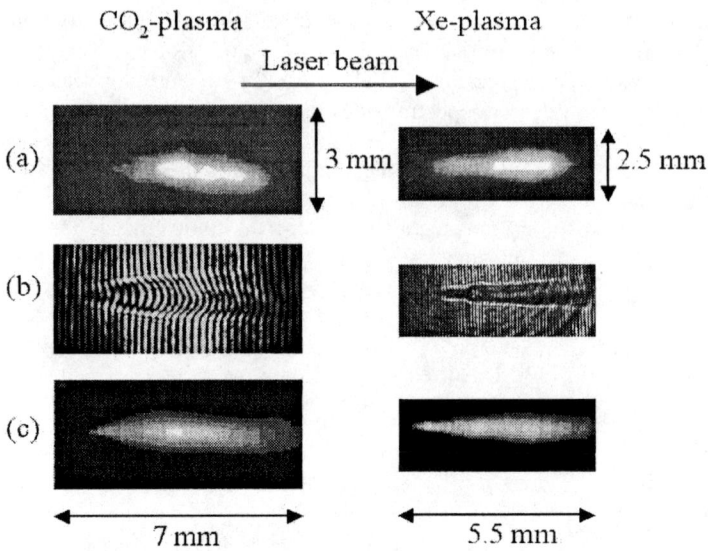

FIGURE 2. Cluster plasma produced by focusing laser radiation with a long-focus lens: (a) x-ray pinhole images; (b) interference patterns; (c) electron density distribution.

Cluster Size Effect on Plasma X-Ray Spectra

The model of the laser–cluster interaction considered above allows us to make some qualitative conclusions about the dependence of plasma parameters on the average size of clusters. According to this model, increasing the cluster size should result in increasing dense plasma region because the laser pre-pulse will destroy a smaller part of the cluster. This should lead, first, to more effective generation of hot electrons by the main fs-pulse, and, second, to increasing of N_{e2} and, to a lesser degree, of N_{e1}. In turn, the increase in the density reduces the time to establish the stationary ionization state, and the ionization temperature will approach the electron temperature.

These qualitative conclusions are confirmed by the results of numerical simulation of the emission spectra of the argon plasma produced upon the interaction of a fs-laser pulse with clusters of different size [21]. To obtain the most uniform cluster target a conical nozzle was used, and the average cluster size was changed by varying the gas pressure in a valve in the range from 15 to 100 bar. The calculations of the cluster formation show that the average number of atoms in a cluster is from $2 \cdot 10^5$ to $2 \cdot 10^7$. The plasma parameters obtained by fitting the experimental spectra to the model spectra are presented in table 1. The experimental spectra at pressure 30 and 100 bar are shown in figure 3 together with the model spectra.

One can see from table 1 that at the initial gas pressure 100 bar there is a large amount $(2 \cdot 10^{19}$ cm$^{-3})$ of hot electrons in dense plasma region. The ionization rate of the He-like Ar ion by hot electrons $(E_0=5$ keV$)$ is $\sim 10^{-11}$ cm^3 s^{-1}, and a noticeable amount of the H-like Ar ions should be formed in the dense plasma region during its lifetime of the order of 1 ps ([H]/[He]=$2 \cdot 10^{-4}$). These ions can be detected by observing the Ly-α line. Because the excitation rates of the $1s^2$ $^1S_0$–$1s2p^1P_1$ and $1s$ $^2S_{1/2}$–$2p$ ^2P_j transitions by hot electrons are of the same order of magnitude, the intensity of the Ly-α line will be extremely low ($\cong 2 \cdot 10^{-4}$ of the He-α line intensity). Nevertheless, it is sufficient for detecting this line with a CCD camera having a broad dynamic range.

TABLE 1. Effect of theAverage Size of Argon Clusters on the Plasma Parameter.

Gas Pressure bar	Average Number of Atoms in a Cluster	Rarefied Plasma			Dense Plasma		
		T_{e1}, eV	N_{e1}, cm^{-3}	$f_1 N_{e1}$, cm^{-3}	T_{e2}, eV	N_{e2}, cm^{-3}	$f_2 N_{e2}$, cm^{-3}
15	$2 \cdot 10^5$	100	$5 \cdot 10^{19}$	$1.5 \cdot 10^{13}$	215	$3 \cdot 10^{20}$	$0.9 \cdot 10^{17}$
30	10^6	160	$5 \cdot 10^{19}$	$5 \cdot 10^{13}$	230	$5 \cdot 10^{20}$	$3.5 \cdot 10^{17}$
70	$7 \cdot 10^6$	140	$2 \cdot 10^{20}$	$2 \cdot 10^{14}$	230	$2 \cdot 10^{21}$	$6 \cdot 10^{17}$
100	$1.6 \cdot 10^7$	195	$1.5 \cdot 10^{21}$	$1.2 \cdot 10^{17}$	400	$2 \cdot 10^{21}$	$2 \cdot 10^{19}$

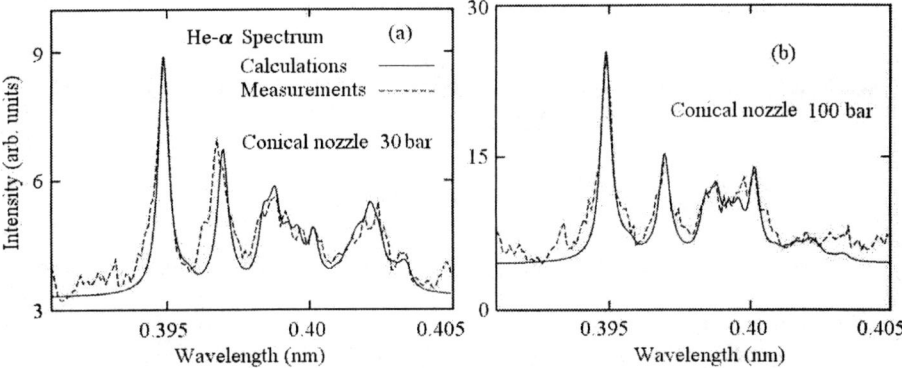

FIGURE 3. X-ray spectra in the region of the He-α line of Ar XVII at gas pressures 30 bar (a) and 100 bar (b). The values of plasma parameters are presented in table 1. The solid curves are model spectra; the dashed curves are experimental spectra.

We have performed special measurements of the emission spectrum in the wavelength region of the Ly-α line [22]. At the initial gas pressure equal to 100 bar we detected a rather weak resonance emission line of the H-like argon (figure 4). The relative intensities of the Ly-α and He-α lines is approximately equal to 10^{-4} that is very close to the expected ratio value of the H- and He-like ions concentrations estimated above. We failed to detect the Ly-α line at lower gas pressures, i.e., for lower cluster sizes.

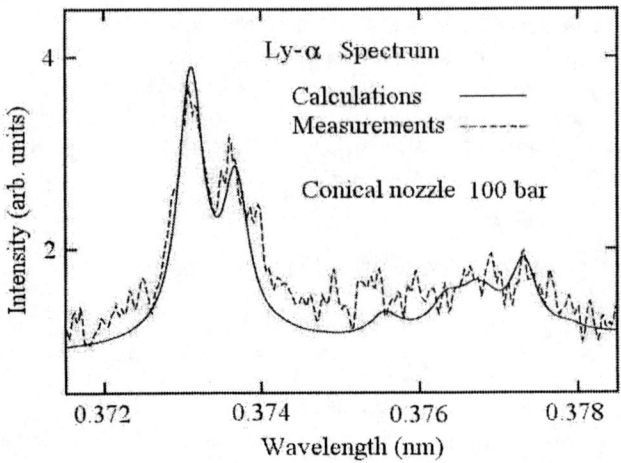

FIGURE 4. The observed Ly-α emission line of the H-like Ar XVIII ion at a gas pressure of 100 atm (dashed curve) and the model spectrum (solid curve) for the plasma parameters presented in table 1.

EFFECT OF THE LASER PULSE PROPERTIES ON PLASMA CHARACTERISTICS

In this section, we consider the influence of the duration and contrast of a laser pulse on the parameters of the laser-induced plasma.

Dependence on the Contrast of Femtosecond Laser Pulse

The simplified model of the laser–cluster interaction predicts that a change in the laser-pulse contrast will affect first of all the plasma temperature and density of the most dense regions (i.e., the value of N_{e2}).

The increase of the main pulse contrast, i.e., the decrease of the pre-pulse intensity, should be accompanied by a decrease of the temperature, whereas the value of N_{e2} should increase because the pre-pulse will destroy a smaller part of the cluster. To verify these predictions, we performed a series of experiments at a fixed energy and fixed duration of the main pulse while keeping the average size of the argon clusters constant. The pulse contrast was varied in a broad range. The duration of the main pulse was 45 fs. The clusters were formed using a conic nozzle with an initial gas pressure of 63 bar. The interaction region was located at a distance of 1.5 mm from the nozzle outlet.

Plasma parameters were defined by fitting the model spectra to the emission data in the spectral region of the He-α line by. One can see from the results presented in table 2 that the plasma temperature monotonically decreases with increasing the pulse contrast, while the density N_{e2} increases and exceeds the critical density $N_{cr}=1.7 \cdot 10^{21}$ cm^{-3} at the pulse contrast $\geq 10^3$. Note that the later result agrees with that in the case of the interaction of ultrashort pulse with solid targets [23–25] where the supercritical density plasma was also observed only at high contrasts of laser pulses.

TABLE 2. Effect of the laser-pulse contrast on the plasma parameters.

Contrast	50	100	350	1000	5000	>>5000
T_{e2} (eV)	250	222	215	215	205	215
N_{e2} (cm^{-3})	$3 \cdot 10^{20}$	$5 \cdot 10^{20}$	$2 \cdot 10^{21}$	$3 \cdot 10^{21}$	$7 \cdot 10^{21}$	$2 \cdot 10^{22}$
$f_2 N_{e2}$ (cm^{-3})	$5.1 \cdot 10^{17}$	$3.5 \cdot 10^{17}$	$1.4 \cdot 10^{18}$	$8.5 \cdot 10^{17}$	$1.2 \cdot 10^{18}$	$3.4 \cdot 10^{18}$

Effect of the Laser-Pulse Duration

We studied the effect of the pulse duration on the plasma parameters by fixing the pulse energy at 15 mJ, so that the change in the pulse duration from 45 fs to 1.1 ps was accompanied by decreasing in the peak intensity from 10^{17} to $3 \cdot 10^{15}$ W/cm^2. As in the previous case, clusters were produced using a conic nozzle with an initial argon pressure of 60 bar.

The plasma parameters determined by fitting the spectra of the He-α line of Ar XVII and its dielectronic satellites are presented in table 3. The relation between the excitation rates of the autoinizing levels by hot electrons and dielectronic capture rates makes even a small amount of hot 5-keV electrons to contribute substantially to excitation of some satellite lines. Figure 5 shows a variation of spectra with the laser pulse duration. The plasma parameters obtained were verified using the independent measurements of the Rydberg resonance lines of He-like ion with $n=5-9$ [26].

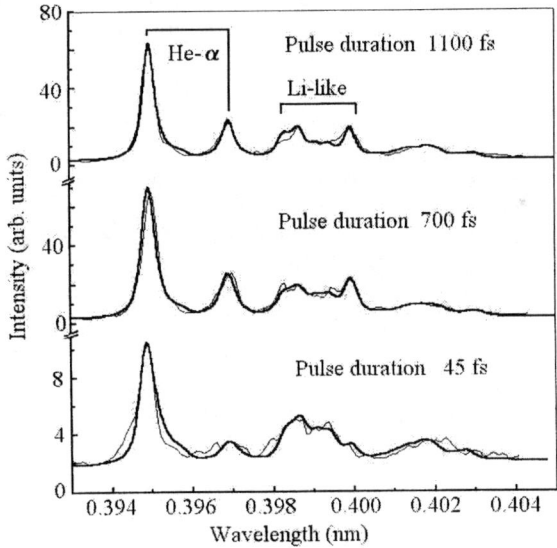

FIGURE 5. The observed He-α emission line of the H-like Ar XVIII ion at a gas pressure of 60 bar and different values of main laser pulse duration (thin curves) and the model spectra (thick curves) for the plasma parameters presented in table 3.

TABLE 3. **Effect of the laser-pulse duration on the plasma parameters.**

Pulse duration (fs)	Rarefied Plasma			Dense Plasma		
	T_{e1} (eV)	N_{e1} (cm^{-3})	f_1N_{e1} (cm^{-3})	T_{e2} (eV)	N_{e2} (cm^{-3})	f_2N_{e2} (cm^{-3})
1100	130	$3 \cdot 10^{19}$	$3 \cdot 10^{11}$	200	$3 \cdot 10^{21}$	$9 \cdot 10^{17}$
700	130	$3 \cdot 10^{19}$	$9 \cdot 10^{12}$	215	$4 \cdot 10^{21}$	$1.2 \cdot 10^{18}$
45	130	$3 \cdot 10^{21}$	$3 \cdot 10^{12}$	200	10^{22}	$7 \cdot 10^{17}$

CONCLUSIONS

The model for describing of the x-ray emission spectra of a plasma produced upon the interaction of femtosecond laser pulses with clusters was considered. It includes a number physical parameters as fitting constants and allows one to reproduce experimental spectra quite accurately. The values of these parameters vary reasonably depending on the laser–cluster interaction conditions. They represent the average values of the plasma parameters in the corresponding space–time regions.

The systematic experimental study performed in this paper for clusters of different sizes and at different durations and contrasts of laser pulses have been performed. The results allow us also to make some conclusions concerning the use of the laser–cluster plasma for several applications. For example, the optimization of heating conditions for enhancing the brightness of an x-ray lithographic source is especially important for the development of these sources. The results of our measurements show that the brightness is a non-monotonic function of the laser pulse duration. The spatial position of the maximum depends on the cluster size and the laser-pulse contrast. For the experimental conditions examined, the optimal pulse duration is about 700 fs.

The laser–cluster interaction in the $\tau_{prep} \sim \tau_{exp}$ regime allows one to obtain strongly ionized plasma. Such a plasma source can be used to obtain lasing at x-ray transitions in multiply charged Ne- and Ni-like ions where the optimal conditions for obtaining large absolute values of the population inversion are naturally satisfied (see, for example, [27–29]). In this case, it is important to produce sufficiently extended and uniform plasma. Uniform extended plasma can be obtained by using a conic nozzle for the formation of a cluster target. Note, that upon heating of clusters by a low-contrast laser pulse, the uniform plasma cannot be produced even using a homogeneous target. This is due to the fact that at low contrast the pre-pulse power is large enough to destroy completely the clusters located in the region of the Rayleigh length near the focal plane. In this case, the produced plasma will be strongly inhomogeneous. Therefore, only high-contrast laser pulses and homogeneous cluster targets can produce the plasma for the x-ray lasing.

ACKNOWLEDGMENTS

This work was supported in part by the U.S. Department of Energy, by Fond European de Development Economique Regional et Conseil Regional d'Aquitaine (France), by NATO grant PST.CLG.977637, and by CRDF grant No. RP1-2328-ME-02.

REFERENCES

1 Ditmire, T., Donnelly, T., Rubenchik, A. M., et al.., *Phys. Rev. A* **53**, 3379 (1996).
2. McPherson, A., Tompson, B. D., Borisov, A. B., et al., *Nature* **370**, 631 (1994).
3. Ditmire, T., Tisch, J. W. G., Springate, E., et al., *Phys. Rev. Lett.* **78**, 2732 (1997).
4. Ditmire, T., Zwelback, J., Yanovsky, V. P., et al., *Nature* **398**, 490 (1999).
5. Zweiback, J., Smith, R.A., Cowan, T.E., et al., *Phys. Rev. Lett.* **84**, 2634 (2000).
6. Parra, E., Alexeev, T., Fan, J., et al., *Phys. Rev. E* **62**, 35931 (2000).
7. Rozet, J.-P., Cornille, M., Dobosz, S., et al., *Phys. Scr.* **T92**, 113 (2001).
8. Auguste, T., D'Oliveira, P., Hulin, S., et al., *JETP Lett.* **72**, 38 (2000).
9. Abdallah, J., Jr., Faenov, A.Ya., Skobelev, I.Yu., et al., *Phys. Rev. A* **63**, 032706 (2001).
10. Boldarev, A.S., Gasilov, V.A., Blasco, F., et al., *JETP Lett.* **73**, 514 (2001).
11. Skobelev, I.Yu., Faenov, A.Ya., Magunov, A.I., et al., *JETP* **94**, 73 (2002).
12. Junkel-Vives, G. C., Abdallah, J., Jr., Auguste, T., et al., *Phys. Rev. E* **65**, 036410 (2002).
13. Skobelev, I. Yu., Faenov, A. Ya., Magunov, A. I., et al., *JETP* **94**, 966 (2002).
14. Stenz, C., Bagnoud, V., Blasco, F., et al., *Quantum Electron.* **30**, 721 (2000).
15. Skobelev, I.Yu., Faenov, A.Ya., Bryunetkin, B.A., et al., *JETP* **81**, 692 (1995).
16. Pikuz, T.A., Faenov, A.Ya., Pikuz, S.A., et al., *J. X-ray Sci. Technol.* **5**, 323 (1995).
17. Young, B.K.F., Osterheld, A.L., Price, D.F., et al., *Rev. Sci. Instrum.* **69**, 4049 (1998).
18. Abdallah, J., Jr., Faenov, A.Ya., Pikuz, T.A., et al., *J. Quant. Spectros. Radiat. Transfer* **62**, 1 (1999).
19. Abdallah, J., Jr., Clark, R.E.H., Faenov, A.Ya., et al., *J. Quant. Spectros. Radiat. Transfer* **62**, 85 (1999).
20. Dobosz, S., Schmitd, M., Perdrix, M., et al., *JETP* **88**, 1122 (1999).
21. Junkel-Vives, G.C., Abdallah, J., Jr., Blasco, F., et al., *J. Quant. Spectros. Radiat. Transfer* **71**, 417 (2001).
22. Junkel-Vives, G.C., Abdallah, J., Jr., Blasco, F., et al., *Phys. Rev. A* **63**, 021201(R) (2001).
23. Faenov, A.Ya., Magunov, A.I., Pikuz, T.A., et al., *Phys. Scr.* **T80**, 536 (1999).
24. Maksimchuk, A., Nantel, M., Ma, G., et al., *J. Quant. Spectros. Radiat. Transfer* **65**, 367 (2000).
25. Rosmej, F.B., Funk, U.N., Geissel, M., et al., *J. Quant. Spectros. Radiat. Transfer* **65**, 477 (2000).
26. Magunov, A.I., Pikuz, T.A., Skobelev, I.Yu., et al., *JETP Lett.* **74**, 375 (2001).
27. Nickles, P.V., Shlyaptsev, V.N., Kalachnikov, M., et al., *Phys. Rev. Lett.* **78**, 2748 (1997).
28. Dunn, J., Osterheld, A.L., Shlyaptsev, V.N., et al., in *Atomic Processes in Plasmas*, edited by E. Oks and M.S. Pindzola, AIP Conference Proceedings 443, New York: American Institute of Physics, 1998, p. 106.
29. Dunn, J., Osterheld, A.L., Shepherd, R., et al., *Phys. Rev. Lett.* **80**, 2825 (1998).

Theoretical Aspects of HULLAC

A. Bar-Shalom*, M. Klapisch¶[1], J. Oreg*

*NRCN, Be'er Sheva 84190 Israel
¶ARTEP, Inc. Columbia, MD 21044

Abstract. HULLAC is an integrated code for calculating atomic structure and cross sections for collisional and radiative atomic processes, aimed at complex spectra of heavy ionized atoms. It is based on relativistic quantum mechanical calculations including configuration interaction. The collisional cross sections are calculated in the distorted wave approximation. The target and continuum wavefunctions are obtained consistently. The theory and algorithms are presented, emphasizing the various novel methods that were developed to obtain accurate results very efficiently. In particular we describe the Parametric Potential method used for both bound and free orbitals, and the NJGRAF graphical method used in the calculation of the angular momentum part of the matrix elements. Collision cross sections are obtained extremely efficiently, thanks to the Factorization-Interpolation method applied in the derivation of collisional rates, and the Phase Amplitude approach for calculating the continuum orbitals. Special effort was devoted to ensure the simplicity of use.

I. INTRODUCTION

The HULLAC [1] code is an efficient generator of cross sections (and rates) for a collisional radiative model (CRM). These data are required for simulating non-LTE plasma emission and absorption spectra. The theory and code were developed over the last two decades, and specific parts were already described in details [2-4]. A general overview was recently published [5]. Various versions of the code are used extensively in several research centers (for a short list, see ref. [5]). The main purpose of the present paper is to introduce to the APIP community the various novel techniques that were developed to obtain a highly efficient tool, that at the same time gives accurate results for complex spectra of heavy and ionized atomic systems. We will focus here on the basic ideas and underlying assumptions in our model.

The CRM amount to solving rate equations for the level populations N_i

$$\frac{dN_i}{dt} = -N_i \sum_{j \neq i} R_{ij} + \sum_{j \neq i} N_j R_{ij} \qquad (1)$$

among all fine structure levels in the model.

---

[1] Contractor to Naval Research Laboratory, Washington, DC 20375.

CP635, Atomic Processes in Plasmas: 13th APS Topical Conference, edited by D. R. Schultz et al.
© 2002 American Institute of Physics 0-7354-0090-3/02/$19.00

The core of the model is the relativistic quantum mechanical calculation of the energies and the rates R_{ij} for all the atomic processes in the plasma i.e. radiative and collisional excitation and de-excitation, radiative and collisional ionizations and their inverse processes - radiative recombination and 3 body recombination- autoionization and electron capture.

Hullac is a coherent set of programs that use the same set of wavefunctions to compute the different processes with the same level of accuracy. Its theoretical framework is first order perturbation theory (PT) with a central field, with configuration interaction (CI). The zero-order wavefunctions are solutions of the Dirac equation. The Hamiltonian is diagonalized on the basis of these wavefunctions, and the resulting eigenvectors are used for all the processes. The domain of best performance is dealing with ionized heavy atoms with several open shells, but it can be applied to any atomic system. The reason for this performance is a combination of theoretical insight and novel numerical techniques.

The typical example is the Factorization-Interpolation method for collisional excitations [4] a cornerstone of the model, which in conjunction with the phase amplitude algorithm [6] shortens the computing time by orders of magnitude with no loss of accuracy. Thanks to this factorization, in effect a decoupling between bound and free electron wavefunctions, the angular matrix elements of the collisional processes can be handled easily as all the other processes by the powerful graphical methods implemented in NJGRAF [3].

For the radial parts, the parametric potential method guarantees solution of the radial wavefunctions, even for configurations with several open shells.

Special efforts were made to insure the ease of use. The input (all in free format) consists only of the configurations of the various ions in the model, and a list of processes described by keywords. When solving the rate equations, populations of levels are also obtained.

The paper contains four sections: Section II focuses on the atomic structure, with the parametric potential and the recoupling coefficients. In section III we describe the collision part, with the factorization-interpolation and the phase amplitude methods. In section IV we briefly present an example demonstrating the efficiency, and we discuss the accuracy of the code.

II. ATOMIC STRUCTURE

1. The Parametric Potential

Using the central field principle [7] the relativistic Hamiltonian can be written as $H = H_0 + H_1$ where, in atomic units:

$$H_0 = \sum_i \left(h_i^D + U(\alpha; r_i) \right) \qquad H_1 = \sum_i \left(-U(\alpha; r_i) - \frac{Z}{r_i} \right) + \sum_{i,j} \frac{1}{|r_i - r_j|} \quad . \quad (2)$$

H_0 is the spherically symmetric zero order Dirac Hamiltonian representing the spherical average of the interaction between the electrons where h_i^D is the single particle kinetic Dirac Hamiltonian and $U(\alpha;r_i)$ is the parametric potential (PP), so that H_1 can be treated as a perturbation.

The parametric potential [2, 8, 9] introduces a screened Coulombic potential

$$U(\underline{\alpha}; r) = -\frac{1}{r}[\Sigma_s q_s f_\ell(\alpha_\ell;r) + z] \quad \text{with} \quad \Sigma_s q_s + z = Z \tag{3}$$

where q_s is the shell's occupation number, Z is the nuclear charge, z is the charge seen by a test electron, and $f_\ell(\alpha;r) = e^{-\alpha r} \Sigma_{j=0}^{2\ell+1}(1 - \frac{j}{2\ell+2})\frac{(\alpha r)^j}{j!}$ describes the screening obtained from a Slater-type "zero order screening orbital". The parameter α is proportional to the radius of the latter orbital.

The first order solutions are obtained by constructing the matrix of $H' = H_0 + H_1$ in the basis of the chosen configurations obtained with the PP, and diagonalizing it, yielding mixed configuration states and first order energies with some correlation corrections. The resultant wave functions are then used to add the Breit and QED corrections to the state energies [10]. These energies thus become algebraic functions of the set of parameters $\underline{\alpha}$. Consequently, contrarily to Hartree-Fock based on integro-differential equations, there are *many* possible ways to choose the parameters according to various equally theoretically valid criteria [2, 8]. One of these, the default option in our code, is the minimization of the first order configuration average energy of any chosen set of configurations. Among other possibilities, which can be used with a simple keyword at run time, one can minimize the *rms* deviation between theoretical and experimental level energies.

Other advantages of the PP are that unlike the HF methods it always converges, even for configurations with many open shells. In addition since first order energies, on which optimization is performed, include the exact exchange contribution, the resulting parameters absorb the effect of exchange, and therefore, there is no need for an explicit additional exchange potential. Moreover the resulting orbital set is orthogonal. Consequently, there can be a complete factorization of radial and angular parts, and all the angular momentum algebra can be handled in one program, as described in the next paragraph.

2. The Graphical Approach for Recoupling Coefficients

We present here a brief discussion of the graphical method, for details see ref. [3]. The Hamiltonian and the various transition rates are represented quantum mechanically by coupled tensor operators of the general form $T = \left[\left[T_1^{(k_1)} T_2^{(k_2)} \right]^{(k_{12})} \cdots \right]^{(K)}$. The reduced matrix elements $\langle \psi(\Gamma J) \| T \| \psi'(\Gamma' J'') \rangle$ are represented as recoupling coefficients (RCs). Any RC can be expressed in terms of 6j symbols, generally including summations over new variables. However this expression is not unique and some expressions can be faster to calculate than others by orders of magnitude. Our approach yields a near optimal decomposition using graphical methods. It is well known that reduced matrix elements correspond to closed diagrams. The decomposition of the diagram is done by graphical rules such as cutting lines or removing loops (closed polygons) For example a loop of two lines introduces a delta function, a "loop" with 3 sides gives a 6j, and a "loop" with 4 sides gives a sum of products of 2 6j's. The simplest decomposition of the graph is achieved by identifying (and taking into account) first the smallest loops.

A simplifying strategy for encoding is to describe the closed diagram as a product of flat diagrams (FD) [11] as shown in Fig. 1.

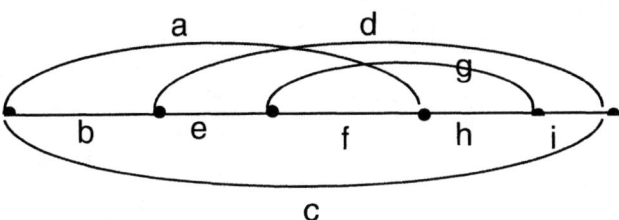

FIGURE 1. A flat closed diagram.

In most cases the closed diagram reduces to a single FD. This FD representation enables locating loops of any order immediately since the FD is one-dimensional where all the nodes appear on the basic line. One needs simply to follow the subsequent nodes from left to right until reaching a node containing a 'j' of a previous node, which graphically closes the loop.

The number of contained nodes (including the edge nodes) defines the order of the loop. It is seen immediately, for instance, that the loop covered by 'g' is of order 3 and is the smallest. An important graphical operation is shown in Fig. 2. It is the interchange of two angular momenta belonging to two neighboring nodes i.e. interchanging b and c in the *lhs* diagram yields the *rhs* diagram. This is one of the rules used to reduce the order of loops. We have used it in the factorization of the collision strength as shown in the next section. This algorithm has been published as a separate program: NJGRAF [3].

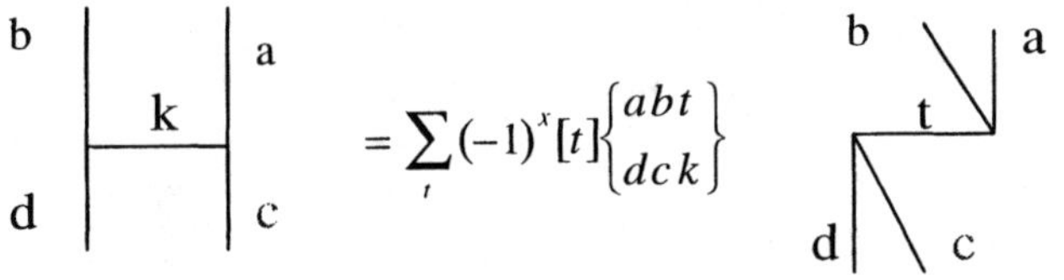

$$= \sum_t (-1)^x [t] \begin{Bmatrix} a\,b\,t \\ d\,c\,k \end{Bmatrix}$$

FIGURE 2. Exchange diagram rule: The phase x is determined by the convention used.

III. ELECTRON COLLISIONS

1. The Factorization Interpolation (FI) Method for Collisional Excitation

For ionized atoms, the distorted wave approximation is probably sufficiently accurate (see next section for a discussion). The collision strength can be therefore be written

$$\Omega_{01} = 8 \sum_{\mathbf{j}_0 \mathbf{j}_1} \sum_{J_T M_T} \left(\psi_0(\gamma_0 J_0), \mathbf{j}_0, J_T M_T \left| \sum_{i<j} \frac{1}{r_{ij}} \right| \psi_1(\gamma_1 J_1), \mathbf{j}_1, J_T M_T \right)^2 . \tag{4}$$

The bra and ket are the initial and final states that are totally anti-symmetric functions of N+1 bound and continuum electrons, coupled together to a total angular momentum $J_T M_T$. The bound parts ψ_0 and ψ_1 are mixed configuration atomic states where γ_0 and γ_1 stand for all the required quantum numbers specifying the states. The lower case $\mathbf{j}_0 \equiv \varepsilon_0 l_0 j_0$ and $\mathbf{j}_1 \equiv \varepsilon_1 l_1 j_1$ represent the incoming and outgoing continuum electrons. Energy conservation requires that

$$E_0 + \varepsilon_0 = E_1 + \varepsilon_1 . \tag{5}$$

We start with the representation of the electrostatic interaction [4]

$$\sum_{i<j} \frac{1}{r_{ij}} = \sum_{abcd} \sum_k X^k(ab,cd) \sum_{i<j} \left(z_i^{(k)}(ac) \bullet z_j^{(k)}(bd) \right) \tag{6}$$

where $a \equiv n_a l_a j_a$ etc., $z^{(k)}$ are the unit operators defined by their reduced matrix elements

$$\left(a' \left| z^{(k)}(ab) \right| b' \right) \equiv \delta_{aa'} \delta_{bb'} \tag{7}$$

and

$$X^k(ab,cd) = \left(j_a \left\| C^{(k)} \right\| j_c \right) \left(j_b \left\| C^{(k)} \right\| j_d \right) R^k(ab,cd) \tag{8}$$

defined in terms of the spherical harmonics operators $C^{(k)}$ and the Slater integrals $R^k(ab,cd)$. In equation (6) each of the indices runs independently over the same complete set of bound and continuum orbitals. Thus ordered sets like $\alpha,\beta,\gamma,\delta$ and $\alpha,\beta,\delta,\gamma$ will appear separately, giving rise to direct and exchange X^k integrals. We can use the exchange diagram rule of Fig. 2 to express the corresponding exchange angular operator as a sum of direct type operators. (see reference [4] for details) We then collect together the corresponding radial parts and obtain the direct type representation of the electrostatic interaction

$$\sum_{i<j} \frac{1}{r_{ij}} = \sum_{ac} \sum_k \phi^k(a\mathbf{j}_0, c\mathbf{j}_1) \sum_{i<j} \left(z_i^{(k)}(a,c) \bullet z_j^{(k)}(\mathbf{j}_0, \mathbf{j}_1) \right) \tag{9}$$

where b and d of equation(6) are now replaced by continuum orbitals as in the matrix element of equation(4). Now only direct angular operators appear, with generalized radial parts

$$\phi^k(a j_0, c j_1) = X^k(a j_0, c j_1) + \sum_t (-1)^{k+t} [t] \begin{Bmatrix} j_a j_1 t \\ j_0 j_c k \end{Bmatrix} X'(a j_0, j_1 c) \quad (10)$$

that include both direct and exchange integrals. This representation allows the use of Racah's rule for separating the two systems [12]. The result, allowing for configuration interaction is:

$$\Omega_{01} = \sum_{ac,a'c'} \sum_k \langle \psi_0(\gamma_0 J_0) | Z^{(k)}(ac) | \psi_1(\gamma_1 J_1) \rangle \langle \psi_0(\gamma_0 J_0) | Z^k(a'c') | \psi_1(\gamma_1 J_1) \rangle Q^k(ac, a'c') \quad (11)$$

where $Q^k(ac, a'c') = \sum_{j_0 j_1} \phi^k(a j_0, c j_1) \phi^k(a' j_0, c' j_1)$ and $Z^{(k)} = \sum_i z_i^{(k)}$. The matrix elements connecting the bound states are decoupled from the continuum part that appear only in the radial integrals Q^k.

Now the radial integrals Q^k depends of each specific bound transition through the energy conservation of equation(5). However, it turns out that these integrals vary very smoothly with energy. Therefore we can calculate the Q^k's with only a few energy points and interpolate for the various transitions.

The same method can be applied to autoionization [13], and to collisional ionization [14].

2. The Phase Amplitude Method for Continuum Calculations

The calculations of the continuum wavefunctions require the solution of equation of the form $P'' + \omega(r)P = 0$. This takes the dominant part of the total computation time [6].

The problem arises from the highly oscillating nature of the solutions. The usual representation (equidistant r grid) requires many points, especially at high energies, to obtain sufficient accuracy and to reach the asymptotic amplitude. Furthermore, the optimal radial grid depends on the energy. The Phase Amplitude (PA) approach speeds up the calculations dramatically. The method is described in details in ref. [6]. Here we only outline the essential points.

Writing the solution as $P = y \sin \varphi$ we obtain two equations for the amplitude and the phase: $y'' + \omega(r)y - a^2/y^3 = 0$, $\phi(r) = \int_0^r a/y^2 dr$ where $a = \sqrt{2/\pi}$ is the normalization that dictates the orthogonality $\int P_\varepsilon(r) P_{\varepsilon'}(r) dr = \delta(\varepsilon - \varepsilon')$. The obvious advantage is that $y(r)$ and $\varphi(r)$ are smooth functions that can be described on the same "*log r*" mesh that is used for bound wave functions.

The PA method is well known [15], but it is seldom actually implemented for collision calculations. There are two main reasons for this.

First, in this case the usual numerical algorithms (e.g. predictor corrector) fail due to the nature of the function $\omega(r)$. The difficulty is that the predictor-corrector scheme uses the differential equation to predict y" from y, i.e: $y" = -\omega(r)y + a^2/y^3$. In a certain region $\omega(r)$ becomes very large while y and y" are very small. Then y" is obtained as a difference of two large almost equal numbers. Thus a minute error in y yields a huge error in the prediction for the small value of y". The problem is solved by exchanging the roles of the y and y" 'in the predictor corrector scheme using $a^2/y^4 = \omega(r) + y"/y$ to obtain y from $y"/y$. Noting that in the problematic region $y"/y$ is a small number, we can see that in this scheme the correction to y is only slightly affected by errors in $y"/y$ and the method becomes remarkably stable.

The second reason is that solving for y and $\varphi(r)$ does not complete the task since the radial integrals still require to follow the high oscillations of the sine functions. We have developed an analytic procedure by changing variables

$$I = \int_0^\infty y_1(r)y_2(r)\sin\varphi_1(r)\sin\varphi_2(r)dr = \int X(\Phi)\cos(\Phi)d\Phi + \int X(\phi)\cos(\phi)d\phi \quad (12)$$

where $\Phi = \varphi_1 + \varphi_2$, $\phi = \varphi_1 - \varphi_2$. The integration now is done on the phases of the cosines. This allows the use of Gauss trigonometric formulas to integrate over half cycles. The result is a sum of an alternating series. Using the Levin transform [16], the sum converges with only a few half-cycles. Although the algorithm and the code are not straightforward, this PA approach yields a dramatic reduction in computation time.

IV. DISCUSSION

1. Efficiency

As an example of the efficiency of the code we have calculated data for Ni-like and Co-like ions of gold. The configurations in the model are listed in Table 1. The total number of levels for these two ions is 3562. The number of rates for the various processes and their run times are listed in Table 2. Thanks to the factorization of angular parts, and to the FI method for collisions, the number of radial integrals to be computed is reduced dramatically. For instance for excitation the FI method reduced the number of required integrals from 3,404,855 to only 875. The *total run time* on a simple PC (AMD XP 1500) was 218 seconds.

TABLE 1. Configurations and levels included in example.

Configurations Included	Type	No. of Levels
•Ni-like		
$(3s\ 3p\ 3d\ )^{18(a)}$	ground	1
$(3s\ 3p\ 3d\ )^{17}\ \ (4s\ 4p\ 4d\ 4f\ )^1$	singly excited	106
$(3s\ 3p)^8\ (3d\ )^8\ (4s\ 4p\ 4d\ 4f\ )^2$	doubly excited	2840
•Co-like		
$(3s\ 3p\ 3d\ )^{17}$	ground	6
$(3s\ 3p\ 3d\ )^{16}(4s\ 4p\ 4d\ 4f\ )^1$	singly excited	609

(a)The notation $(nl_1\ nl_2\ nl_3...)^N$ means distributing N electrons in all possible ways among the shells included in the parenthesis.

TABLE 2. Breakdown of number of rates and computing time.

Process	Number of Elements	Computing Time (seconds)
Angular recoupling for all processes	N/A	41
Levels	3562	⎫
Radiative rates	66,246	⎬27
Autoionization	1,492	⎭
Collisional excitation	225,878	64
Photo-ionization	442,224	25
Collisional ionization	1,414,308	61

2. Accuracy

As mentioned in the introduction, HULLAC is based on solving homogeneous Dirac equations in a central potential. Consequently, the accuracy of the energy levels is very good for highly ionized atoms when exchange is not very large. This is the case for systems isoelectronic to alkalis or rare gases. Conversely, for levels involving large contributions of G^1 integrals — such as Ni-like $3d^9\ 4f\ ^1S_0$ — the disagreement with experiment can be 2-3 times larger than the average deviation. This problem can only partially be compensated by configuration interactions, and it affects obviously the quality of the various transitions probabilities.

For collisions, we have the well known limits of the distorted wave approximation (DWA), without unitarization or resonances. Nevertheless, a recent comparison [17] shows that for ionized atoms, such as Mg^{10+}, and resonant transitions, the disagreement with much more sophisticated methods is not more than 10-20%. Weaker transitions are more susceptible to the effects of resonances. For lowly ionized atoms (e.g $z \leq 5$), unitarization could be important.

We will conclude with two remarks: (1) In the framework of the CRM, most of the weak transitions are not observed, except at extremely low density. Moreover, part of the resonances can be simulated in the "physical picture" as transitions through doubly excited states. (2) Because the algorithms are so efficient, it is worthwhile considering adding modules to HULLAC for the evaluation of second order effects, resonances in collisions, etc. Work is in progress in this direction.

ACKNOWLEDGMENTS

A.B-S and J.O. thank the Naval Research Laboratory, Laser Plasma Branch for hospitality and support during the year 2000-2001.

The Parametric Potential was developed was developed by M.K. while he was affiliated with Laboratoire Aimé Cotton (LAC), Orsay, France, and the relativistic version of it by E. Luc-Koenig, of LAC. The overall concept was developed at the Hebrew University of Jerusalem, Israel, by M.K. and A.B-S. The majority of the algorithms and of the coding are by A.B-S, with some parts by M.K. Other contributors include J.O. and W.H. Goldstein. During 1985-1988, the project benefitted from a grant from Lawrence Livermore National Laboratory, hence the name Hullac. Recently, the code has been completely overhauled and many parts added without this support, but the name remained.

REFERENCES

1. Hebrew University Lawrence Livermore Atomic Code – see Acknowledgments.
2. Klapisch, M., *Comput. Phys. Comm.* **2**, 239 (1971).
3. Bar-Shalom, A., and Klapisch, M., *Comput. Phys. Comm.* **50**, 375 (1988).
4. Bar-Shalom, A., Klapisch, M., and Oreg, J., *Phys. Rev. A* **38**, 1773 (1988).
5. Bar-Shalom, A., Klapisch, M., and Oreg, J., *J. Quant. Spectrosc. Radiat. Transfer* **71**, 169 (2001).
6. Bar-Shalom, A., Klapisch, M., and Oreg, J., *Comput. Phys. Comm.* **93**, 21 (1996).
7. Slater, J. C., *Phys. Rev.* **34**, 1293 (1929).
8. Klapisch, M., Ph. D. Dissertation, *Une Nouvelle Methode pour ie calcul des fonctions d'onde atomique.* University of Paris, France (1969).
9. Koenig, E., *Physica* (Utrecht) **62**, 393 (1972).
10. Grant, I.P., *Adv. Phys.* **19**, 747 (1970).
11. Brink, D. M., and Satchler, G. R., *Angular momentum* (Clarendon Press, Oxford, 1968).
12. Judd, B. R., *Operator Techniques in Atomic Spectroscopy* (McGraw-Hill, New York, 1963).
13. Oreg, J., Goldstein, W. H., Klapisch, M., and Bar-Shalom, A., *Phys. Rev. A* **44**, 1750 (1991).
14. Sampson, D. H., and Zhang, H. L., *Phys. Rev. A* **45**, 1657 (1992).
15. Milne, W. E., *Phys. Rev.* **35**, 863 (1930); Wheeler, J. A., *Phys. Rev. A* **52**, 1123 (1937).
16. Levin, D., *Intern. J. Computer Maths.* **3B**, 371 (1973); Blakemore, M., Evans, G.A., and Hyslop, J., *J. Comp. Phys.* **22**, 352 (1976).
17. Mitnik, D. M., and Pindzola, M. S., *Phys. Rev. A* **62**, 062711 (2000).

Non-Maxwellian Laser Produced Plasmas

F. B. Rosmej[†*], A. Calisti[†], B. Talin[†], R. Stamm[†], D. H. H. Hoffmann[*],
M. Geißel[*], Ya. Faenov[¶], T.A. Pikuz[¶], I.Yu. Skobelev[¶]

[†]*Laboratoire de Physique des Interactions Ioniques et Moléculaires, Université de Provence,
13397 Marseille cedex 20, France*
[*] *Gesellschaft für Schwerionenforschung GSI, Plasmaphysik,
Planckstr. 1, D-64291 Darmstadt, Germany*
[¶] *Multicharged Ions Spectra Data Center of VNIIFTRI, Mendeleevo 141570, Russia*

Abstract. Dense non-Maxwellian plasmas have been created at the high energy nhelix-laser test
bed facility under well defined hydrodynamic conditions. Space resolved hot electron induced
K_α-emission from the target surface has been observed simultaneously with the x-ray radiation
emission of highly charged target ions. Two-electron inner-shell transitions are introduced as
reference lines to provide a unique parameter determination. Non-Maxwellian non-LTE kinetics
simulations for opaque plasmas as well as Stark broadening calculations of complex
autoionizing levels provide a proof of the proposed methods. Excellent agreements with the data
are demonstrated. Space resolved x-ray Doppler spectroscopy provides fast ion energy
distribution.

INTRODUCTION

The interaction of high intensity laser beams with matter generates suprathermal
electrons as well as fast ions. A continuously growing interest in these non-
Maxwellian particles is due to their importance for both, basic physics studies and
numerous technical and diagnostic applications.

The development of short pulse x-ray sources driven by hot electron induced K_α-
emission from the target material for the study of ultrafast phenomena undergoes
intense studies in, e.g., chemistry, biology and physics [e.g. 1]. Further interest in the
hot-electron generation is derived from the proposal of the fast ignitor for inertial
confinement fusion [2]. In laser driven ICF hohlraums, hot electron production
prevents final compression and burn [3] and the control of the hot electron fraction
and the characterization of instability mechanisms is essential for a high performance
[4]. This has strongly stimulated the development of spectroscopic methods for the
analysis of plasmas containing hot electrons [e.g., 4-12].

The generation of fast ions in intense laser field is of equally growing interest.
Applications range from new accelerator concepts [e.g. 13] and material processing
until the use of accelerated protons for ultrafast proton radiography [14]. Traditionally
characterization of the fast ions and protons has been performed with particle
analyzers located several meters away from the interaction region. Useful scaling laws
have been derived [15, 16]. However, recent experiments [17] have demonstrated
order of magnitude deviations showing that MeV-energy Target ions can be generated

even for rather moderate laser intensities. Understanding of these phenomena is still going on. In order to receive data about the ion energy distribution inside the plasma rather than outside (by means of particle detectors) space resolved Doppler x-ray spectroscopy of target ions at various laser installations has been conducted [18].

In almost all cases, a space resolved parameter information is requested. For these purposes, we employ high-resolution x-ray optics at the high energy **nhelix**-laser facility at GSI-Darmstadt to receive simultaneously high spectral and spatial resolution while maintaining high luminosity. Under rather well defined hydrodynamic conditions we report about the investigation of non-Maxwellian particles (electrons, highly charged target ions).

EXPERIMENTAL SETUP

Experiments were carried out at the *"nhelix-laser"* installation (nano second **high** energy laser for **h**eavy **i**on e**x**periments) at GSI in Darmstadt, Germany. *"nhelix"* is a Nd:Glass/YAG laser (λ_{las}=1.06 μm) with a pulse duration of 15 ns and an energy up to 100 J. The laser radiation is focused with a plane-convex lens (diameter 100 mm, focal length f = 130 mm) onto solid F, Mg, Al and Si targets. The focal spot diameter varied between 100 and 1000 μm giving an intensity up to 10^{13} W/cm^2 onto the target. In order to obtain different laser intensities, the distance between lens and target was changed (movement Δz of lens).

Soft X-ray emission from highly charged target ions were recorded with X-ray optical methods employing spherically bent mica crystals [19] and Kodak DEF-5 X-ray film (grain size less than 2 μm). Two foils of 1 μm polypropylene covered from each side with 0.1 μm Al protected the film from visible light. The distances between target and crystal were 200-300 mm. The crystals had a curvature radius of R = 150 mm and lattice spacing of d_1 = 9.95745 Å, d_2 = 9.96885 Å and d_3 = 9.97095 Å for the first (F), second (Mg, Al) and third (Si) order of reflection, respectively. X-ray images have been scanned with a 10.000 dpi (1 pixel corresponds to 2.5 μm) drum scanner (EUROCORE). The spectral resolution was up to $\lambda/\Delta\lambda$ = 7000 and the total spatial resolution was up to about δx = 7 μm. Spectra have been corrected for non-linear dispersion curves, filter transmission, crystal reflectivity and film response using the calibration code SCALE.

FAST IONS

Until now the main part of the experimental information on fast ion production in laser plasmas have been obtained with the help of mass spectrometry methods [15]. Such methods are based on the direct observation of ions over large distances: performed usually several meters away from the place of plasma creation by charged particle detectors (e.g., Faraday detector). In this case, the results depend strongly on the recombination processes which occur during the plasma expansion to large distances. Thus, these methods are suitable to investigate laser-produced plasmas as possible sources of multi-charged fast ions for some practical applications but have serious limitations for studies of the mechanisms of fast ion production inside the laser produced plasma. For this purpose the indirect spectroscopic methods are more suitable: observation of photons emitted by fast ions rather than observing directly fast ions.

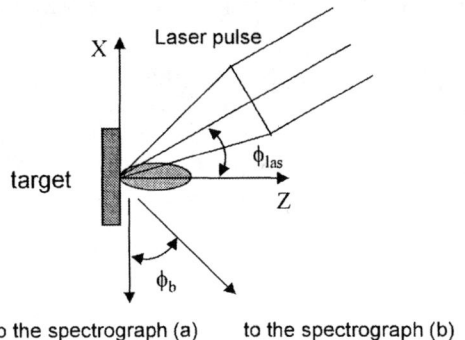

FIGURE 1. Scheme of the experiment and geometry of Doppler spectroscopy.

The main idea to use X-ray spectroscopy for the observation of fast ions is the following. Suppose that we have a plasma expansion which is predominantly in the direction normal to the target surface (z-axis). In this case the observed emission spectra depend on the angle between the direction of observation and z-axis due to the Doppler effect. By using several spectrographs simultaneously it is possible to derive from the observed spectra the direction of the predominant plasma expansion and its velocity distribution. E.g., if the plasma is cylindrically symmetric (the usual case for the interaction of a laser pulse with a flat target) the use of even two spectrographs (as shown schematically in Fig. 1) provides the necessary information. In this case spectrograph (a) observes the plasma in a direction perpendicular to the z-axis and the line profiles are expected to be essentially symmetric. The width of the lines (apart the random walk characterized by the ion temperature T_i) is due to the transverse component $V_{x,y}$ of the plasma expansion velocity:

$$\frac{\Delta \lambda(a)}{\lambda} \approx \frac{V_{x,y}}{c}. \tag{1}$$

For the second spectrograph (b) the situation will be different. In this case we have no symmetry for the plasma motion as the plasma moves only in positive z-direction. The spectral line will be shifted to shorter wavelengths by

$$\Delta \lambda = \lambda_0 \frac{V_z}{c} \sin \varphi_b. \tag{2}$$

Because different ions may have different expansion velocities, the observed spectra will show corresponding wings on the blue side of all spectral lines. The line profile is strongly asymmetric. Due to the relation

$$I\left(\lambda_0 + \lambda_0 \frac{V_z}{c} \sin \varphi_b\right) \propto N(V_z) \tag{3}$$

it is possible to determine the number $N(V_z)$ of fast ions for a given velocity, that means the velocity distribution. It should be noted that usually $V_z \gg V_{x,y}$ and consequently the asymmetry observed by spectrograph (b) must be much more pronounced than the broadening observed by spectrograph (a).

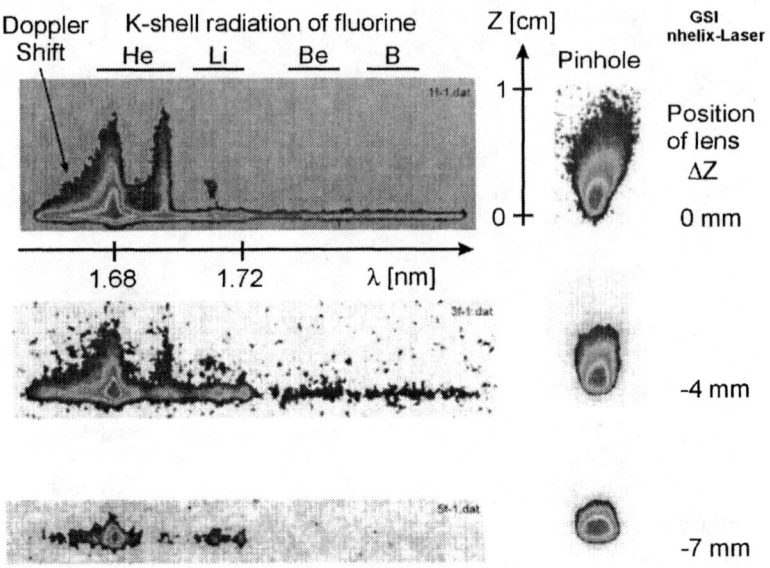

FIGURE 2. X-ray images of fluorine (space resolved K-shell spectra and x-ray pinholes) for various positions of the focussing lens. At optimum focus ($\Delta z = 0$), a strong Doppler shift at the blue wing of the He-like resonance line is seen which gradually decreases for larger spot sizes ($\Delta z = -4, -7$ mm). Spatial extension of the He-like lines correlates with the extension of the x-ray pinholes.

Figure 2 shows the x-ray images of fluorine for various positions of lens: $\Delta z = 0$ corresponds to the optimum focus (spot size about 100 µm). At $\Delta z = 0$, the He-like resonance line shows a strong blue wing due to the Doppler shift (indicated with arrow in Fig. 2) and a large spatial extension up to cm. This extension correlates with the x-ray pinhole images. Simultaneous observation of the H-, He-, Li-, Be-, B-, C-, N-, O-, F- like x-ray radiation as well as the K_α-radiation indicates, that only the He-like resonance line $1s2p\ {}^1P_1 - 1s^2\ {}^1S_0$ and the intercombination line $1s2p\ {}^1P_1 - 1s^2\ {}^1S_0$ radiate up to cm far away from the target. This indicates that the image sizes of the pinhole camera pictures are predominantly determined by the He-like resonance and intercombination lines.

Figure 3 shows the relative intensity in the blue wings of the He_β line of fluorine versus the Doppler shift measured in terms of ion energy (directed motion) near the target surface. The smooth line represents a Maxwellian distribution fit (with the temperature defined by the angle to the abscise axis in log-plots). The results show that the experimental uncertainty is sufficiently small to determine the average energy values for the fast ions.

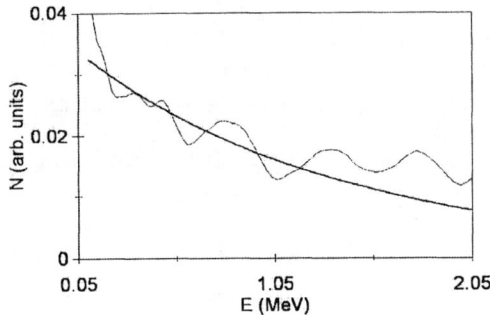

FIGURE 3. Ion energy distribution function of fluorine determined from the Doppler shifted blue wing of the He$_\beta$-line near the target surface.

SPACE RESOLVED X-RAY SPECTRA

Parameter analysis strongly depends on the spatial variation of temperature, density and hot the electron fraction. We are therefore highly interested in the spatial emissivity profiles of various transitions while keeping the extremely high spectral resolution. Fig. 4 shows the X-ray images of the He$_\beta$, Ly$_\alpha$, He$_\alpha$ and its dielectronic satellite structure (irradiating a solid Mg target with E = 17 J).

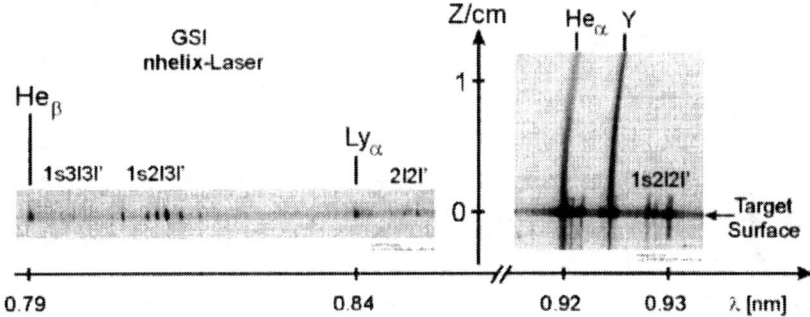

FIGURE 4. Space resolved X-ray images (observed at 45° to the target) and He$_\alpha$ (observed at 20° to the target) of Mg. The Z-coordinate is the direction of the expanding plasma.

Figure 4 shows the characteristic resonance lines He$_\beta$ = $1s^2 - 1s3p\ ^1P_1$, Ly$_\alpha$ = $1s -$ 2p, He$_\alpha$ = $1s^2 - 1s2p\ ^1P_1$, Y = $1s^2 - 1s2p\ ^3P_1$ as well as numerous satellite transitions originating from the states 1s3l3l', 1s2l3l', 2l2l', 1s2l2l' of hollow ions. It can clearly be seen, that satellite emission is confined to the target surface, whereas He$_\alpha$ and Y show strong emissions even up to cm-distances indicating a large inhomogeneity (different emissions in Z-direction) between different types of line emissions. The emission of He$_\beta$ and Ly$_\alpha$ and in particular the dielectronic satellites are well located near the target surface: their emission spatially extends to only some 100 μm (corresponding to the laser spot size). These results suggest the way towards the observation of a local emission profile: the use of various satellite transitions for diagnostics.

In this case, the emissivity peaks in a very restricted spatial area and line of sight integration effects are small. This provides a localized emission source of the highest density plasma region and is due to the functional energy dependence of the satellite excitation mechanism: dielectronic recombination is a resonance process. Satellite transitions therefore work as a "local probe" of plasma parameters. Abel-type inversion techniques which are based on rather uncertain assumptions (either test-functions or hydro-grids) are avoided.

HOT ELECTRONS

Figure 5 shows the influence of hot electrons on the ion abundance's. It can be seen clearly, that hot electrons lead to a larger population of higher charge states. This is due to the increased effective ionization rates. However, rather than shifting only the ion abundance towards higher charge states, also qualitative deformation shows up. This qualitative deformation is a great challenge for the characterization of hot electrons [5] and has lead to important applications, e.g., the determination of the hot electron fraction in NOVA-Hohlraums and the identification of responsible instability mechanisms [4].

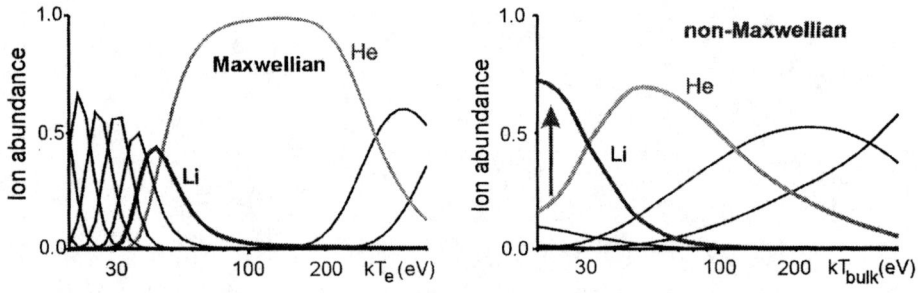

FIGURE 5. MARIA-simulation of ion abundances of non-Maxwellian Mg plasma. a) $n_e = 10^{21}$ cm^{-3}, b) $n_e = 10^{21}$ cm^{-3}, $kT_{beam} = 3$ keV, $f_{hot} = 0.1$.

The qualitative distortion, however, results in a non-unique parameter solution when applying standard methods. Its basic origin is the lost of stable reference lines [20]. Under non-Maxwellian parameter conditions of dense plasmas, the requirements for stable reference lines: a) Insensitivity to density, i.e., negligible redistribution effects between the autoionizing levels, b) Insensitivity to hot electrons, i.e., small inner-shell contributions, c) Insensitivity to electric field effects, i.e., no serious changes in transition probabilities, d) Negligible opacity, and e) Spatially restricted. In the present work we show, that two-electron satellite transitions, β_6 and β_7 (1s2s3s,3d $-$ 1s^{2}2p, for designations see [20]) near He$_\beta$ and op $=$ 1s2s^2 $-$ 1s^{2}2p fulfill these requirements.

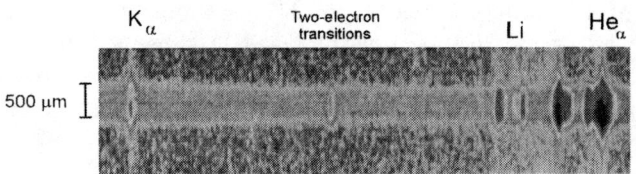

FIGURE 6. Space resolved high resolution x-ray image of Al.

The essential properties of the two-electron satellite transitions are: first, very high autoionizing rates and, second, transition probabilities that are 1-2 orders of magnitude smaller than the largest in their complexes. Large autoionizing rates give rise to stability versus collisional redistribution between the autoionizing levels, because population transfer cannot compete with the large dielectronic capture rates. Moreover, large autoionizing rates result in small sensitivity to inner-shell collisional excitation, so that effects of hot electrons are likewise small. Due to the relatively low transition probabilities, photo absorption effects are also negligible even for large-scale dense plasmas; see [21] for an example where opacity is an issue. Figure 6 demonstrates, that despite the low transition probabilities, two-electrons transitions can be well observed.

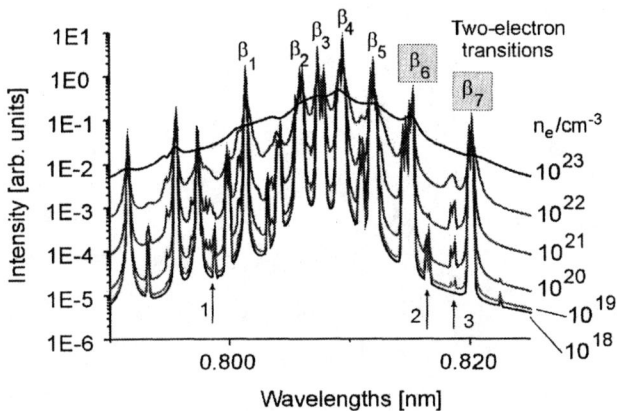

FIGURE 7. "PPP-code" Stark broadening calculations of Mg He_β-satellites, $1s2l3l' - 1s^2 2l''$ for various electron densities, $kT_e = kT_i = 100$ eV. The arrows indicate changes in the intensity patterns.

For low transition probabilities, electric field effects in dense plasmas may alter remarkably the intensity. In order to investigate these effects in detail, extended Stark broadening calculations (including Doppler, electron broadening and Stark broadening) have been performed with the PPP, a code based on the Frequency Fluctuation Model, [22] including the entire complex $1s2l3l'$. Fig. 7 shows the results for the He_β-satellites for various electron densities. A statistical population of the $1s2l3l'$ levels was assumed in order not to mask electric field effects with population redistribution driven by electron collisions. The arrows indicate the changes in the intensity pattern. For the He_β-satellites the absolute intensity level of the affected transitions is of the order of two-electron intercombination transitions, e.g., the $1s2s[^3S]3d\ ^4D$ – $1s^23p\ ^2P$ transition (arrow 2). However, doublet two-electron transitions, i.e. β_6 and β_7, show stable intensity patterns. These investigations using PPP indicate that for n_e from 10^{18} to 10^{22} cm^{-3} the electric field does not change the intensity pattern of the two electron transitions β_6 and β_7 near He$_\beta$ (and also not those of the intercombination transitions $1s2p[^3P]3d\ ^2D$ – $1s^23d\ ^2D$ and $1s2p[^3P]3p\ ^2P$ – $1s^23p\ ^2P$ near He$_\alpha$).

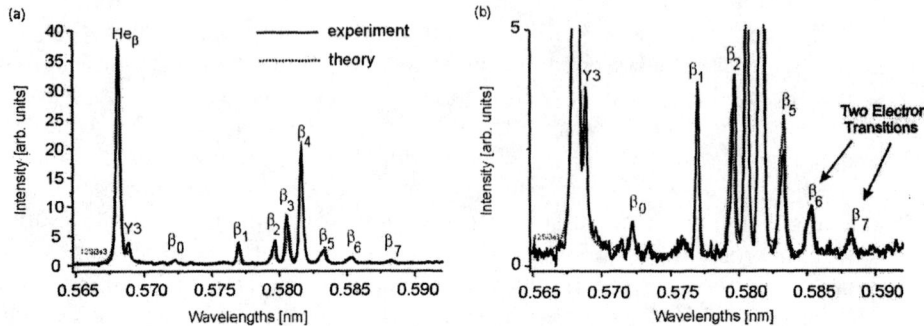

FIGURE 8. (a) MARIA-simulation of the He$_\beta$ spectral range, $n_e = 3\cdot10^{20}$ cm^{-3}, kT$_e$ = 330 eV, L$_{eff}$ = 100 μm. (b) enhanced scale (linear).

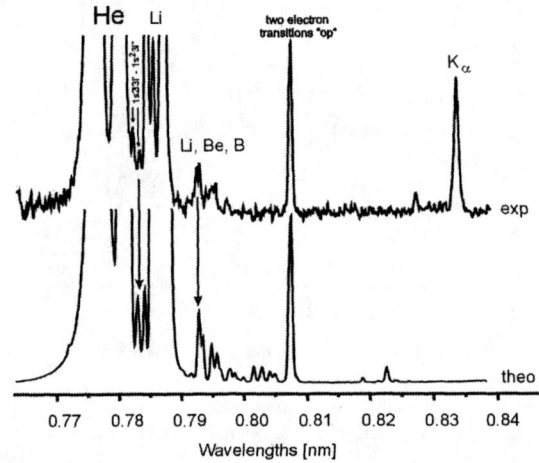

FIGURE 9. (a) MARIA-simulation of the He$_\beta$ spectral range, $n_e = 3\cdot10^{20}$ cm^{-3}, kT$_e$ = 330 eV, L$_{eff}$ = 100 μm, (b) enhanced scale (linear).

Let us now consider spectral simulations including two-electron satellite transitions as reference lines. Figure 8 shows the He$_\beta$ spectral range of silicon obtained at the GSI *nhelix* laser and the MARIA-simulations using the two-electron transitions β_6 and β_7 as reference lines. Overall agreement is found even on the enhanced linear scale: for the He_β line and its intercombination line $Y3 = 1s3p\ ^3P_1 - 1s^2\ ^1S_0$, the 1s3l3l'-satellites and almost all details for transitions β_1 to β_7. The fit also includes the opacity sensitivity of He_β and $Y3$ yielding information about the source size in dense optically thick plasmas [23].

The two-electron transitions β_6 and β_7 also serve as reference lines for hot electrons. Simulations carried out separately for the inner-shell and dielectronic capture contributions of the 1s2l3l'-satellites indicate a strong inner-shell sensitivity for β_1, β_3 and β_5 and also partly for β_2 and β_4. The two-electron transitions β_6 and β_7, however, have very low inner-shell contributions. Similar considerations hold true for the two-electron transitions near He_α, namely op $= 1s2s^2 - 1s^22p$. Therefore, a good fit that takes into account the two-electron transitions provides confidence in a predicted the hot electron fraction.

Figure 6 shows an example where the hot electron contribution is important. The observation of Al II K_α-lines originates from hot electrons returning from the expanding plasma and penetrating into the massive Al target. Figure 9 shows the corresponding non-Maxwellian MARIA-simulation. Very good agreement with the data is obtained.

CHARGE-EXCHANGE IN INTERPENETRATING PLASMAS

The two-electron transitions $1s2s^2\ ^2S_{1/2} \rightarrow 1s^22p\ ^2P_{1/2,3/2} + h\nu$ (Fig. 6 and 9) have anomalous high intensities (factor of 10) compared to the dipole transitions $1s2p^2 \rightarrow 1s^22p + h\nu$. Photo absorption of strong dipole transitions in a charge-exchange perturbed dense plasma was proposed [24] to explain the experimental findings. In high intensity laser fields high energy plasma jets are generated with interact with the background target plasma. Due to the interpenetration of plasmas with different charge states charge-transfer processes between highly charged and low charged ions may result.

Figure 10 shows several relevant line intensity ratios (OP $= 2s^2\ ^1S_0 - 1s2p\ ^{1,3}P_1$, J $= 2p^2\ ^1D_2 - 1s2p\ ^1P_1$, op $= 1s2s^2\ ^2S_{1/2} - 1s^22p\ ^2P_{3/2,1/2}$, j $= 1s2p^2\ ^2D_{5/2} - 1s^22p\ ^2P_{3/2}$) in dependence of the charge exchange fraction $f_x = n_i(donor)/n_e$. Single and double electron capture processes have been consistently included in a collisional radiative modeling. Two-electron satellite ratios are raised by about an order of magnitude, the He_β-emission by about a factor of two. This is in agreement with several experimental observations made at different laser installations [24].

We note that the good agreement between the data and the simulation presented in Fig. 9 was obtained considering low opacity transitions only; therefore, the issue of anomalous high intensity two-electron transitions was avoided.

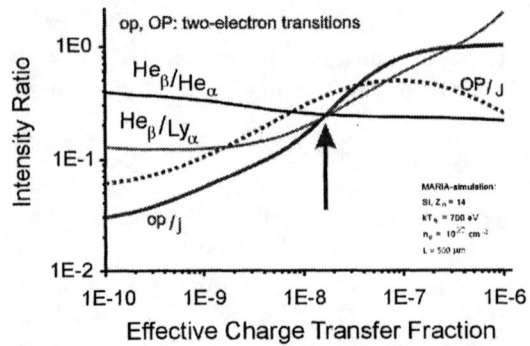

FIGURE 10. (a) MARIA-simulation of line intensity ratios in charge-transfer-coupled plasmas, as functions of the cold donor-ion fraction (relative to the electron density). The interpenetration velocity $V = 10^8$ cm/s, the electron density $n_e = 10^{22}$ cm^{-3}, the electron temperature $kT_e = 700$ eV, $L_{eff} = 500$ μm.

CONCLUSION

Hot electrons and fast ions have been characterized in dense laser produced plasmas by means of space resolved non-Maxwellian x-ray spectroscopic methods. Hot electron fractions, ion velocity profiles, temperature and density have been obtained.

REFERENCES

1. Rose-Petruck, C. et al., *Nature* **398**, 310 (1999).
2. Tabak, M. et al., *Phys. Plasmas* **1**, 1626 (1994).
3. Lindl, J. D, *Phys. Plasmas* **2**, 3933 (1995).
4. Glenzer, S.H., Rosmej, F.B., Lee, R.W. et al., *Phys. Rev. Lett.* **81**, 365 (1998).
5. Rosmej, F.B., *J. Phys. B Lett.: At. Mol. Opt. Phys.* **30**, L819 (1997).
6. Rosmej, F.B., *J. Phys. B Lett.: At. Mol. Opt. Phys.* **28**, L747 (1995).
7. Rosmej, F.B. et al., *Phys. Rev. A* **63**, 032716 (2001).
8. Inal, M.K, Dubau, J., *J. Phys. B : At. Mol. Opt. Phys.* **20**, 4221 (1987).
9. Kieffer, J.C. et al., *Phys. Rev. Lett.* **68**, 480 (1992).
10. Peyrusse, O., *JQSRT* **51**, 281 (1994).
11. Abdallah, J. Jr. et al., *Physica Scripta* **53**, 705 (1996).
12. Abdallah, J. Jr. et al., *AIP* **381**, 131 (1996).
13. *PHELIX, Petawatt High-Energy Laser for Heavy-Ion Experiments*, GSI-98-10, Report (1998).
14. Ruhl, H. et al., *Plasma Phys. Reports* **27**, 363 (2000).
15. Gitomer, S.J. et al., *Phys. Fluids* **29**, 2679 (1986).
16. Gibbon, P., Förster, E., *Plasma Phys. Controlled Fusion* **38**, 769 (1996).
17. Rosmej, F.B. et al., *JETP Lett.* **70**, 270 (1999).
18. Rosmej, F.B. et al., *JETP* **94**, 60 (2002).
19. Skobelev, I. Yu. et al., *JETP* **81**, 692 (1995).
20. Rosmej, F.B. et al., *JQSRT* **71**, 639 (2001).
21. Rosmej, F.B. et al., *JQSRT* **58**, 859 (1997).
22. Talin, B., Calisti, A., Godberg, L., Stamm, R., Lee, R.W., Klein, L., *Phys. Rev. A* **51**, 1918 (1995).
23. Rosmej, F.B., *J. Phys. B Lett.: At. Mol. Opt. Phys.* **33**, L1 (2000).
24. Rosmej, F.B., H.R. Griem, R. Elton et al., *Phys. Rev. E*, submitted.

ATOMIC AND MOLECULAR PHYSICS

Atomic Processes in Plasmas – An Overview

Hans R. Griem

Institute for Research in Electronics and Applied Physics
University of Maryland
College Park, Maryland 20742-3511

Abstract. Natural and man-made plasmas almost always not only contain free electrons and completely-stripped ions, but also bound-state atoms and positive, or even negative, ions, not to mention molecules, clusters, dust particles, etc. All of these constituents interact more or less strongly with electro-magnetic fields, and with each other. The often rather complex atomic processes or interactions to be discussed here are essential for our understanding of basic plasma properties (e.g., of the equation of state) and for many plasma applications, including materials processing. Special emphasis is placed on diagnostics and modeling under extreme plasma conditions and on the interpretation of unusual measurements, e.g., polarization spectroscopy, spectral line broadening, charge-exchange and beam-emission spectroscopy.

I. INTRODUCTION

Atoms and ions containing one or more bound electrons, immersed in plasmas, continue to play important roles in astrophysics and laboratory plasma physics. By means of spectroscopy, atomic physics enables us to study the composition and densities, temperatures and even electric and magnetic fields of and in these plasmas. Perhaps less appreciated, but equally important for such and other diagnostic applications, are the requirements for reasonably complete and accurate atomic data in the calculations of plasma equations of state, radiative opacities and kinetic models for non-equilibrium (non-LTE) situations. Much progress has been made in all these areas by the development of new experimental approaches and advanced computing, and in at least one case, also by analytic theory.

II. ELECTRON-ATOM (OR ION) COLLISIONS

Assuming that all atomic structure and radiative transition probabilities are available, the next common requirement is usually for the cross sections of electron collisions causing excitation and ionization, because cross sections and rates for the inverse processes can usually be inferred from them using the principle of detailed balancing. As shown, e.g., for Ar I [1] and Kr I [2] groundstate excitation, agreement between (electron loss or fluorescence) measurements and calculated cross sections for neutral atoms is usually within 25% for Ar I, but less satisfactory for Kr I, an even more complicated system. Especially serious are the disagreements between various Kr I excitation cross sections at high electron energies.

The energy resolution in excitation cross section measurements for ions has been much improved by the MEIBEL (merged electron-ion beam electron loss) method, e.g., for the 2s-2p excitation of C IV [3] and 3s-3p of Si III [4]. In both cases, agreement with close-coupling R-matrix calculations is excellent (after convolution with 0.17 or 0.24 eV FWHM Gaussians representing the experimental resolutions, and notwithstanding the importance of dielectronic resonances in case of Si III). Very similar agreement [5] was obtained between a recent [6] measurement of the Al III groundstate ionization cross section and R-matrix (with pseudo states) and CCC (converging close coupling) calculations. While reliable inelastic cross sections thus have become rather accessible, not much can be said about elastic cross sections, except that they are needed for the interpretation of spectral line broadening and polarization spectroscopy experiments (see Sec. IV and VI, respectively).

III. DIELECTRONIC AND THREE-BODY RECOMBINATION

Since most plasmas containing multiply-ionized atoms are far from LTE conditions, thus limiting the utility of the principle of detailed balancing, direct measurements and calculations of recombination rates continue to be of great interest. Examples for such situations are photo-ionized cosmic plasmas surrounding accretion-powered x-ray binaries or active galactic nuclei. In such plasmas the electron temperature can be as much as a factor ten lower for a given charge-state distribution compared with electron-ionized plasmas, necessitating measurements and calculations of dielectronic recombination, e.g., of Fe XX at $kT \gtrsim 1$ eV electron energies [7]. Such measurements can and have been performed using the heavy-ion test storage ring at Heidelberg, yielding energy-resolved dielectronic and radiative recombination rates. They clearly show the important role of $\Delta n = 0$ core excitations and allow detailed comparisons with various codes. The corresponding Maxwell-averaged rate coefficients are significantly larger than previous predictions for $kT_e < 100$ eV.

Another mechanism for accelerated recombination at high electron densities but also relatively low electron temperatures has been demonstrated recently by incorporating doubly-excited states and their resonant collisional deexcitation into time-dependent CR (collisional radiative) model calculations for carbon ions [8]. Considering, e.g., a beam of fully ionized carbon ions injected into a $N_e = 3 \times 10^{19}$ cm^{-3}, $kT_e = 3$ eV, background plasma, the CR code is used to predict the recombination under conditions where dielectronic recombination would normally be assumed not to be important. However, it turns out that the inclusion of C V and C IV doubly-excited states and their resonant deexcitation does have a significant effect on the calculated time histories, suggesting again that dielectronic rate coefficients may have been underestimated at low temperatures.

Also of great current interest is three-body recombination at extremely low temperatures and densities [9]. Although collisional capture into principal quantum number $n \approx 200$ levels is then very rapid, the captured electrons almost entirely go into the highest angular momentum ℓ states. Their radiative cascade decay is much too slow to be effective. However, ion-(Rydberg) atom collisions have large cross sections for elastic transfer to lower ℓ states, which then complete the recombination by radiative decay. Because of the invariance of total angular momentum **L** and of the Runge-Lentz vector **A**, elegant expressions can be derived for the $n\ell \to n\ell'$ transfer probabilities [10,11] for applications in future kinetic models. They should also be useful in spectral line broadening calculations for hydrogen and Rydberg atoms in general at low densities, at which the impact approximation [12] is valid even for ions as perturbers.

IV. COLLISIONAL BROADENING OF SPECTRAL LINES

In most sufficiently dense plasmas for Stark broadening to be important, only the electron effects can be described in the impact approximation just mentioned. However, besides inelastic (and super elastic) cross sections, one now also needs elastic cross sections. If both upper and lower levels of the line of interest are significantly perturbed, the various elastic cross sections enter the expressions for such line widths in a rather delicate way [13], making it difficult to estimate corresponding theoretical errors. For most (isolated) lines, the resulting line shapes are slightly shifted Lorentzians, the widths and shifts increasing linearly with electron density, until lines from different levels begin to overlap. Many measurements and (mostly semiclassical) calculations are listed in the NIST Physics Laboratory Physical Reference Data and have been critically reviewed since over about 25 years.

In contrast to semiclassical calculations, which tend to agree with measurements on lines from neutral atoms and low charge state ions, often to within 20%, most fully quantum-mechanical calculations are rather recent, especially those based on the CCC (convergent close-coupling) method [14]. A very recent example are the calculations [15] for 3s-3p transitions in Li-like ions. Two trends seem evident in the comparisons with measurements and semiclassical calculations: for low charge states there is reasonable agreement with measurements and some previous calculations. Two of the semiclassical calculations, however, yield overestimates by as much as 40%. For high charge state ions, on the other hand, both measurements and semiclassical calculations give significantly larger widths than the CCC calculations, by as much as a factor of 3. This may be due to motional Doppler effects in the experimental light source for these lines, which are relatively insensitive to Stark broadening.

V.   HEAVY PARTICLE COLLISIONS; CHARGE EXCHANGE

Besides the elastic ℓ-changing ion-atom (or ion) collisions just mentioned, electron transfer from donor atoms (or ions) to receptor ions may also have large cross sections. These are, however, very state-specific, whereas the former cross sections tend to increase rapidly with principal quantum number [10,11]. Interest in ion-ion charge-transfer into specific receptor states was generated recently by observations of anomalous intensities of normally weak He- and Li-like satellite lines to the Ly_α and He_α resonance lines of Al in a laser-produced plasma [16]. These observations were inconsistent with kinetic models allowing only for electron collisions; but charge transfer into the particular upper states involved here was soon recognized as a likely cause for the discrepancies [17,18]. This possibility was supported by spatial resolution in these measurements, indicating that these satellites were especially pronounced near the target surface where highly ionized ions might interact with lower charge states and even neutrals and where K_α lines were observed. Furthermore, charge exchange cross sections between ions can perhaps surprisingly be rather large, e.g., for the C^{3+} and He^{2+} system as measured in the ion-ion cross-beam experiments [19] in Giessen. Two-electron transfer cross sections may be assumed of comparable magnitude, e.g., for fully-ionized Si ions interacting with L-shell ions below the target surface, leading to doubly-excited He-like Si, etc. Including such reactions, opacities, and nonthermal electrons in the kinetic model then yields much improved agreement with the observed satellite intensities at reasonable ion beam intensities [17,18].

VI. POLARIZATION SPECTROSCOPY

Most plasma spectroscopy measurements are still made without much attention to polarization, i.e., assuming implicitly that the radiation is unpolarized. However, it has long been known that, e.g., macroscopic magnetic fields can lead to polarization. As in astronomy, one can therefore infer the strength of magnetic fields along the line of sight by measuring more or less subtle differences between (mostly Doppler broadened) Zeeman patterns for parallel (π) and perpendicular (σ) polarizations. For laser-produced plasmas, early measurements on the C V $2s^3S$-$2p^3P$ transitions of this kind were especially successful for ring-shaped laser focal spots [20], in part because the stronger radial gradients would drive larger currents but also because the cooler plasma core would absorb the line emission from the receding portion of the radially- and axially-expanding plasmas.

To interpret the observed line profiles, it was necessary to calculate the Zeeman shifts for the magnetic sublevels J',M' of 2^3P_3 for arbitrary fields and also the relative intensities of the π and σ components (J',M',M), M being the magnetic quantum number of the 2^3S_1 states. The observed differences in widths and intensities of the Doppler-smeared π and σ emissions indicated an azimuthal magnetic fieldstrength of about 20 T.

Besides such polarization due to anisotropy caused by macroscopic fields (termed polarization of the first kind [21]), there may also be polarization due to alignment of atoms or ions in particular M-levels due to beam-like, rather than isotropic, electron velocity distributions. Such beams are especially strong in pico-second (high intensity) laser-produced plasmas. This polarization of the second kind was the interpretation [22] for a significant dependence of the Al XII resonance line intensity on polarization, while the intercombination line was not polarized, presumably because its upper level is mostly populated by recombination with low-energy, isotropic thermal electrons.

Returning to polarization-dependent line shapes, Al XIII Ly-β and Ly-γ lines emitted from a high current z-pinch where assumed [23] to be mainly Stark-broadened by electric fields in electron-plasma waves generated by currents in the axial direction. Polarization was reported only for the γ line, for which Doppler and opacity effects are less important than for the β line.

116

VII. CHARGE-EXCHANGE RECOMBINATION SPECTROSCOPY (CERS)

Two practical difficulties in diagnosing high temperature laboratory plasmas spectroscopically can be overcome by injecting diagnostic neutral beams, usually hydrogen or deuterium atoms. These firstly provide electrons to completely ionized atoms into selected states, which then radiate experimentally convenient lines through strong $\Delta n = 1$ transitions. Secondly, the emission volume is reasonably well defined by the intersection of the beam with the sight line of the collection optics, in contrast to the usual integration along the line of sight in standard emission spectroscopy. Any shift of the CERS emission lines should thus indicate the flow velocity component along the sight line at the point of intersection.

However, since the radiating ions execute gyro-motion in a confining magnetic field in a tokamak device, for example, and since the transfer cross sections are dependent on the relative beam-atom, ion velocities, shifts associated with this effect must be carefully evaluated and subtracted [24], accounting for the delay between electron capture and line emission. These considerations not only require realistic atom-ion charge transfer cross sections, but also a rather detailed CR model, with results again being sensitive to ℓ mixing, elastic electron-ion (C IV, $n = 8$) cross sections. Because the poloidal velocities v_θ are much smaller than the toroidal velocities, v_ϕ, such corrections are especially important for v_θ and had therefore to be tested by comparing shifts obtained from upward and downward poloidal views in the torus.

VIII. BEAM EMISSION SPECTROSCOPY (BES)

Line emission from excited neutral beam atoms is also of considerable interest for the measurements of magnetic fields via the motional Stark effect in the $\underset{\sim}{v}_B \times \underset{\sim}{B}$ $\underset{\sim}{E}$-field in the atoms' frame of reference. The ensuing spectra [25] resemble classical Stark spectra, except that the individual components overlap because of Doppler effects due to the beam energy spread. At the relatively low fields in this reversed-field pinch experiment, $B \approx 0.5$ T, polarization effects are not very important, and fine structure of the Balmer-α line observed here may not be entirely negligible. Statistical distribution of populations in the sublevels was assumed, but may of course not be entirely achieved at the relatively low electron densities.

IX. CONCLUSIONS

In the over 50 years of research, the author has enjoyed fruitful interactions with many colleagues from his own and other physics and astronomy subfields and many students. Also, in the present case, several colleagues were very helpful with their comments and contributions. Clearly, much improved measurements and computations for many processes have been achieved, facilitating astrophysical and laboratory plasma applications over very large parameter ranges in electron density from $0.1\text{-}10^{25}\text{cm}^{-3}$. Still, there is a lot more work to be done on electron-neutral atom collisions, kinetic models for recombining plasmas and on elastic collisions for line broadening and polarization spectroscopy. Last but not least, cross sections for charge transfer and double charge transfer between ions will hopefully be measured and calculated in the near future.

ACKNOWLEDGMENTS

Partial support by the National Science Foundation and the U.S. Department of Energy is gratefully acknowledged.

REFERENCES

1. Dasgupta, A., Blaha, M., and Giuliani, J. L., *Phys. Rev. A* **61**, 012703 (2000).
2. Dasgupta, A., Bartschat, K., Vaid, D., Grune-Grzhimaito, A.N., Madison, D. H., Blaha, M., and Guiliani, J. L., *Phys. Rev. A* **64**, 052710 (2001).
3. Bannister, M. E., Durić, Woitke, O, Dunn, G. H., Chung, Y.-S., Smith, A. C. H., and Wallbank, B, 11th APS Topical Conference on Atomic Processes in Plasmas, edited by E. Oks and M. S. Pindzola, AIP Conference Proceedings 443, New York, 1998, p. 149. See also Bannister, M. E, et al., *Phys. Rev. A* **57** 278 (1998) and Janzen, P. H. et al., *Phys. Rev. A* **59**, 4821 (1999).
4. Wallbank, B., N. Durić, N., Woitke, O., Zhou, S., Dunn, G. H., Smith, A.C.H., and Bannister, M. E., *Phys. Rev. A* **56**, 3714 (1997). See also Reisenfeld, D. B. et al., *Phys. Rev. A* **60**, 1153 (1999).
5. Bartschat, K., AIP Conference Proceedings 443, 11th APS Topical Conference on Atomic Processes in Plasmas, edited by E. Oks and M. S. Pindzola, New York, 1998, p. 121.
6. Thomason, J.W.G., and Peart, B., *J. Phys. B* **31**, L201 (1998).
7. Savin, D.W. et al., *Astrophys. J. Suppl.* **138**, 377 (2002).
8. Ralchenko, Yu. V., and Maron, Y., *J. Quant. Spectrosc. Radiat. Transfer* **71**, 601 (2001).
9. Flannery, M. R., and Vrinzeanu, D., 11th APS Topical Conference on Atomic Processes in Plasmas, edited by E. Oks and M. S. Pindzola, AIP Conference Proceedings 443, New York, 1998, p. 317.
10. Vrinzeanu, D., and Flannery, M. R., *Phys. Rev. A* **63**, 03270 (2001).
11. Vrinzeanu, D., and Flannery, M. R., *J. Phys. B* **34**, L1 (2001).
12. Griem, H. R., *Principles of Plasma Spectroscopy*, Cambridge University Press, 1997.
13. Baranger, M., *Phys. Rev.* **112**, 855 (1958).
14. Bray, I., *Phys. Rev. A* **49**, 1066 (1994).
15. Ralchenko, Yu. V., Griem, H. R., and Bray, I., to be published.
16. Elton, R. C., Cobble, J. A., Griem, H. R. Montgomery D. S., Mancini, R. C., Jacobs V. L., and Behar, E., *J. Quant. Spectrosc. Radiat. Transfer* **65**, 185 (2000).
17. Rosmej, F. B. et al., to be published.
18. Rosmej, F. B., these proceedings.
19. Trassl, R., Bräuning, H., Diemar, K. V., Melchert, F., Salzborn, E., and Hofmann, I., 12th APS Topical Conference on Atomic Processes in Plasmas, edited by R. C. Mancini and R. A. Phaneuf, AIP Conference Proceedings 547, New York, 2000, p. 157.
20. McLean, E. A., Stamper, J. A., Manka, C. K., Griem, H. R., Droemer, D. W., and Ripin, B. H., *Phys. Fluids* **27**, 1327 (1984).
21. Fujimoto, T., Proceedings of the 3rd Japan-US Workshop on Plasma Polarization Spectroscopy, edited by P. Beiersdorfer, Livermore, 2002, p. 1.
22. Kieffer, J. C., Matte, J. P., Pépin, H., Chaker, M., Beaudoin, Y., and Johnston, T. W., *Phys. Rev. Lett.* **64**, 480 (1992).

23. Oks, E., 13th International Conference on Spectral Line Shapes, edited by M. Zoppi and L. Ulivi, AIP Conference Proceedings 386, New York, 1997, p. 3.
24. Bell R., and Synakowski, E., 12th APS Topical Conference on Atomic Processes in Plasmas, edited by R. C. Mancini and R. A. Phaneuf, AIP Conference Proceedings 547, New York, 2000, p. 39.
25. Den Hertog, D. J., Craig, D., Fiksel, G. and the MST Group, Davydenko, V. I., Ivanov, A. A., and Lizunoy, A. A., Proceedings of the 3rd Japan-US Workshop on Plasma Polarization Spectroscopy, edited by P. Beierdorfer, Livermore, 2002, p. 205.

Fermi Molecular Dynamics Applied to Problems in the Atomic Physics of Plasmas

Ken LaGattuta

Plasma Physics Group, Applied Physics Division, Los Alamos National laboratory,
Los Alamos, NM 87545

Abstract. A report on work in progress: The approach to steady state of a partially ionized fluid, in the regime of strong ion-ion coupling, is being simulated with the quasi-classical method known as Fermi Molecular Dynamics (FMD). We begin with a description of FMD, a statement of its advantages and disadvantages, and an overview of past work. We have continued to develop the FMD method as a tool for simulating a variety of inhomogeneous, partially ionized, dense plasma systems.

INTRODUCTION

Fermi Molecular Dynamics (FMD) is a "quasiclassical" approach to the molecular dynamic description of an assembly of electrons, partially stripped ions, atoms and molecules arising from a solution of Hamilton's equations. The equations themselves have a conventional form. However, to the otherwise classical Hamiltonian for this system of electrically charged point particles are added two sets of momentum dependent pair-potentials. The aim of these pair-potentials is to simulate certain effects of the Heisenberg uncertainty principle and the Pauli exclusion principle [1].

To the classical Hamiltonian H_0, describing a system of electrically charged point electrons and point nuclei, we add the Heisenberg "potential" V_H and the Pauli "potential" V_P:

$$H = H_0 + V_H + V_P \tag{1}$$

where the classical Hamiltonian is given by,

$$H_0 = \Sigma_i[p_i^2/2m_i + Z_i\Sigma_{j<i}(Z_j/r_{ij})] \tag{2}$$

and the Heisenberg and Pauli potentials, operating between pairs of point particles, and depending on their relative coordinates and relative momenta, are defined as:

$$V_H = \Sigma_i\Sigma_{j<i}(A_H/r_{ij}^2)exp(-B_H.r_{ij}^4.p_{ij}^4) \tag{3}$$

$$V_P = \Sigma_i\Sigma_{j<i}(A_P/r_{ij}^2)exp(-B_P.r_{ij}^4.p_{ij}^4) \tag{4}$$

CP635, *Atomic Processes in Plasmas: 13th APS Topical Conference*, edited by D. R. Schultz et al.
2002 American Institute of Physics 0-7354-0090-3

As seen, V_H and V_P are chosen to have the same algebraic form. The relative coordinates and relative momenta have the standard definitions:

$$r_{ij} = |\vec{r}_i - \vec{r}_j| \tag{5}$$

$$p_{ij} = |m_j \vec{p}_i - m_i \vec{p}_j|/(m_i + m_j) \tag{6}$$

Then, with these additions of V_H and V_P to the classical Hamiltonian, the equations of motion of individual point particles are the familiar ones:

$$d\vec{r}_i/dt = \nabla_{p_i} H \tag{7}$$

$$d\vec{p}_i/dt = -\nabla_{r_i} H \tag{8}$$

Note that the purely repulsive Heisenberg potential V_H (eq'n 3) is intended to operate only between electrons and nuclei. It acts to stabilize the ground, and singly excited states, of multielectron atoms or ions against unphysical autoionization by providing these systems with absolute ground states. Otherwise, in a purely classical mechanics, an initially bound multielectron atom or ion would rapidly autoionize until only a bound hydrogen-like ion remained; typically, the remaining electron would be bound into an unphysically deep state.

V_H contains two positive constants, A_H and B_H, which can be selected to match physical constraints; e.g., the electronic energies of the g.s. of atomic hydrogen, and the g.s. of H_2.

The purely repulsive Pauli potential V_P (eq'n 4) acts to distribute electrons bound in an atom or ion into shells. It acts only between electrons of the same "spin". Electrons in FMD carry a two-state spin attribute, which is fixed at the beginning of a calculation. There are no spin changing collisions in this version of FMD.

V_P also contains two positive constants, A_P and B_P, which can be selected to match physical constraints; e.g., the average spatial separation of the n=1 and n=2 bound levels of four electron ions.

For work in which high precision is not required, it has sufficed to pick the four constants A_H, B_H, A_P, and B_P such that the FMD computed first ionization potential agrees, on the average, to within $\sim 10\%$, with its corresponding physical value, for all atoms throughout the Periodic Table; i.e., for $1 \leq Z \leq 94$ [2].

In summary, some of the advantages of the FMD approach to mechanics are:

a) Multielectron atoms and ions, in ground or singly excited states, are stabilized against unphysical autoionization by the Heisenberg potential V_H.

b) It can be arranged that atoms and ions possess good values of their ionization potentials.

c) Multielectron atoms and ions possess a shell structure, arising from the Pauli potential V_P.

d) It can be arranged that the radii of individual electronic shells of atoms are physically reasonable.

It is also important to note that, with the FMD approach, the behavior of a system of electrons, partially stripped ions, atoms and molecules, all interacting, can be simulated successfully by solving a set of coupled ordinary differential equations.

This extended system of charged particles may be inhomogeneous spatially, and these inhomogeneities may themselves exhibit time-dependent behavior.

Fluctuations about local averages, both spatial and temporal, are also accessible by the FMD method. For instance, transport coefficients can be computed directly from data generated in the course of an FMD simulation for a system at, or close to, equilibrium; e.g., either by invoking the fluctuation-dissipation theorem, or by watching the rate at which a deliberately installed gradient relaxes.

CONNECTION BETWEEN FMD AND QUANTUM MECHANICS

As an approach to the precise description of real systems, FMD has an obvious shortcoming. It offers only an approximate description of the inherently quantum phenomena arising from the Heisenberg and Pauli principles, and it accomplishes this in an apparently adhoc manner. Moreover, the degree of approximation is not easy to assess, nor is it clear how to systematically improve upon the approximation.

In this context, we refer to some work which has appeared recently [3], based upon earlier work of Bohm [4]. Starting from the time-dependent Schrodinger equation, for a system of point particles described by a Hamiltonian H, which is

$$i\partial\psi/\partial t = H\psi \tag{9}$$

we rewrite the wavefunction as

$$\psi \equiv R\exp(iS) \tag{10}$$

where R and S are both real functions of the coordinates and the time. Following several algebraic manipulations, the Schrodinger equation can be transformed into the following pair of coupled ordinary differential equations:

$$m d\vec{v}/dt = -\nabla_r(V + Q) \tag{11}$$

$$dR/dt = -\frac{1}{2}R\nabla_r \cdot \vec{v} \tag{12}$$

where the velocity is

$$\vec{v} = \nabla_r S \tag{13}$$

the potential V is the purely classical potential, contained in H, and Q is the so-called quantum potential:

$$Q = -\frac{1}{2}\nabla_r^2 R/mR \tag{14}$$

122

To the extent that FMD provides a good description of real systems, then effects arising from Q, in the equation of motion (eq'n 11), must be incorporated by the Heisenberg potential V_H and the Pauli potential V_P, in the equation of motion (eq'n 8).

For Hamiltonians which do not depend explicitly on the time, R (eq'n 10) is time-independent and, therefore, Q is also time-independent. For such cases, it is straightforward to show [5] that V_H and V_P do indeed approximate well to Q, if the constants A_H, B_H, A_P, and B_P are well-chosen. If, instead, H depends explicitly on the time, then Q also depends explicitly on time, and the form of V_H and V_P becomes time-dependent. At the least, this means that the constants $A_H, ...B_P$ should be time-dependent, too.

EXAMPLES OF FMD BASED SIMULATION

I. The dynamics of the ionization of multielectron atoms by short pulse, long wavelength, laser beams has been simulated. In particular, "nonsequential" multiple ionization of isolated atoms has been visualized [6]. Similar phenomena occurring for atoms bound into small clusters has also been simulated [7]. In these cases, photoionized electrons driven to high energies by the laser field are seen to recollide with either the parent ion, or to collide with a neighbor atom. (This is an example of a system for which the Hamiltonian is explicitly time-dependent, but FMD simulations performed with values of $A_H, ...$ fixed are apparently satisfactory [5].)

II. The equilibrium net positive charge on a fission fragment (FF) arising from the fissioning of an unstable nucleus has been simulated, both for fission occurring in a vacuum and in an extended medium. As expected, the charge on a FF propagating through a dense plasma is found to be higher than that for a FF propagating through a cold material of a similar density. (In these calculations, a steady-state condition was attained, and the FMD parameters were fixed.)

Propagation of FFs through dilute hot plasmas has not been simulated since one expects that [8], for such systems, the dominant recombination mechanism will be dielectronic recombination (DR). But, if DR is to operate successfully then some model for bound-bound radiative decay must be included in FMD. This is perhaps an area for future work.

III. The behavior of the EOS for dense hydrogen, in the vicinity of the Hugoniot curve, has been simulated; i.e., both $U = U(n, T)$ and $P = P(n, T)$, where U and P, are internal energy and pressure, and n and T are density and temperature, have been determined. Not surprisingly, when departures from the Hugoniot are allowed, FMD calculations for these systems show values of the compressibility which are larger than those obtained when the hydrogen is fully shocked. (The problem is time-independent, unless propagation of the shock itself is to be traced.) An approach to the simulation of an extended electron gas, similar to FMD, has been described previously [9].

The Rankine-Hugoniot equation (RHE) must be satisfied for fully shocked material:

$$[U_0 - U] + 0.5[V_0 - V][P_0 + P] = 0 \qquad (15)$$

123

where, for example, the unshocked hydrogen is characterized by the quantities:

$$U_0 = -1.15Ry$$

$$V_0 = (5.09a.u.)^3 \qquad (16)$$

$$P_0 \approx 0$$

Off the Hugoniot, and for values of P and V determined by the simulation, the RHE may be solved for $U \equiv U_{solve}$. Comparing this value with the corresponding value of the internal energy U determined by the simulation one finds that $U < U_{solve}$, implying that the system is more bound, in this regime, than for the fully shocked material.

IV. In FMD simulations, we tracked the equilibration of the average kinetic energies of free electrons and ions, following a period during which the ion average kinetic energy was deliberately lowered. The observed time history of the difference of these two average kinetic energies was then compared with a history obtained by solving the equations

$$d < KE_e > /dt = -(\Xi_{eq}/\tau_{eq})[< KE_e > - < KE_i >] \qquad (17)$$

$$d < KE_i > /dt = -d < KE_e > /dt \qquad (18)$$

where τ_{eq} is a characteristic equilibration time, as given by Lee and More [10], and Ξ_{eq} is an empirical constant selected by us to produce agreement between the FMD history and that obtained from eq'ns 17 and 18. The values of Ξ_{eq} appearing in the Table suggest a slowing down of the equilibration process, as compared with the analytic theory. (In the Table, the column labeled $< Z >^{(SB)}$ refers to values computed by a reliable Saha-Boltzmann code [11], and is included here as a check on the FMD result.)

$Z_{nuclear}$	$< Z >^{(FMD)}$	$< Z >^{(SB)}$	$\rho(g/cc)$	$k_BT(eV)$	Ξ_{eq}
1	0.5	0.4	0.4	2.8	0.07
6	3.5	2.5	2.2	28.	0.20

REFERENCES

1. Kirschbaum, C., and Wilets, L., *Phys. Rev. A* **21**, 834 (1980).
2. Cohen, J. S., *Phys. Rev. A* **57**, 4964 (1998).
3. Lopreore, C., and Wyatt, R., *Phys. Rev. Lett.* **82**, 5190 (1999).
4. Bohm, D., *Phys. Rev.* **85**, 166 (1952).
5. LaGattuta, K., *Opt. Exp.* **8**, 401 (2001).
6. LaGattuta, K., and Cohen, J. S., *J. Phys. B* **31**, 5281 (1998).
7. LaGattuta, K., *Euro. Phys. J. D* **2**, 267 (1998).
8. Peter, T., and Meyer-ter-Vehn, J., *Phys. Rev. A* **43**, 2015 (1991).
9. Ebeling, W., and Schautz, F., *Phys. Rev. E* **56**, 3498 (1997).
10. Lee, Y. T., and More, R., *Phys. Fluids* **27**, 1273 (1984).
11. LaGattuta, K., and Winske, D., *JQSRT* **58**, 793 (1997).

Ion-Implantation-Related Atomic Collision Studies at the ORNL Multicharged Ion Research Facility

F. W. Meyer, M. E. Bannister, C. C. Havener, H. F. Krause,
P. Krstic, D. R. Schultz, A. Agarwal[1], D. Swenson[1*], F. Yan[2]

Physics Division, Oak Ridge National Laboratory, Oak Ridge, TN 37831-6372
[1]Axcelis Technologies, Beverly, MA
[2]Plasmaco Inc., Highland, NY

Abstract. In this article, some atomic collision data needs in the ion-implantation industry are discussed, and illustrated by specific examples of electron impact and heavy particle cross sections measured or calculated in our atomic collisions group for ion source modeling, beam energy contamination determination, and wafer dose error corrections. An example is also provided of how ion implantation has been used in our laboratory to investigate ways of improving the operating characteristics of ac plasma display panels.

INTRODUCTION

Ion implantation [1] is a critical technology in forming all doped regions of modern integrated circuit (IC) structures. B or In ion beams are used for creation of p-type doped regions, while P, As, or Sb beams are used to create regions of n-type doping. Required beam energies cover an extremely wide range. On one extreme, the continuing decrease of the lateral dimensions of complimentary metal oxide semiconductor (CMOS) devices to achieve speed improvements requires a corresponding decrease of vertical dimensions. Thus, ultra-shallow doping for gate profiling and creating source/drain regions requires beam energies down to about 100 eV. There is also increasing use of higher implantation energies for, e.g. modulated wells for CMOS formation, requiring ~200 keV, retrograde wells utilizing implant energies up to 1.5 MeV, and triple wells for flash memory and high dose buried layers, performed at implant energies extending to several MeV.

While traditionally three types of implantation machines have been required to cover this broad energy range, new generations of ion sources and implanters are under development with wider energy range capabilities, to reduce the number of re-quired implantation systems and thus overall production cost. Wafer biasing, either decel or accel, is being explored to extend the energy range of implanter machines. In addition, the use of multicharged dopant ions is contemplated to extend high energy

---

* Present address: Epion Corp., Billerica, MA

capabilities, while dimer, trimer, and cluster dopant ions (the latter based, e.g., on $B_{10}H_{14}$ source gas [2]) are being increasingly used for low energy applications.

As is well documented by now, the performance improvement of semiconductor devices continues to proceed at an incredible pace [3,4]. These improvements impose ever-increasing constraints on all aspects of the implantation process. Of particular relevance to the present article are the tolerances on ion beam energy contamination and wafer dose errors, which are now less than 1%.

Continuation of performance improvements on the present pace will be possible only with a significant evolution in process control methodology (see the *International Technology Roadmap for Semiconductors: 2001*, Ref. [5]). Whereas a largely empirically based (i.e., trial and error) approach to optimization of semiconductor processing technology has sufficed in the early days, a gradual change to a more phenomenological approach has occurred in recent years. Continuing along this trend, future optimization will have to rely increasingly on model-based approaches that require basic physics understanding of the relevant processes in order to enable real time process control, and to permit predictive simulation and exploration of new, as yet unexplored, operating regimes and conditions.

Efforts have been undertaken to make the transition to basic physics understanding in a number of semiconductor processing areas. Among them are implantation ion source modeling, estimation of implantation beam energy contamination, and wafer dose correction. In the following three sections, some of the atomic collision data requirements arising from such endeavors are highlighted using cross sections measured at the ORNL Multicharged Ion Research Facility (MIRF) [6] or calculated in-house by CFADC/Theoretical Atomic physics for Fusion [7] personnel. In the final section, work carried out at ORNL MIRF is described illustrating an alternate application of ion implantation, namely in the improvement and optimization of plasma display panels. The work also highlights an alternate approach to device optimization via a small test bench where the global effects of parameter changes can be conveniently evaluated. The work described below has in large part been carried out under the auspices of the US DOE Laboratory Technology Research Program, and under a Work for Others (WFO) contract with Eaton Semiconductor Corp. (now Axcelis Technologies, Inc.), who own the intellectual property rights to some of the data highlighted below.

IMPLANTATION ION SOURCE MODELING DATA NEEDS

For the reasons mentioned in the previous section, there is appreciable interest in assessment of possible optimization for multicharged ion production of Bernas and ELS hot cathode ion sources [8] presently used for singly charged ion production. The use of multicharged ions for higher energy ion implantation operations presents significant cost savings since the same acceleration facilities already in place for singly charged ion implantation could be used, while extending the maximum implantation energy by factors of two or three, depending on extracted charge state. Whether the same ion source can be used as well, will depend on the beam intensities that can be produced for the desired multiply charged ion. Knowledge of the relevant electron and heavy particle impact inelastic cross sections is required to guide the range of arc voltages and source pressures to explore, since those two parameters are the main determinants of the electron energy distributions in the above ion sources.

Some of the atomic collision cross sections relevant to an implantation ion source plasma based on BF_3 source gas were determined as part of a WFO project for Eaton Semiconductor Corp. (now Axcelis Technologies, Inc). The goal of the latter was to lay the groundwork for quantitative ion source plasma modeling which would enable prediction of the source conditions giving the required boron beam properties as well to facilitate source optimization of multicharged B ion production for use in next generation micro-electronic devices. Most of the collision cross sections for a BF_3 plasma were until recently unknown.

At the ORNL Physics Division MIRF, unique apparatus exists for the measurement of such collision cross sections. This includes a state-of-the art electron cyclotron resonance (ECR) [6] ion source for producing singly and multiply charged B and F ions, as well as the many singly charged radicals formed from neutral BF_3; an electron crossed beam apparatus for measuring electron impact ionization and dissociation cross sections; and a beam - gas cell set-up for measuring many of the heavy particle charge exchange and dissociation cross sections of interest. To complement and extend the experimental work, theoretical estimates were made of the relevant electron-neutral collision cross sections, as well as of the heavy particle cross sections from the ~ 200 eV minimum energies experimentally accessible with the present set-up down to the 5-30 eV energy required for the source plasma modeling.

Selected Cross Section Results

One of most important class of collisions in low temperature plasmas is collisions with electrons. Using known scalings, comparisons with similar systems, additivity theorems, and semi-empirical relations, theoretical cross section estimates were made for electron impact ionization of BF_3, as well as the major dissociation channels. In addition, measurements were performed of electron impact ionization of B^{q+} and F^{q+}, ions (q = 1-3) in the energy range 0-200 eV, and of the following electron fragmentation cross sections in the energy region 1-100 eV:

$$e^- + BF_3^+ \rightarrow BF_2^+$$
$$e^- + BF_2^+ \rightarrow BF^+$$
$$e^- + BF^+ \rightarrow B^+, F^+ \ .$$

Another important class of collisions taking place in low temperature plasmas are heavy particle charge exchange and dissociation. For example, charge exchange collisions provide the dominant recombination process that determines mean charge states of extracted ions. In-house experimental apparatus for measuring heavy particle collision cross sections was used for the determination of

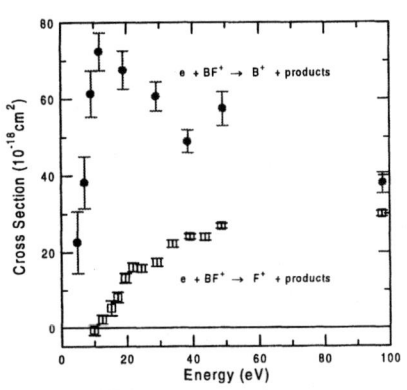

Figure 1. Electron impact fragmentation of BF^+.

charge exchange and total attenuation cross sections at collision energies down to about 200 eV in energy:

$$B^{q+} + BF_3 \rightarrow B^{(q-1)+} \quad (q=1\text{-}3)$$
$$F^{q+} + BF_3 \rightarrow F^{(q-1)+} \quad (q=1\text{-}3)$$
$$BF_3^+ + BF_3 \rightarrow \Sigma \text{ all channels}$$
$$BF_2^+ + BF_3 \rightarrow \Sigma \text{ all channels}$$
$$BF^+ + BF_3 \rightarrow \Sigma \text{ all channels } .$$

By "Σ all channels" is meant the sum of charge exchange and overall dissociation cross sections, i.e. total attenuation. In conjunction with the heavy particle collision

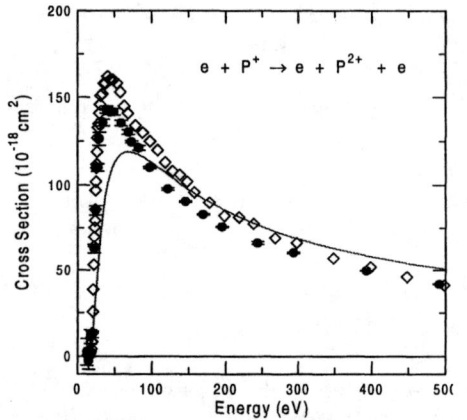

Figure 2. Electron impact ionization of P^+: open symbols - Ref. 9; solid line - Lotz [10], solid symbols – present data.

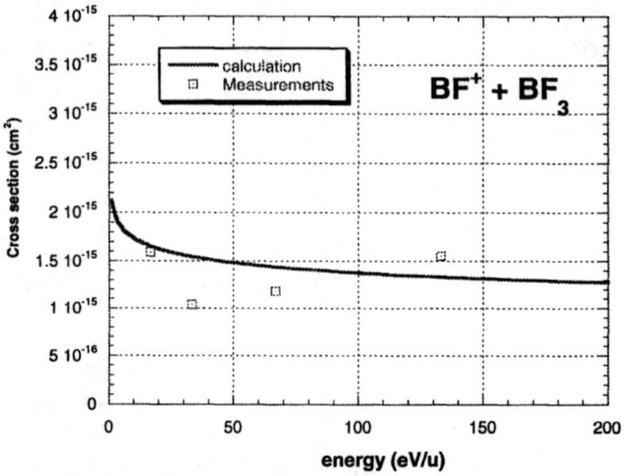

Figure 3. Experimental attenuation cross section compared with estimated single electron capture cross section for BF^+ in BF_3.

cross section measurements, theoretical estimates based on Demkov or Landau-Zener models of non-adiabatic coupling were made to guide extrapolation down to the energies relevant for ion source modeling. The total experimental uncertainties of the cross section measurements are estimated to be at the 20% level . A measurement of a previously published electron capture cross section [11] for $O^{3+} + H_2 \rightarrow O^{2+}$ obtained with completely different apparatus agreed to within 3% of the published value. Figures 1-3 show sample cross section results for electron impact ionization, fragmentation, and heavy particle charge exchange.

Boron Beam Development With The MIRF ECR Source

As already mentioned earlier, the development of intense multicharged B ion beams is of great interest to the ion implantation industry because it cheaply extends the energy range of current ion implanters required for certain applications. The ORNL MIRF ECR ion source was used to investigate plasma conditions that optimized multicharged B ion production. For comparison, both BF_3 source gas and a solid B sample biased at negative voltage to enhance sputtering were investigated. The effect of varying microwave power levels, source magnetic field configurations, source gas feed rates, and mix gas species on extracted beam charge state distributions was investigated and documented. Results for the biased B sample are summarized in the table below. Although not providing nearly as intense beams of multicharged ions as was achievable with BF_3 source gas, the biased B sample test provided an interesting insight into the mechanism whereby the solid B sample is converted to vapor. From the roughly proportional dependence of the extracted beam currents on sputter current under low microwave power conditions (i.e. B^+ ion production), the dominant mechanism is inferred to be simple sputtering. In contrast, at the high rf power conditions required for B^{2+} and B^{3+} production, the more exponential increase of the extracted currents with the power to the B sputter sample indicates a transition to sample evaporation, as the sample vapor pressure increases close to exponentially with temperature.

Although not documented here, ECR sources may prove to be better than the present Bernas ion implantation source for the production of B^{2+} and B^{3+} ion beams from BF_3. For example, an all-permanent-magnet ECR ion source developed by Pantechnik [12] in Caen, France is currently producing almost a mA of B^{2+}, and more than 100μA of B^{3+}.

TABLE 1. Boron single and multicharged ion generation in the ORNL MIRF ECR source using a biased B sputter sample; 10 kV source high voltage; 5×10^{-6} Torr gauge reading for He sputter gas; I(mA) is sputter sample current; Beam(nA) is extracted ion beam current.

Bias voltage(V)	B^+ (54 W)		B^{2+} (254 W)		B^{3+} (254W)	
	I(mA)	Beam(nA)	I(mA)	Beam(nA)	I(mA)	Beam(nA)
250	10	15			0	0
500	12	40				
750	12	50				
1000	15	60	32	35	34	3
1250	15	70	36	90	37	50
1500	17	80	40	200	45	300
1750	18	93	45	650	49	750
2000	20	-	54	2000	51	1800

ENERGY-CONTAMINATION-RELATED CROSS SECTIONS

A common scheme for extending the range of energies and penetration depths obtained from any implantation equipment is to use multiply charged monatomic, or singly charged polyatomic ion species. However, the use of such species introduces the risk of having some energy contamination [13] in the beam, which ultimately results in a broadening of the implantation depth distribution. This risk arises because the other species are simultaneously emitted from the ion source. As an example, in a beam of doubly charged atomic ions, accelerated at a potential V, there can be a spurious component of singly charged monatomic ions which has 1/4 the kinetic energy per atom of the doubly charged ion, and therefore penetrate to a correspondingly smaller depth in the target material. The singly charged monatomic ion beam results from dissociation of singly charged diatomic ions with the background gas in the implanter beamline between ion source and analyzing magnet.

For instance, when implanting P^{2+} at 2 keV [1 kV/(unit charge)], the beam may be contaminated with P^+ at 0.5 keV, the latter having been accelerated through the 1-kV potential as P_2^+ and subsequently undergoing dissociation in the beamline residual gas preceding the analyzing magnet. Both species have the identical momentum per charge ratio, and so are not separated in the analyzing magnet. To address this issue, cross section measurements and calculations were made of simple dissociation and total attenuation of As_2^+ and P_2^+ colliding with H_2, and CH_4 gases at keV energies. Experimental results for As_2^+ are shown in Table 2.

TABLE 2. Dissociation and total attenuation cross sections for As_2^+ ions in H_2 and CH_4 gases.

Energy (keV)	H_2		CH_4	
	σ_{diss} (10^{-16} cm^2)	σ_{att} (10^{-16} cm^2)	σ_{diss} (10^{-16} cm^2)	σ_{att} (10^{-16} cm^2)
3.0	1.8	3.3	2.1	11
5.0	3.4	4.1	3.0	13

As an additional example, in a beam of singly-charged diatomic ions, accelerated across a potential V, there can be a spurious component of singly-charged atomic ions which have 4 times the kinetic energy per atom of the diatomic ion, and therefore penetrate to a correspondingly greater depth in the target material. The singly charged atomic ion beam results from charge exchange of doubly charged atomic ions with the background gas molecules in the implanter beamline. For instance, when implanting P_2^+ at 1 keV (0.5 keV per atom), the beam may be contaminated with P^+ at 2 keV, produced from extracted P^{2+} that has undergone single electron capture prior to magnetic analysis. Ion implantation machine vendors must be able to specify for their customers the amount of energy contamination which will be present in any beam as a

TABLE 3. Data summary of single electron capture cross section of B^+, P^{2+}, and As^{2+} incident on various gases; (*) denotes total attenuation cross section.

Projectile	Energy (keV)	Cross Section (10^{-16} cm^2)					
		H_2	CO	N_2	O_2	CH_4	CO_2
B^+	2.0		6.4			10.2	10.0
B^+	5.0	3.7	7.2	4.3		7.3	7.9
P^{2+}	4.0	24.1		26.1	18.4		
	6.0	22.7		23.6	18.3		
	10.0	17.4		20.9	18.0		
	20.0	16.9*		26.2*	21.0*		
	36.0	11.0*		18.7*	16.0*		
As^{2+}	4.0	11.8		11.6	13.7		
	6.0	11.5		11.8	15.1		
	10.0	10.7		11.4	17.9		
	20.0	7.9*		13.6*	15.6*		
	36.0	7.0*		12.8	17.8*		

function of known or measurable equipment parameters such as the ion energy or the pressure and composition of the background gas in the beamline. The energy contamination can be predicted from the probability of charge exchange, characterized by the charge changing or dissociation cross section of the various ions with gases present in the beamline. Until recently, there was little such data applicable to the ion species, energies, and gases used today in the microelectronics field. Most available data had been obtained in support of fusion energy research. Table 3 itemizes some sample cross section results.

Below is a more complete list of the processes needed to address energy contamination as well as the dose correction issues to be discussed in the next section.

Electron capture: $B^{2+}, P^{2+}, As^{2+} + BF_3, AsH_3, PH_3, H_2, CO, CO_2, C_nH_m \rightarrow B^+, P^+, As^+$

Dissociation: $B_2^+, P_2^+, As_2^+ + BF_3, AsH_3, PH_3, H_2, CO, CO_2, C_nH_m \rightarrow B^+, P^+, As^+$

Neutralization: $B^+, P^+, As^+ + BF_3, AsH_3, PH_3, H_2, CO, CO_2, C_nH_m \rightarrow B^0, P^0, As^0$

Stripping: $B^+, P^+, As^+ + BF_3, AsH_3, PH_3, H_2, CO, CO_2, C_n H_m \rightarrow B^{2+}, P^{2+}, As^{2+}$.

The energies for which these cross sections are needed range from a hundred eV to a few MeV! With exception of selected cases in the range 3 – 36 keV, and the B^+ stripping cross section shown in the next section, these cross sections are for the most part not known.

DOSE CORRECTION RELATED CROSS SECTIONS

An additional problem that arises from charge exchange is accuracy of implant-dose measurement, which relies on integration of the beam current at the target. In this case, the concern is charge exchange that occurs in the beamline section between the analyzing magnet and target. For instance, singly ionized species that undergo single electron capture in this section of beamline will

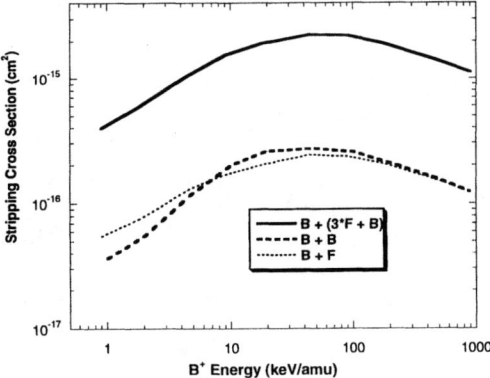

Figure 4. Stripping cross section for B^+ on BF_3 leading to B^{2+}.

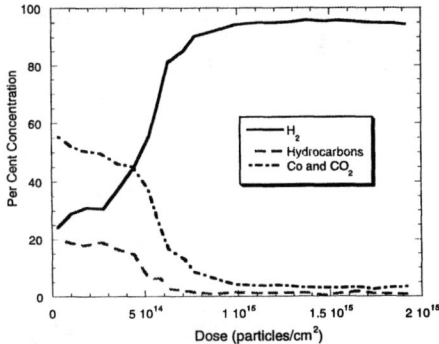

Figure 5. Evolution of the concentration of various outgassing species with accumulated dose for a 150 keV P implant.

not be detected at all when they reach the target, leading to an undercounting of the implantation dose. The cross sections of Table 2 and 3 are relevant for this issue as well, in the case of low energy implantation. In the case of high-energy implantation, stripping collisions prior to wafer impact but after magnetic analysis stage can similarly result in errors in dose estimation, leading to overestimation rather than underestimation of dose. A relevant cross section for this case is illustrated in Figure 4. The cross section shown was determined theoretically using a classical trajectory Monte Carlo (CTMC) approach under the assumption that the additivity rule applies, whereby the cross section for a molecular target is determined by a weighted sum of the cross sections of the constituent atoms, also shown in Fig. 4.

It is important to keep in mind that ion implantation occurs usually onto wafers with patterned photoresist coatings $1.5 - 4$ μm thick. These coatings are usually organic in origin and respond to mA ion beam bombardment with significant out-gassing, releasing a broad spectrum of gases ranging from H_2, CO, and CO_2, to a variety of hydrocarbons (CH_2, CH_4, and C_2H_2). This outgassing can raise the pressure in the wafer vicinity to as high as 10^{-3} Torr! Furthermore, as illustrated in Figure 5, the species released vary with accumulated dose, i.e. are time dependent [14]. These two features together make total dose estimation at the less than 1% level extremely challenging. Efforts are underway to implement real time partial pressure monitoring and species specific cross sections to extend the range of energies and pressures over which the robustness of present dose control algorithms can be maintained. The potential impact of photoresist outgassing on energy contamination is obviously most pronounced if wafer biasing is used, but may affect upstream (i.e. before magnetic analysis) energy contaminating collisions as well. Thus implanters almost universally require high pumping speed in the wafer vicinity and excellent differential pumping with respect to the upstream injection line.

OTHER IMPLANTATION APPLICATIONS

While dopant ion implantation has by far the biggest commercial usage, we highlight here briefly another application in the area of plasma display panels, a cutaway sketch of which is shown in Figure 6. Optimization of the top insulating coatings used in the construction of ac type Plasma Display Panels (PDP's) is presently an area of intense research. Choice of the coating material is critical to the operating voltage and lifetime of the plasma display panel. MgO has been empirically found to be an excellent coating material, resulting in desirable low operating voltages and long panel lifetimes [15]. This is presumably due to the high secondary electron

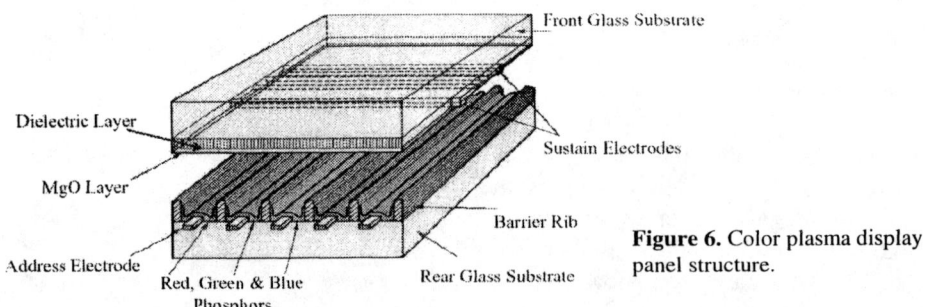

Figure 6. Color plasma display panel structure.

emission and low sputtering rate properties of this material. As a possible aid in further improvements of device design, we have investigated the feasibility of using slow highly charged ions to modify MgO thin films. Any increases in secondary electron emission can help reduce the operating voltage of a PDP and thus reduce the cost of the electronics (almost half of the cost of a PDP) and boost the luminous efficiency of the device. Reduced sputtering will increase panel lifetime, and thereby its competitiveness relative to other display types.

One possible mechanism that might enhance the secondary electron emission properties of the MgO thin film, is the creation of surface states close to the top of the 8.7 eV bandgap of the material by near-surface implantation of highly charged ions (HCI). If the implantation sufficiently distorts the local lattice, those local lattice defects may trap electronic excitations sufficiently long to provide a stepping stone for electrons leaving the surface, effectively enhancing the secondary electron emission yields. Alternatively, by suitable choice of species, the near-surface implanted HCI may themselves provide electronic states lying energetically between the MgO valence band and the vacuum, thus effectively lowering the threshold energy required for secondary electron emission, and thereby increasing electron yield.

The overall effect of near-surface implantation of MgO is of course extremely complex and difficult to disentangle in terms of the affected atomic and solid state processes. For this reason an approach using a mini-plasma display test panel developed at Plasmaco was employed which incorporates extensive diagnostics of the relevant display parameters and permitted almost real-time testing of the

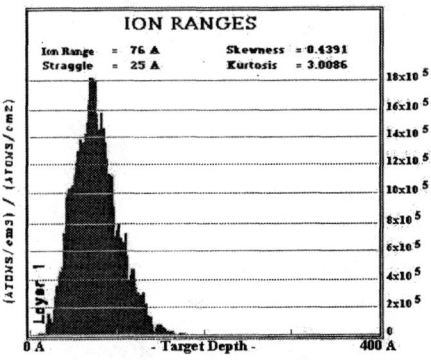

Figure 7. SRIM 2000 simulation of implanation depth distribution for 10 keV Ni incident on MgO.

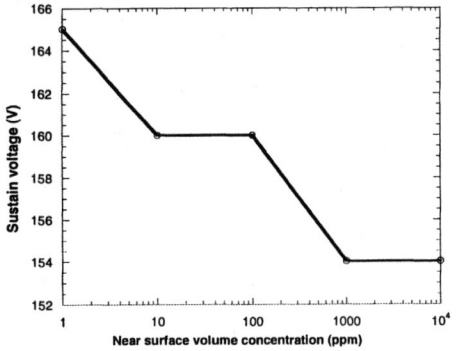

Figure 8. Dependence of minimum sustain voltage on near-surface concentration of implanted Al.

implantation regimen. This test stand employs easily removable microscope-slide-size lead glass substrates onto which mini-electrode arrays were deposited, which in turn were coated with a thin MgO layer. A set of such slides was implanted by various doses (representing near surface concentrations of 10 to 10^4 ppm) of Al, Fe, Ni, and Cl ion beams at the ORNL MIRF, and then evaluated in the test stand for possible performance improvements after transport back to Plasmaco. Energies were in the 10 keV range to result in depth distributions that peaked (see Fig. 7) at less than 10nm according to SRIM 2000 [16]. The exposed samples were specifically evaluated for changes in minimum sustain voltage, luminosity, and priming time (related to the discharge formation time delay).

In this manner conditions (i.e. beam species, energy, and dose) that gave maximum display panel performance improvements could be immediately identified.

The MgO samples were mounted on a sample stage that could be heated up to 300° C to outgass the samples prior to implantation. The sample stage was designed to accommodate up to 4 15x50mm MgO samples to permit maximum throughput. A low energy electron flood gun was used to prevent sample charging during HCI beam exposure. In addition, a set of horizontal and vertical sweep plates was used to assure uniform illumination of the entire 7x20 mm active area of the MgO thin film.

Figure 8 summarizes the change in minimum sustain voltage observed as function of the near-surface concentration of implanted 10 keV Al. As can be seen from the figure, an almost 10% decrease was observed. A 50% improvement in priming performance was noted for 10 keV Al implantation as well. The improvement of both parameters appears to peak at concentrations near the 1000-ppm level. These results show sufficient promise that further implantation runs are planned in the near future.

ACKNOWLEDGMENTS

This research was sponsored in part by the Office of Fusion Energy Sciences and by the Office of Science Laboratory Technology Research Program of the U.S. Department of Energy under contract No. DE-AC05-00OR22725 with UT-Battelle, LLC.

REFERENCES

1. McKenna, C. M., "History of Ion Implantation Equipment," in *Ion Implantation Science and Technology, 2000 Edition,* edited by J. F. Ziegler, Edgewater, Maryland: Ion Implantation Technology Co., 2000, pp. 1-45.
2. Sosnowski, M., et al., *Mat. Res. Soc. Symp. Proc.* **568**, 49 (1999).
3. Moore, G. E., *Electron.* **38**, 114-117 (April 19, 1965).
4. Isaac, R. D., *IBM J. Res. Develop.* **44**, 369 (2000).
5. see, e.g., http://public.itrs.net/
6. Meyer, F. W., "ECR-based Atomic Collision research at the ORNL MIRF," in *Trapping Highly Charged ions: Fundamentals and Applications,* edited by J. Gillaspy, Huntington, New York: Nova Science Publishers, Inc., 1997, pp. 117-165.
7. See, e.g. http://www-cfadc.phy.ornl.gov
8. Stephens, K. G., "Ion Source Physics," *op.cit. ref* [1], pp. 413-456.
9. Yamada et al., *J. Phys. Soc. Japan* **58**, 1585 (1999)
10. Lotz W., *Astrophys. J. Suppl.* **14**, 207 (1967); *Z. Phys.* **216**, 241 (1968); *Z. Phys.* **220**, 466 (1969).
11. Phaneuf, R. A. et al., *Phys. Rev. A* **26**, 1892 (1982).
12. See, e.g. http://www.pantechnik.net
13. Freeman, J. H., Chivers, D. J., and Gard, G. A., *Nucl. Instr. Meth.* **143**, 99 (1977).
14. Horsky, T. N., "Photoresist Outgassing in High Energy and High Current Ion Implantation," in *1998 Int. Conf. on Ion Implantation Proceedings,* Kyoto, Japan, June 22-26, 1998, p. 654.
15. Weber, L., "Color Plasma Displays", in *The Electrical Engineering Handbook*, CRC Press, pp. 1939 – 1950, 1997.
16. See, e.g. http://www.srim.org/SRIM/SRIMLEGL.htm

New Insights into the X-ray Spectra of Heliumlike and Neonlike Ions

P. Beiersdorfer*, H. Chen*, D. Hey*, M. J. May*, A. L. Osterheld*, K. J. Reed*, J. H. Scofield*, D. B. Thorn*, M. Bitter[†], K. R. Boyce**, G. V. Brown**, K. C. Gendreau**, R. L. Kelley**, F. S. Porter**, C. K. Stahle**, A. E. Szymkowiak**, R. E. Olson[‡], J. K. Lepson[§], M.-F. Gu[¶], E. Behar[‖] and S. M. Kahn[‖]

*Lawrence Livermore National Laboratory, Livermore, CA
[†]Princeton Plasma Physics Laboratory, Princeton, NJ
**Goddard Space Flight Center, Greenbelt, MD
[‡]University of Missouri, Rolla, MO
[§]University of California, Berkeley, CA
[¶]Massachusettes Institute of Technology, Cambridge, MA
[‖]Columbia University, New York, NY

Abstract. Recent measurements of the K-shell and L-shell x-ray spectra of highly charged heliumlike and neonlike ions are presented that were performed on the Livermore electron beam ion traps and the Princeton tokamaks. These measurements provide new insights into collisional and indirect line formation processes, identifications of forbidden lines, and a new plasma line diagnostic of magnetic field strength.

INTRODUCTION

The x-ray spectra of heliumlike and neonlike ions provide excellent diagnostic opportunities of high-temperature plasmas, including those found in magnetic and inertial confinement fusion research, the Sun, stellar coronae, supernova remnants, galaxy clusters, and cometery comae. These spectra are now commonly used to measure such parameters as the plasma density, temperature, bulk motion, opacity, and ionization balance. Despite this widespread use and the ensuing familiarity with the spectral emission, continued investigation of the spectra and the underlying atomic physics still offers surprises.

In the following we present several examples of new insights gained in recent studies of the x-ray spectra of heliumlike and neonlike ions. The studies were carried out at the EBIT-I and EBIT-II electron beam ion traps at the Lawrence Livermore National Laboratory and at the National Spherical Tokamak Experiment (NSTX) facility at the Princeton Plasma Physics Laboratory. Our studies of heliumlike ions include the identification of the lithiumlike $1s2s2p\,^4P_{5/2} \to 1s^22s\,^2S_{1/2}$ magnetic quadrupole line in the K-shell satellite spectrum of heliumlike Ar^{16+}, which has generally been overlooked in past analyses of tokamak spectra. We also present high-resolution measurements of the K-shell x-ray emission produced by charge exchange between plasma ions and cold neutrals. These measurements were carried out with the Goddard Space Flight Cen-

CP635, *Atomic Processes in Plasmas: 13th APS Topical Conference*, edited by D. R. Schultz et al.
© 2002 American Institute of Physics 0-7354-0090-3/02/$19.00

ter high-resolution x-ray microcalorimeter. The measurements show that a description of the resulting x-ray emission requires capture cross sections into individual atomic levels and that statistical assumptions for populating angular momentum states used in cometary x-ray models are insufficent for proper description.

Our measurements of the L-shell spectra of neonlike ions have revealed what we believe are the first x-ray lines that are sensitive to the magnetic field. These lines can be used as magnetic field diagnostic of low-density high-temperature plasmas. Moreover, our measurements resolved many questions surrounding the relative intensities of $3d \rightarrow 2p$ and $3s \rightarrow 2p$ transitions in neonlike ions. The puzzles surrounding these lines ratios had resulted in wide speculation on possible physical processes left out in the modeling calculations. For example, the discrepancy between solar observations and models had been attributed to opacity effects. This, however, could be ruled out when our measurements on the opacity-free Princeton tokamaks found the same ratios as in the Sun and other stellar coronae. Combined with measurements on EBIT-I and EBIT-II, our measurements demonstrated instead that most of the discrepancy was due to a systematic overestimate or underestimate of the relative excitation rates in the available calculations. Our electron beam ion trap measurements showed that the remainder was due to blending with lines whose existence had been missed in many recent calculations.

The present measurements should encourage continued investigations on tokamaks and electron beam ion traps to resolve further puzzles in the x-ray spectra of highly charged ions in high-temperature plasmas.

K-SHELL SPECTRA OF HELIUMLIKE IONS

Beginning with early measurements on tokamaks and the Sun [1, 2, 3, 4, 5, 6], the K-shell spectra of heliumlike ions are probably by now the most studied high-resolution x-ray spectra from high-temperature collisional plasmas. The following examples highlight two recent advances in our understanding these spectra.

Identification of the $1s2s2p\ ^4P_{5/2} \rightarrow 1s^22s\ ^2S_{1/2}$ magnetic quadrupole line among the heliumlike K-shell satellite lines

Gabriel, in his fundamental work on the K-shell satellite lines to the x-ray emission lines of heliumlike ions, reserved 22 letters of the alphabet (a–v) for labeling all electric-dipole allowed lithiumlike transitions [7]. He used the remaining four letters (w–z) to label the heliumlike lines. One possible transition in lithiumlike ions was not labeled. This was the dipole-forbidden magnetic quadrupole transitions $1s2s2p\ ^4P_{5/2} \rightarrow 1s^22s\ ^2S_{1/2}$.

Many, though not all, lithiumlike satellite lines have been identified in the K-shell spectra of heliumlike ions from tokamaks or the Sun. Others have been observed in high-density laser-produced plasmas. The $1s2s2p\ ^4P_{5/2} \rightarrow 1s^22s\ ^2S_{1/2}$ magnetic quadrupole (M2) line has eluded identification, however. Although it was observed in time-delayed beam-foil spectroscopy [8, 9], resulting in the measurements of the lifetime of its upper level, it was thought to be too weak to be seen in collisional plasmas.

136

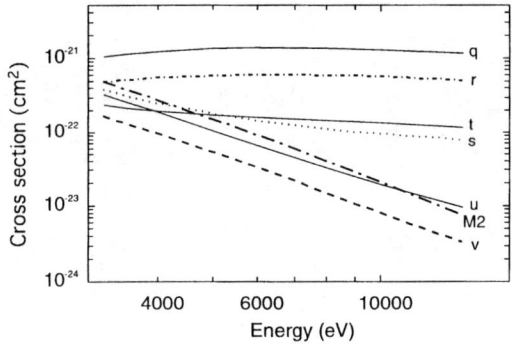

FIGURE 1. Electron-impact excitation cross sections for populating the upper levels of the lithiumlike innershell satellite lines q, r, s, t, u, v, and the $1s2s2p\ {}^4P_{5/2} \rightarrow 1s^22s\ {}^2S_{1/2}$ quadrupole line, labeled $M2$. (From [11].)

Excitation by dielectronic recombination is essentially non-existent because of the small radiative and Auger rates associated with the forbidden decay of the $1s2s2p\ {}^4P_{5/2}$ level. Morever, the level decays predominantly by Auger decay. However, it turns out that electron-impact excitation is very effective in populating the $1s2s2p\ {}^4P_{5/2}$ level close to threshold [10]. In fact, in lithiumlike Ar^{15+} it is second only to the excitation of the resonance line q, intimating that the line may be seen in low-temperature, low-density plasmas where electron collisions near threshold dominate the excitation process, as illustrated by a distorted wave calculation of the electron-impact excitation cross sections shown in Fig. 1.

By analyzing relatively cold spectral data from the Princeton NSTX tokamak in comparison with measurements on the Livermore EBIT-II electron beam ion trap where electron-impact excitation near threshold was the only excitation mechanism, we were able to identify the line in the K-shell spectrum of Ar^{16+} [11], as shown in Fig. 2.

Heliumlike x-ray spectra excited by charge exchange

The heliumlike K-shell spectra from non-collisional plasmas have received much less attention in the laboratory, including charge-exchange produced spectra which have now found an important application in understanding cometary x-ray emission [12]. Crossed-beam experiments generated extensive charge-exchange cross sections using particle counting techniques, but they did not record the x-ray emission. By contrast, some x-ray measurements were reported from tokamaks where charge exchange contributed to various extent to the excitation of K-shell spectra [13, 14, 15, 16]. K-shell x-ray spectra produced purely by charge-exchange, however, have not been reported until recently, when we started using the Livermore electron beam ion traps to measure charge-exchange produced spectra [17, 18, 19].

High-resolution measurements of charge-exchange produced x-ray emission were

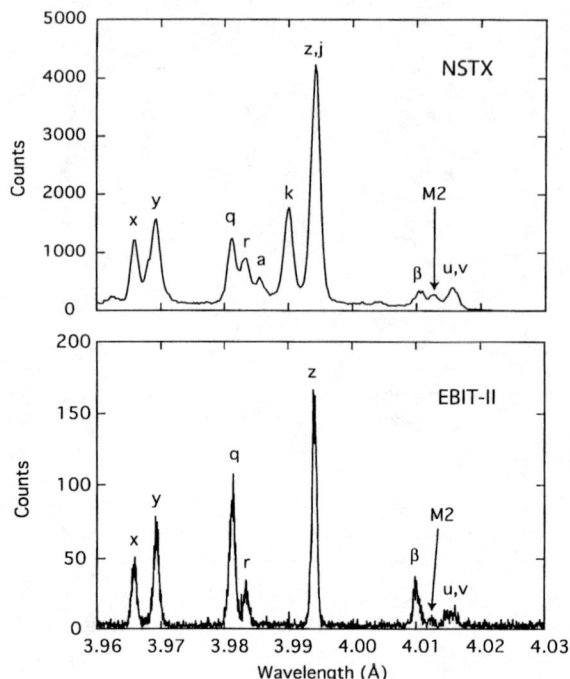

FIGURE 2. K-shell emission spectrum of argon. Top: obtained with the von Johann crystal spectrometer on NSTX. This spectrum shows the Ar^{16+} lines x, y, and z, the Ar^{15+} lines a, j, k, q, r, u, v, and the $M2$ line, and the Ar^{14+} line β. The spectrum was obtained at a plasma temperature of about 700 eV. Bottom: obtained with the von Hámos crystal spectrometer on EBIT-II. Only collisionally excited lines are seen; the dielectronic satellite lines (a, j, and k) are absent. The spectrum was obtained at a constant electron beam energy 200 eV above threshold for electron-impact excitation where dielectronic resonances cannot be excited. (From [11].)

enabled on the Livermore EBIT-I and EBIT-II electron beam ion traps through the use of the 32-channel Goddard Space Flight Center high-resolution x-ray microcalorimeter [20]. The x-ray emission from Ne^{8+} following charge exchange between Ne^{9+} and neutral neon is shown in Fig. 3.

The $np \rightarrow 1s$ emission from levels with principal quantum number $n \geq 3$ is not very bright. It is, however, stronger than predicted by early models developed for describing cometary x rays [21]. In fact, the early x-ray models predicted the $1s2p \, ^1P_1 \rightarrow 1s^2 \, ^1S_0$ resonance transition, labeled w by Gabriel [7], to be the only emission. Our measurements show that this is not correct. This line is neither the only line, nor is it the strongest. Instead, the $1s2s \, ^3S_1 \rightarrow 1s^2 \, ^1S_0$ forbidden transition (line z) is the strongest line produced. A subsequent model based on equipartition of the x-ray emission [22] is just as poor in predicting the correct x-ray emission.

Recently, we were also able to record charge-exchange produced spectra on the NSTX tokamak. This was possible because the line of sight of the NSTX high-resolution crystal

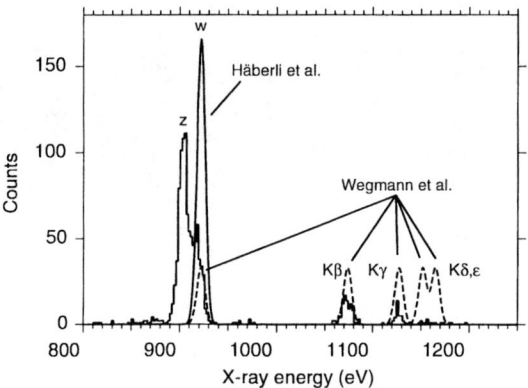

FIGURE 3. K-shell emission spectrum of Ne^{8+} produced by charge exchange between neutral neon and Ne^{9+} ions. The measurements were made with the Goddard microcalorimeter on EBIT-II. Also shown are the predictions from [21] and [22]. The ion-neutral interaction energy was about $150 \pm 100\,eV$.

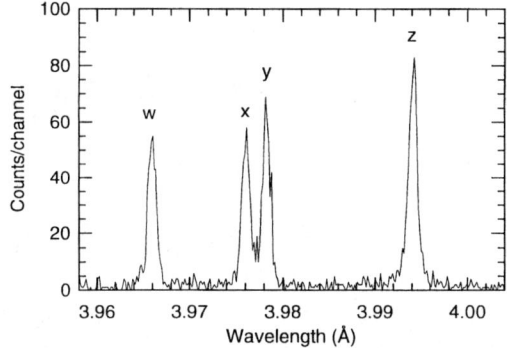

FIGURE 4. K-shell emission spectrum of Ar^{16+} produced by charge exchange between atomic hydrogen and Ar^{17+} ions. The measurements were made with the Johann crystal spectrometer on NSTX. The ion-neutral interaction energy is 80 keV for the dominant hydrogen component in the neutral beam.

spectrometer crosses the path of the diagnostic neutral beam injector [23]. The spectrum in Fig. 4 shows the emission of Ar^{16+} produced by the reaction of Ar^{17+} with H. The interaction energy was 80 keV (full energy). This contrasts with a (thermal) interaction energy of about $150 \pm 100\,eV$ in the EBIT-I experiments above involving neon.

L-SHELL SPECTRA OF NEONLIKE IONS

L-shell x-ray spectra are much more complex than the K-shell spectra of heliumlike ions and therefore offer the possibility of greater diagnostic utility. However, these spectra have not been as intensely studied, and therefore their diagnostic utility is still rather

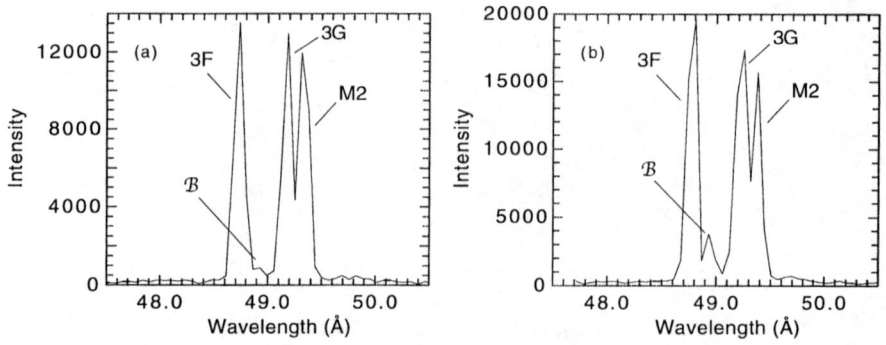

FIGURE 5. Spectra of the Ar^{8+} $3s \rightarrow 2p$ emission for different magnetic field values: (a) $B = 1$ T, (b) $B = 3$ T. The spectra were obtained on EBIT-II. (From [24].)

limited. New discoveries are readily made, as we illustrate by showing that neonlike lines can be used as a magnetic field diagnostic. Moreover, our measurements illustrate that theory has not been able to give a reliable model of the observed spectra.

Magnetic field x-ray diagnostic for high-temperature plasmas

Measurements of magnetic fields embedded in astrophysical plasmas have relied on measuring the line splitting in the optical caused by the Zeeman effect. X-ray lines typically split much less than optical lines, and this technique cannot be used. To our knowledge there are currently no x-ray line diagnostics of magnetic field strength.

Using the Livermore electron beam ion trap we showed that the $(2p_{1/2}^5 3s)_{J=0}$ level can decay to the $2p_{J=0}^6$ neonlike ground state provided there is a sufficiently strong magnetic field. In the absence of a magnetic field this decay channel is strictly forbidden. In the presence of a field, the $(2p_{1/2}^5 3s)_{J=0}$ level mixes with the neighboring $(2p_{1/2}^5 3s)_{J=1}$ level, making decay to the ground state possible [24]. A comparison of the L-shell emission of neonlike Ar^{8+} recorded in a high (3 T) and low (1 T) field is shown in Fig. 5. The magnetic field induced line, labeled $\mathscr{B}$ is clearly seen in the high field case.

The ratio of this line to that of the $(2p_{1/2}^5 3s)_{J=1} \rightarrow 2p_{J=0}^6$ resonance line (labeled $3F$ in Fig. 5) is sensitive to the magnetic field. This ratio is shown for Ar^{8+} in Fig. 6. The principle behind this line diagnostic applies to all neonlike ions, and thus higher magnetic fields can be measured with higher-Z neonlike ions.

Resolving puzzles in L-shell diagnostic line ratios

The L-shell x-ray spectra of neonlike ions have posed several puzzles over the years that have not been resolved by theory. One such puzzle concerned the ratio of the $(2p_{1/2}^5 3d_{3/2})_{J=1} \rightarrow 2p_{J=0}^6$ resonance line, labeled $3C$, to that of the $(2p_{3/2}^5 3d_{5/2})_{J=1} \rightarrow$

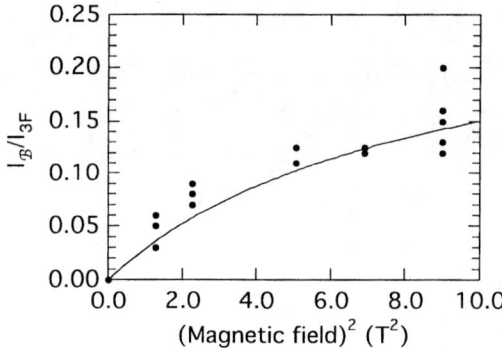

FIGURE 6. Intensity of the magnetic field induced line $\mathscr{B}$ relative to that of the $3F$ line as a function of the square of the applied magnetic field in EBIT-II. The line is drawn to guide the eye. The scatter in the data represents the statistical uncertainty of each measurement. (From [24].)

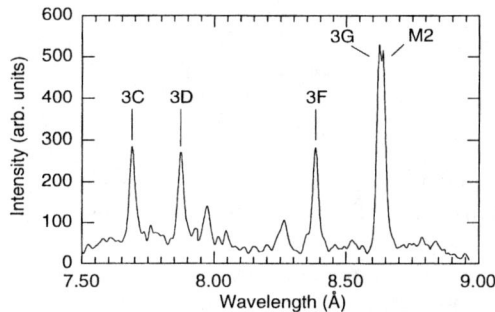

FIGURE 7. L-shell emission spectrum of Se XXV recorded with the vacuum flat-crystal spectrometer on the PLT tokamak. The spectrum represents the sum of 17 similar discharges. (From [26].)

$2p^6_{J=0}$ intercombination line, labeled $3D$. This ratio was observed to be consistently smaller than predicted when studying astrophysical plasmas and the Sun. This led to the hypothesis that resonant scattering of the $3C$ line causes a reduction in its intensity. In other words, it was assumed that this line was optically thick compared to the $3D$ line.

Systematic studies of this line ratio along the isoelectronic sequence both with our electron beam ion traps and with the Princeton Large Torus tokamak have shown that opacity does not need to be invoked to explain the solar and astrophysical observations [25, 26]. A typical spectrum of neonlike Se^{24+} recorded on the Princeton Large Torus is shown in Fig. 7.

Opacity effects play neither a role in an ion trap nor in a tokamak. Yet the observed ratios reproduced those of astrophysical and solar plasmas. The reason is that collisonal excitation calculations clearly overestimated this ratio. This is demonstrated in Fig. 8,

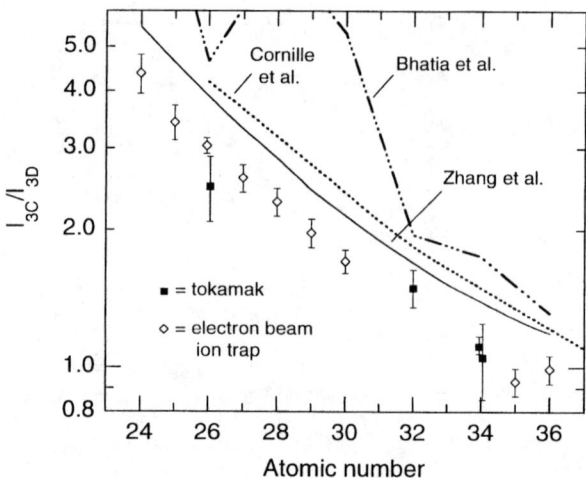

FIGURE 8. Comparison of the ratios measured on the EBIT-II electron beam ion trap (open diamonds) [25] and the PLT tokamak (closed squares) [26] with theory values from Zhang *et al.* (solid line) [29], Cornille *et al.* (dotted line) [30], and Bhatia *et al.* (dot-dashed line) [31]. (From [26].)

which compares the ratios measured in the laboratory to various calculations. In cases where the solar or astrophysical line ratio is even smaller than the laboratory ratio, the differences could be attributed to line blending with a sodiumlike transition that perfectly coincides with line $3D$ [27, 28].

The ratio of the intensity of the $3s \rightarrow 2p$ transitions to those of the $3d \rightarrow 2p$ transitions in neonlike Fe^{16+} ions was found to be much larger in astrophysical plasmas than predicted by model calculations. The difference was roughly a factor of two. A measurement of this ratio in the laboratory using the NIST electron beam ion trap and a single-channel microcalorimeter claimed to demonstrate that the modeling calculations were correct [32]. The authors, therefore, invoked yet-to-be-determined mechanisms that would bring the calculations to the level observed in the astrophysical sources. Needless to say, finding such mechanisms should be paramount to any spectral studies of the Sun and deep-space plasmas.

Using the Livermore electron beam ion trap, we have made careful studies that disprove the NIST claims. Our laboratory measurements of this ratio utilized not only crystal spectrometers, but also the Goddard microcalorimeter and a grating spectrometer. All three instruments provided measurements that agreed very well with the ratios observed in the Sun and other astrophysical objects, such as the stars Capella and HR 1099 as well as the galaxy NGC 4636 [28, 33, 34, 35, 36, 37, 38]. The comparison is shown in Fig. 9. Following our laboratory measurements, no additional mechanisms are needed to explain the extraterrestrial observations.

Measurements of this ratio in Fe^{16+} are now planned on the NSTX tokamak with a new high-resolution crystal spectrometer.

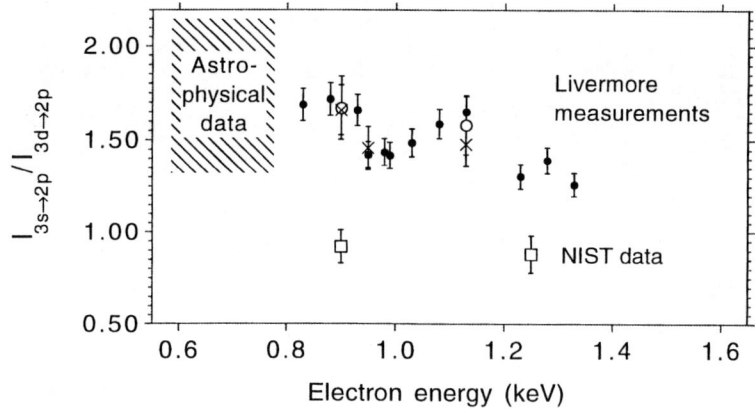

FIGURE 9. Measured $3s \rightarrow 2p$ to $3d \rightarrow 2p$ line intensities versus electron energy. Livermore data [39]: solid circles – Goddard X-ray calorimeter; crosses – crystal spectrometer; open circles – grating spectrometer. NIST results [32]: open squares – Harvard-Smithsonian calorimeter. Also shown are observational values. The energy scale on the x-axis should be disregarded for these. In collisional ionization equilibrium, the Fe XVII emission is dominated by collisions with electrons near excitation threshold, i.e., the region shown. (From [39].)

ACKNOWLEDGMENTS

This work was performed by the University of California Lawrence Livermore National Laboratory under the auspices of the Department of Energy under Contract No. W-7405-Eng-48. This work was supported in part by the DOE Office of Fusion Energy Sciences as part of the Basic and Applied Plasma Science Initiative and NASA Space Astrophysics Research and Analysis and Planetary Atmospheres programs.

REFERENCES

1. Grineva, Yu. I., Karev, V. I., Korneev, V. V., Krutov, V. V., Mandelstam, S. L., Vainshtein, L. A., Vasilyev, B. N., and Zhitnik, I. A., *Sol. Phys.*, **29**, 441 (1973).
2. Doschek, G. A., Kreplin, R. W., and Feldman, U., *Astrophys. J.*, **233**, L157 (1979).
3. Culhane, J. L., Gabriel, A. H., Acton, L. W., Rapley, C. G., Phillips, K. J., Wolfson, C. J., Antonucci, E., Bentley, R. D., Catura, R. C., Jordan, C., Kayat, M. A., Kent, B. J., Leibacher, J. W., Parmar, A. N., Sherman, J. C., Springer, L. A., Strong, K. T., and Veck, N. J., *Astrophys. J. (Lett.)*, **244**, L141 (1981).
4. Hill, K. W., von Goeler, S., Bitter, M., Campbell, L., Cowan, R. D., Fraenkel, B., Greenberger, A., Horton, R., Hovey, J., Roney, W., Sauthoff, N. R., and Stodiek, W., *Phys. Rev. A*, **19**, 1770 (1979).
5. Bitter, M., von Goeler, S., Hill, K. W., Horton, R., Johnson, D., Roney, W., Sauthoff, N., Silver, E., and Stodiek, W., **47**, 921 (1981).
6. Platz, P., Ramette, J., Belin, E., Bonnelle, C., and Gabriel, A., *J. Phys. E*, **14**, 448 (1981).
7. Gabriel, A. H., Mon. Not. R. Astron. Soc. **160**, 99 (1972).
8. Dohmann, H. D., and Mann, R., *Z. Physik A*, **291**, 15 (1979).
9. Pegg, D. J., Haselton, H. H., Griffin, P. M., Laubert, R., Mowat, J. R., Peterson, R., and Sellin, I. A., *Phys. Rev. A*, **9**, 1112 (1974).

10. Bely-Dubau, F., Dubau, J., Faucher, P., and Gabriel, A. H., *Mon. Not. R. Astron. Soc.*, **198**, 239 (1982).
11. Beiersdorfer, P., Bitter, M., Hey, D., and Reed, K. J., *Phys. Rev. A*, in press (2002).
12. Lisse, C. M., Christian, D. J., Dennerl, K., Meech, K. J., R. Petre, H. A. W., and Wolk, S. J., *Science*, **292**, 1343 (2001).
13. Källne, E., Källne, J., Dalgarno, A., E. S. Marmar, J. E. R., and Pradhan, A. K., *Phys. Rev. Lett.*, **52**, 2245 (1984).
14. Kato, T., Morita, S., Masai, K., and Hayakawa, S., *Phys. Rev. A*, **36**, 795 (1987).
15. Mattioli, M., Peacock, N. J., Summers, H. P., Denne, B., and Hawkes, N. C., *Phys. Rev. A*, **40**, 3886 (1989).
16. Kato, T., Masai, K., Fujimoto, T., Koike, F., Källne, E., Marmar, E. S., and Rice, J. E., *Phys. Rev. A*, **44**, 6776 (1991).
17. Beiersdorfer, P., Schweikhard, L., López-Urrutia, J. C., and Widmann, K., *Rev. Sci. Instrum.*, **67**, 3818 (1996).
18. Beiersdorfer, P., Olson, R. E., Brown, G. V., Chen, H., Harris, C. L., Neill, P. A., Schweikhard, L., Utter, S. B., and Widmann, K., *Phys. Rev. Lett.*, **85**, 5090 (2000).
19. Beiersdorfer, P., Lisse, C. M., Olson, R. E., Brown, G. V., and Chen, H., *Astrophys. J. (Lett.)*, **549**, L147 (2001).
20. Porter, F. S., Audley, M. D., Beiersdorfer, P., Boyce, K. R., Brekosky, R. P., Brown, G. V., Gendreau, K. C., Gygax, J., Kahn, S., Kelley, R. L., Stahle, C. K., and Szymkowiak, A. E., *Proc. SPIE*, **4140**, 407 (2000).
21. Häberli, R. M., Gombosi, T. I., Zeeuw, D. L. D., Combi, M. R., and Powell, K. G., *Science*, **276**, 939 (1997).
22. Wegmann, R., Schmidt, H. U., Lisse, C. M., Dennerl, K., and Englhauser, J., *Planet. Space Sci.*, **46**, 603 (1998).
23. Bitter, M., Hill, K. W., Roquemore, A. L., Beiersdorfer, P., Kahn, S. M., Elliott, S. R., and Fraenkel, B., *Rev. Sci. Instrum.*, **70**, 292 (1999).
24. Beiersdorfer, P., Scofield, J. H., and Osterheld, A. L., *Phys. Rev. Lett.*, submitted (2002).
25. Brown, G. V., Beiersdorfer, P., and Widmann, K., *Phys. Rev. A*, **63**, 032719 (2001).
26. Beiersdorfer, P., von Goeler, S., Bitter, M., and Thorn, D. B., *Phys. Rev. A*, **64**, 032705 (2001).
27. Brown, G. V., Beiersdorfer, P., Chen, H., Chen, M. H., and Reed, K. J., *Astrophys. J.*, **557**, L75 (2001).
28. Behar, E., Cottam, J., and Kahn, S. M., *Astrophys. J.*, **548**, 966 (2001).
29. Zhang, H., Sampson, D. H., Clark, R. E. H., and Mann, J. B., *At. Data Nucl. Data Tables*, **37**, 17 (1987).
30. Cornille, M., Dubau, J., and Jacquemots, S., *At. Data Nucl. Data Tables*, **58**, 1 (1994).
31. Bhatia, A. K., Feldman, U., and Seely, J. F., *At. Data Nucl. Data Tables*, **32**, 435 (1985).
32. Laming, J. M., Kink, I., Takacs, E., Porto, J. V., Gillaspy, J. D., Silver, E. H., Schopper, H. W., Bandler, S. R., Brickhouse, N. S., Murray, S. S., Barbera, M., Bhatia, A. K., Doschek, G. A., Madden, N., Landis, D., Beeman, J., and Haller, E. E., *Astrophys. J.*, **545**, L161 (2000).
33. Raassen, A. J. J., Mewe, R., Audard, M., Güdel, M., Behar, E., Kaastra, J. S., van der Meer, R. L. J., Foley, C. R., and Ness, J.-U., *Astron. Astrophys.*, in press (2002).
34. McKenzie, B. J., Grant, I. P., and Norrington, P. H., *Comput. Phys. Commun.*, **21**, 233 (1980).
35. R. J. Hutcheon, J. P. Pye, and K. D. Evans, *Mon. Not. R. Astron. Soc.*, **175**, 489 (1976).
36. Canizares, C. R., Huenemoerder, D. P., Davis, D. S., Dewey, D., Flanagan, K. A., Houck, J., Markert, T. H., Marshall, H. L., Schattenburg, M. L., Schulz, N. S., Wise, M., Drake, J. J., and Brickhouse, N. S., *Astrophys. J.*, **539**, L41 (2000).
37. Xu, H., Kahn, S. M., Peterson, J. R., Behar, E., Paerels, F. B. S., Mushotzky, R. F., Jernigan, J. G., and Makishima, K., *Astrophys. J.*, in press (2002).
38. Brinkman, A. C., Gunsing, C. J. T., Kaastra, J. S., van der Meer, R. L. J., Mewe, R., Paerels, F., Raassen, A. J. J., van Rooijen, J. J., Bräuninger, H., Burkert, W., Burwitz, V., Hartner, G., Predehl, P., Ness, J.-U., Schmitt, J. H. M. M., Drake, J. J., Johnson, O., Juda, M., Kashyap, V., Murray, S. S., Pease, D., Ratzlaff, P., and Wargelin, B. J., *Astrophys. J. (Lett.)*, **530**, L111 (2000).
39. Beiersdorfer, P., E. Behar, K. R. B., Brown, G. V., Chen, H., Gendreau, K. C., Gu, M.-F., Gygax, J., Kahn, S. M., Kelley, R. L., F. S. Porter, C. K. S., and Szymkowiak, A. E., *Astrophys. J. (Lett.)*, submitted (2002).

Benchmark Calculations of Electron-Lithium Scattering for Beam Diagnostics on Tokamaks

J. Colgan*, M. S. Pindzola*, S. D. Loch*, F. Robicheaux* and D. C. Griffin[†]

*Department of Physics, Auburn University, Alabama 36849
[†]Department of Physics, Rollins College, Winter Park, Florida 32789

Abstract.
   Beams of light elements, such as lithium, are important diagnostics in controlled plasma fusion devices. Consequently there is a great need for accurate atomic data for excitation and ionization processes in neutral lithium and its ions.
   In recent years, much progress has been made in calculating accurate cross sections for electron-impact excitation and ionization processes using non-perturbative techniques. These methods can now provide benchmark accuracy for the electron-impact excitation and ionization of simple atomic systems. For example, for the electron-impact ionization of hydrogen, all the non-perturbative methods yield total cross section results that are in excellent agreement with experimental measurements. These techniques have recently been extended to calculate excitation and ionization cross sections for neutral lithium and its ions. Excellent agreement is found between the three non-perturbative methods employed, and in some cases, large disagreements are found with previous existing experimental measurements.

INTRODUCTION

There are several main beam diagnostics in use today in the study of controlled plasma fusion devices. Beams of neutral hydrogen, helium, and lithium atoms all have a role to play in extracting detailed information about the plasma. The original purpose of injecting neutral beams into plasmas was as a means of heating the plasma [1]. However, it was quickly noted that active beams could be used as a local plasma diagnostic since no line of sight integration is required, and so allowed measurements with spatial resolution in the intersection region.

   Atomic hydrogen beams are generally used as a charge exchange diagnostic of the densities and temperatures of impurities within the plasma. Beam emission spectroscopy can also be used to examine the current profile. Thermal helium atomic beams are used to study electron density and temperature near the plasma edge using a line intensity ratio technique. Fast helium beams have a higher penetration depth and can be used to diagnose electron densities and temperatures closer to the plasma core. Thermal lithium beams are used as an edge diagnostic of the plasma to study the temporal variation of the electron density [2, 3]. Fast lithium beams, which can carry a lot of energy, can be used to study charge exchange for impurities in the plasma as well as the current profile via Zeeman polimetry. These beam diagnostic techniques have been used extensively by fusion groups at JET, UK, and in Garching, Germany, among others.

CP635, *Atomic Processes in Plasmas: 13th APS Topical Conference*, edited by D. R. Schultz et al.

It is clear that accurate atomic data is necessary so that the atomic processes which occur in fusion plasmas can be modeled with some degree of confidence. Therefore much effort has gone into the provision of an atomic database, known as ADAS (Atomic Data and Analysis Structure) [4, 5] which aims to provide accurate data for atomic excitation, ionization, charge exchange, recombination, and collisional radiative processes for many atomic species.

Up until around 10 years ago, the most accurate theoretical methods for calculations of electron-impact direct ionization processes of atoms were based on perturbative methods [6]. These were known to have limited accuracy, especially for neutral species, due to the incomplete treatment of long-range Coulomb interactions. For example, perturbative distorted-wave theoretical calculations give ionization cross sections for neutral lithium that are too high at the peak of the cross section by over 50 %. Recently however, several non-perturbative methods have been developed which can provide benchmark accuracy for electron-impact excitation and ionization of light atomic species. For the simplest case of the electron-impact ionization of hydrogen, the convergent close-coupling[7], the hyperspherical close-coupling[8], the R-matrix with pseudostates[9], the time-dependent close-coupling[10], and the exterior complex scaling[11] methods yield results over a wide range of incident energies that are all within the error bars of the total cross section measurements of Shah *et al* [12].

Very recently, the convergent close-coupling, the R-matrix with pseudostates, and the time-dependent close-coupling method have also calculated benchmark cross sections for the ionization of neutral lithium from its ground (2s) and first excited (2p) states [13]. This work follows previous calculations of electron-impact excitation of neutral lithium [14] and a series of calculations on excitation and ionization of neutral lithium using the convergent close-coupling method [15]. Studies have also been made of electron-impact ionization of ions in the lithium isonuclear sequence [16, 17]. Here we give a short review of the non-perturbative theoretical techniques used in our work and present some of the results obtained from the calculations of excitation and ionization of neutral lithium and its ions. We conclude with a brief summary.

THEORY

In this section we discuss chiefly the time-dependent close-coupling method which has been used in all of the calculations presented here. We also briefly give an overview of other non-perturbative techniques, the R-matrix with pseudostates and the convergent close-coupling methods, which have been compared to some of the time-dependent calculations here.

Time-dependent Close-Coupling Method

The time-dependent close-coupling (TDCC) theory used to determine excitation and ionization for lithium and lithium ions has been set out in detail in earlier publications [16, 17, 18]. Here we review the salient points of the calculations.

146

For the excitation and ionization of neutral lithium the $1s^2$ ground state of Li^+ was calculated in the Hartree-Fock approximation. A set of bound $\bar{n}l$ and continuum $\bar{k}l$ radial orbitals were then obtained by diagonalization of the one-dimensional hamiltonian given by

$$h(r) = -\frac{1}{2}\frac{\partial^2}{\partial r^2} + \frac{l(l+1)}{2r^2} - \frac{Z}{r} + V_D(r) + V_X(r), \qquad (1)$$

where $V_D(r)$ and $V_X(r)$ are the direct Hartree and local exchange potentials respectively, Z is the nuclear charge of the target, and atomic units are used throughout. These potentials were calculated using the $1s$ orbital, and a parameter in the exchange term was adjusted so that the single particle energies for each angular momentum were in good agreement with the experimental term energies. A pseudopotential was used to generate a $\bar{2}s$ orbital in order to eliminate the inner node of the wave function and avoid problems associated with core superelastic scattering[19]. With the exception of the missing node, the $\bar{2}s$ pseudo-orbital is very similar to the $2s$ orbital found from a Hartree-Fock calculation for the $1s^2 2s$ ground term of lithium.

For ionization of Li^+ a frozen-core $1s$ orbital is calculated as the hydrogenic ground state of Li^{2+}. A set of radial orbitals is then obtained by diagonalization of Eq. (1). Again, a parameter in the exchange term is adjusted so that the single particle energies are in good agreement with experiment. Similarly for ionization of Li^{2+} a set of radial orbitals is found by diagonalization of Eq. (1), except of course that in this case there are no direct or exchange potentials.

The total wave function $\Psi^{LS}(\vec{r}_1, \vec{r}_2, t)$ for the valence and continuum electrons is expanded in coupled spherical harmonics

$$
\begin{aligned}
\Psi^{LS}(\vec{r}_1, \vec{r}_2, t) = &\sum_{l_1 l_2} \frac{P^{LS}_{l_1 l_2}(r_1, r_2, t)}{r_1 r_2} \\
&\times \sum_{m_1 m_2} C^{l_1 l_2 L}_{m_1 m_2 0} Y_{l_1 m_1}(\hat{r}_1) Y_{l_2 m_2}(\hat{r}_2),
\end{aligned}
\qquad (2)
$$

where L and S are the total orbital and spin angular momentum of the system; (l_1, l_2) are the angular momenta for the target valence and initial scattered electrons, and later, the excited valence (or ejected) and final scattered electrons; $Y_{lm}(\hat{r})$ is a spherical harmonic; and $C^{l_1 l_2 l_3}_{m_1 m_2 0}$ is a Clebsch-Gordan coefficient. At a time $t = 0$ before the collision, the two-electron radial wave functions $P^{LS}_{l_1 l_2}(r_1, r_2, t)$ are given by antisymmetrized or symmetrized spatial products of the bound radial orbital of the ion under consideration and an incoming radial wavepacket. The time propagation is governed by the time-dependent Schrödinger equation which takes the form

$$
\begin{aligned}
i\frac{\partial P^{LS}_{l_1 l_2}(r_1, r_2, t)}{\partial t} = &\; T_{l_1 l_2}(r_1, r_2) P^{LS}_{l_1 l_2}(r_1, r_2, t) \\
&+ \sum_{l'_1, l'_2} U^L_{l_1 l_2, l'_1 l'_2}(r_1, r_2) P^{LS}_{l'_1 l'_2}(r_1, r_2, t),
\end{aligned}
\qquad (3)
$$

where $T_{l_1 l_2}(r_1, r_2)$ contains kinetic energy, centrifugal barrier, nuclear, direct Hartree, and local exchange operators; and $U^L_{l_1 l_2, l'_1 l'_2}(r_1, r_2)$ couples the various $(l_1 l_2)$ scattering channels. At a suitable time $t = T$ following the collision, excitation and ionization cross sections may be extracted. The ionization cross section is given by

$$\sigma_{ion} = \frac{\pi}{4k^2} \sum_S \sum_L (2L+1)(2S+1) \wp^{LS}_{ion} \tag{4}$$

where $\wp^{LS}_{ion}$ is the probability for ionization given by

$$\wp^{LS}_{ion} = \sum_{l_1 l_2} \int_0^\infty dk_1 \int_0^\infty dk_2 |P^{LS}_{l_1 l_2}(k_1, k_2, t)|^2, \tag{5}$$

where $P^{LS}_{l_1 l_2}(k_1, k_2, t)$ is the time-propagated wavefunction in momentum space found by projecting the propagated wavefunction $P^{LS}_{l_1 l_2}(r_1, r_2, t)$ onto products of continuum state radial orbitals. Equivalently we can write the probability for ionization as

$$\wp^{LS}_{ion} = 1 - \sum_{nlm} \wp^{LS}_{nlm} - \sum_{nlm} \sum_{n'l'm'} | < \Psi^{LS}(\vec{r}_1, \vec{r}_2, t) | \phi_{n'l'm'}(\vec{r}_1)\phi_{nlm}(\vec{r}_2) > |^2, \tag{6}$$

where $\wp^{LS}_{nlm}$ is the probability of finding one electron in a bound state and the other in the continuum. The last part of Eq. (6) is the probability of finding both electrons in bound states. These two methods of obtaining the ionization probability, by directly summing over the continuum and indirectly by subtracting the population in bound states from unity, have been found to give very similar results. We note that the first expression allows a straightforward definition of angular differential cross sections [18].

Similarly the excitation cross section is found by

$$\sigma_{nlm} = \frac{\pi}{4k^2} \sum_S \sum_L (2L+1)(2S+1) \wp^{LS}_{nlm}. \tag{7}$$

The time-dependent close-coupling equations (3) for the two-electron radial wavefunctions are solved on a numerical lattice of constant grid spacing. The mesh spacing is usually determined by the accuracy required of the energies of the bound orbitals found by the diagonalization in Eq.(1), and the box size is determined by the incident electron energy and the convergence of the collision probabilities. The radial wavefunctions are propagated in time until these collision probabilities are converged. The projection onto bound and/or continuum states is then made as described.

Other Non-Perturbative Techniques

The R-matrix with pseudostates and convergent close-coupling theories describing electron ionization of atoms have been well described in the original papers [7, 9, 20]. In the R-matrix with pseudostates method, the bound radial orbitals are obtained by solution of the radial frozen-core Hartree-Fock equation; i.e., the $V_X(r)$ of Eq.(1) becomes a non-local operator. The remaining bound and continuum radial orbitals begin as a set of non-orthogonal Laguerre orbitals of the form:

$$P_{nl}(r) = N_{nl}(\lambda_l r)^{l+1} e^{-\lambda_l r/2} L_{n+l}^{2l+1}(\lambda_l r) , \tag{8}$$

which are subsequently orthogonalized to the Hartree-Fock orbitals and each other in the process of diagonalizing the target Hamiltonian. The screening parameters, λ_l, are adjusted so that the ionization limit for lithium (say) is roughly midway between two term energies of the same symmetry. In the convergent close-coupling method, all bound radial orbitals begin as a set of orthogonal Laguerre orbitals similar to those of Eq.(8). The screening parameters, λ_l, are adjusted so that diagonalization of the target Hamiltonian yields accurate low-lying eigenvalues and a reasonable spread of continuum eigenvalues in the energy range of interest for ionization.

In the R-matrix with pseudostates method, the time-independent (N+1)-electron Hamiltonian is diagonalized in a basis of antisymmetrized products of N-electron target states and a complete set of bound and continuum radial orbitals whose derivatives vanish at the surface of an internal region box. By matching this solution with the solution in the external region a K-matrix may be extracted which yields excitation cross sections to positive-energy pseudostates that may be interpreted as ionization of the target. The pseudostate expansion used is very similar to that employed in [20] where more details can be found. Just below the ionization limit, the negative-energy pseudostates contain some continuum character, and just above the ionization limit, the positive-energy pseudostates contain some bound character. Thus, to calculate a more accurate total cross section, we determine the sum of partial cross sections to those pseudostates just below the ionization limit plus all pseudostates above the ionization limit, while projecting out the contributions from the bound portions of these pseudostates (see Eq. (1) of [20]).

In the convergent close-coupling method, the time-independent (N+1)-electron Lippman-Schwinger equation for the K-matrix is solved directly in momentum space. The resulting K-matrix again yields excitation cross sections to positive-energy pseudostates that may be interpreted as ionization of the target. The quality of this pseudostate representation of a true continuum state may be measured by the overlap of the two states. We note also that these pseudostates are very similar to those used in the R-matrix with pseudostates method, with the main difference being that, in the R-matrix with pseudostates case the Laguerre orbitals are orthogonalised to Hartree-Fock orbitals and themselves, where in the convergent close-coupling case an orthogonal Laguerre basis is used for all radial target-space orbitals. The equations may be solved for every partial wave treated, with the higher partial waves requiring less computational effort since exchange incorporated in the potential matrix elements dies off gradually with increasing angular momentum. The number of partial waves treated increases with the incident electron energy, and enough are taken to ensure convergence.

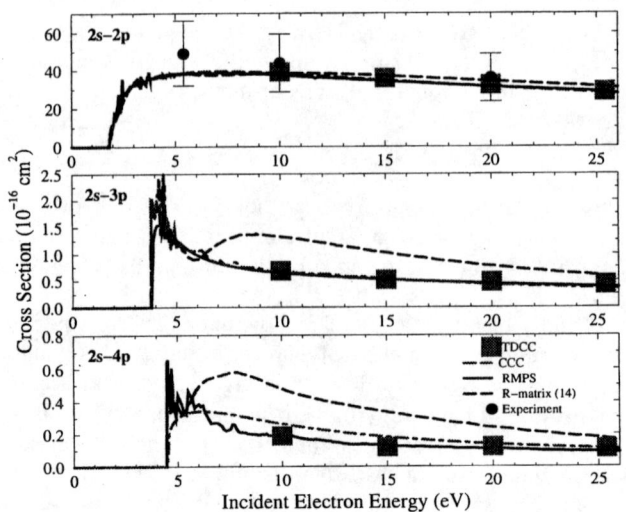

FIGURE 1. Total electron-impact excitation cross sections for the $2s \rightarrow 2p$, $2s \rightarrow 3p$, and $2s \rightarrow 4p$ transitions in lithium. We compare the time-dependent close-coupling calculations (TDCC), the convergent close-coupling calculations (CCC), the 55-state R-matrix with pseudostates (RMPS), a 14-state R-matrix calculation, and an experimental measurement of Williams *et al* [21].

RESULTS

We now turn to some results for electron-impact excitation and ionization of the lithium isonuclear sequence.

Electron-Impact Excitation of Neutral Lithium

In Fig. (1) we present electron-impact excitation cross sections for lithium for the $2s \rightarrow 2p$, $2s \rightarrow 3p$, and $2s \rightarrow 4p$ transitions. We compare the time-dependent close-coupling calculations, the convergent close-coupling calculations, the 55-state R-matrix with pseudostates calculations, and a 14-state non-pseudostate R-matrix calculation. For the $2s \rightarrow 2p$ transition we can compare with experimental measurements of Williams *et al* [21]. For this transition, the non-perturbative theoretical calculations are all in excellent agreement with each other and with the experimental measurement. The 14-state non-pseudostate R-matrix calculation, is also in good agreement with the other calculations, indicating that for this transition the effects of the target continuum on excitation to $1s^2 2p$ are relatively small. This is not the case for the $2s \rightarrow 3p$, and $2s \rightarrow 4p$ transitions where now the 14-state R-matrix calculations are larger than the other methods, indicating, as expected, increasing effects of coupling to the target continuum with increasing principal quantum number of the excited state.

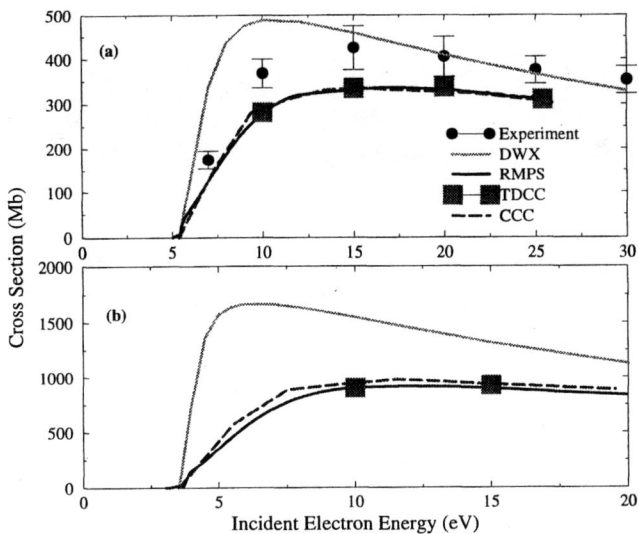

FIGURE 2. Total electron-impact ionization cross sections for (a) Li(2s) and (b) Li(2p). We compare the time-dependent close-coupling calculations (TDCC), the convergent close-coupling calculations (CCC), the *R*-matrix with pseudostates (RMPS), and distorted-wave with exchange (DWX) calculations. For Li(2s) we compare with the experimental measurements of Zapesochnyi and Aleksakhin [22].

Electron-Impact Ionization of Lithium (2*s*) and (2*p*)

Total integral cross sections for the electron-impact ionization of Li(2s) and Li(2p) are presented in Fig. (2) (a) and (b) respectively. Again we show three non-perturbative calculations; the time-dependent close-coupling, the convergent close-coupling, and the *R*-matrix with pseudostates, as well as perturbative distorted-wave with exchange calculations. For Li(2s) we also compare with experimental measurements of Zapesochnyi and Aleksakhin [22]. The non-perturbative calculations are all in excellent agreement with each other, whereas the perturbative distorted-wave calculations are up to 50% higher than the non-perturbative peak near 15 eV. The experimental measurements [22] are also higher than the non-perturbative calculations and we believe that the apparent agreement between experiment and the perturbative distorted-wave calculations is fortuitous.

For Li(2p) there are no experimental measurements with which to compare. Again, we see that the non-perturbative methods are all in excellent agreement with each other, and that perturbative distorted-wave calculations are much higher than the non-perturbative calculations (up to 90% at the peak of the cross section). We also remark that the Li(2p) ionization cross section is almost three times larger than the Li(2s) ionization cross section. The very good agreement between the non-perturbative calculations also extends to the spin-resolved partial wave cross sections, as is discussed in more detail in [13].

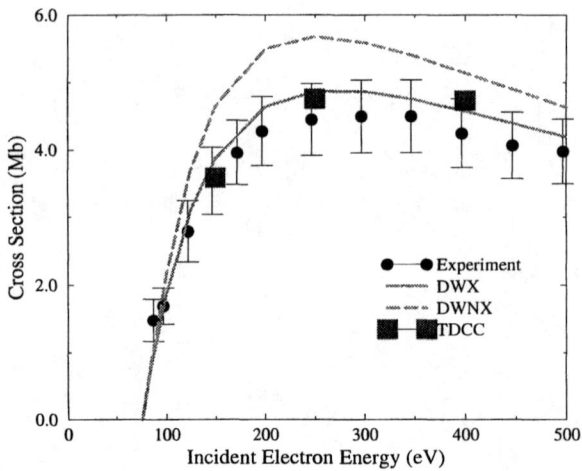

FIGURE 3. Total electron-impact ionization cross section for Li$^+$. We compare the time-dependent close-coupling calculations (TDCC) with distorted-wave with exchange (DWX) calculations, distorted-wave with no exchange (DWNX) calculations, and with experimental measurements of Lineberger *et al* [24].

Electron-Impact Ionization of Li$^+$

We now turn to studies of electron-impact ionization of ions in the lithium isonuclear sequence. In Fig. (3) we show total ionization cross sections for electron scattering from Li$^+$. In this case we compare the time-dependent close-coupling method with two distorted-wave calculations; one where an exchange term is included in the calculation and one without. There are also the experimental measurements of Lineberger *et al* [24] with which to compare. We see that the time-dependent close-coupling calculations are within the error bars of [24] and are generally in good agreement with the distorted-wave calculations which include exchange. The distorted-wave calculations where exchange is excluded are higher than all the other calculations over the range of incident electron energies.

Electron-Impact Ionization of Li^{2+}

Finally, in Fig. (4) we present total ionization cross sections for electron-impact ionization of Li^{2+}. In this case we have used a modified form of the time-dependent close-coupling technique [17] which allows us to extract the total ionization cross section over a wide range of incident electron energies for only one time propagation, giving much more information for very little increase in computer time. This is the

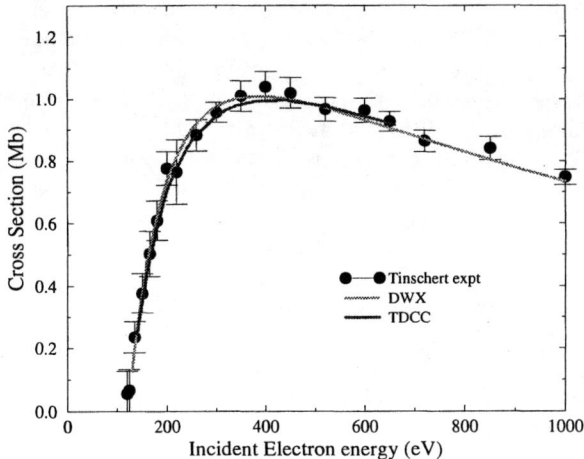

FIGURE 4. Total electron-impact ionization cross section for Li^{2+}. We compare the time-dependent close-coupling method (TDCC) with distorted-wave calculations (DWX) and experimental measurements of Tinschert *et al* [23].

first atomic system to which this new method has been applied. Here we compare these time-dependent results with distorted-wave calculations and with experimental measurements of Tinschert *et al* [23]. We see that both theoretical calculations are in excellent agreement with experiment over a wide range of incident electron energies.

As expected, the distorted-wave with exchange method gives results that are much closer to non-perturbative calculations as one goes up the lithium isonuclear sequence. By Li^{2+}, the distorted-wave calculations are extremely close to non-perturbative time-dependent calculations. This reflects the fact that distorted-wave theory does much better when calculating electron scattering from more highly charged Coulombic targets.

SUMMARY

In this review we have presented a selection of benchmark non-perturbative calculations for electron-impact excitation and ionization of neutral lithium and its ions, and compared to available experimental measurements. This atomic data is important for lithium beam spectroscopy which is becoming increasingly used in controlled fusion plasma devices [2, 3]. Time-dependent calculations for electron-impact excitation of Li^+ and Li^{2+} have already been made (as they are produced in the ionization calculations) and comparisons with similar RMPS calculations are underway.

The non-perturbative techniques discussed here can also be used to give more information about the scattering process. For example, single differential cross sections for the electron-impact ionization of lithium have also been calculated using non-perturbative techniques [13, 18]. This provides cross section data as a function of the energy of one of the outgoing electrons for a given incident electron energy. Even more detailed calculations of triple differential cross sections have been made for the electron-impact ionization of hydogen [25], and the time-dependent method is expected to be applied to the study of triple differential cross sections for lithium in the near future. Similar calculations have already been made for lithium using the convergent close-coupling method [26].

ACKNOWLEDGMENTS

This work is supported in part by the U. S. Department of Energy, including a SciDAC award. Computational work was carried out at the National Energy Research Supercomputer Center in Oakland, CA.

REFERENCES

1. J. Wesson, in *Tokamaks* (2nd Edition) (Clarendon Press, Oxford, UK, 1997).
2. Z. A. Pietrzyk, P. Breger, and D. D. R. Summers, Plasma Phys. Control. Fusion **35**, 1725 (1993).
3. J. Schweinzer *et al* Plasma Phys. Control. Fusion **34**, 1173 (1992).
4. H. P. Summers, JET Joint Undertaking Report JET-IR(94)96, (1994).
5. H. P. Summers, ADAS User Manual (2nd Edition), available from `http://adas.phys.strath.ac.uk/adas/docs/manual` (1999).
6. S. M. Younger, Phys. Rev. A **22**, 111 (1980).
7. I. Bray and A. T. Stelbovics, Phys. Rev. Lett. **70**, 746 (1993).
8. D. Kato and S. Watanabe, Phys. Rev. Lett. **74**, 2443 (1995).
9. K. Bartschat and I. Bray, J. Phys. B **29**, L577 (1996).
10. M. S. Pindzola and F. Robicheaux, Phys. Rev. A **54**, 2142 (1996).
11. M. Baertschy *et al*, Phys. Rev. A **63**, 022712 (2001).
12. M. B. Shah, D. S. Elliott, and H. B. Gilbody, J. Phys. B **20**, 3501 (1987).
13. J. Colgan, M. S. Pindzola, D. M. Mitnik, D. C. Griffin, and I. Bray, Phys. Rev. Letts. **87**, 213201 (2001).
14. D. C. Griffin, D. M. Mitnik, J. Colgan, and M. S. Pindzola, Phys. Rev. A **64**, 032718 (2001).
15. J. Schweinzer *et al*, At. Data Nucl. Data Tables **72**, 239 (1999).
16. M. S. Pindzola, D. M. Mitnik, J. Colgan, and D. C. Griffin, Phys. Rev. A **61**, 052712 (2000).
17. J. Colgan, M. S. Pindzola, and F. Robicheaux, Phys. Rev. A, submitted (2002).
18. J. Colgan, D. M. Mitnik, M. S. Pindzola, and D. C. Griffin, Phys. Rev. A **63**, 062709 (2001).
19. M. S. Pindzola, F. Robicheaux, I. Bray, and D. C. Griffin, J. Phys. B **32**, 911 (1999).
20. N. R. Badnell, M. S. Pindzola, I. Bray, and D. C. Griffin, J. Phys. B **31**, 911 (1998).
21. W. Williams, S. Trajmar, and D. Bozinis, J. Phys. B **9**, 1529 (1976).
22. I. P. Zapesochnyi and I. S. Aleksakhin, Zh. Eksp. Teor. Fiz. **55**, 76 (1968) [Sov. Phys. JETP **28**, 41 (1969)].
23. K. Tinschert *et al*, J. Phys. B **22**, 531 (1989).
24. W. C. Lineberger, J. W. Hooper, and E. W. McDaniel, Phys. Rev. **141**, 151 (1966).
25. J. Colgan, M. S. Pindzola, F. Robicheaux, D. C. Griffin, and M. Baertschy, Phys. Rev. A **65**, 042721 (2002).
26. I. Bray, J. Beck, and C. Plottke, J. Phys. B **32**, 4309 (1999).

Using Quantum Defect Theory in the (e,2e) Ionization of Argon

S. Mazevet*, K. Fakhreddine[†], G. Nguyen Vien[†], R.J. Tweed[†] and
J. Langlois[†]

*Theoretical Division, Los Alamos National Laboratory, Los Alamos, New Mexico 87545
[†]Laboratoire des Collisions Electroniques et Atomiques, Université de Bretagne Occidentale,
6 Avenue le Gorgeu - BP: 809 - 29285 Brest Cedex, France

Abstract. Quantum Defect theory is a well established theoretical concept in modern spectroscopy
that was found particularly powerful in the study of Rydberg states and photoionization of various
atomic and molecular species. We show that this approach can also be useful in electron impact
ionization problems where state of the art theoretical methods are presently restricted mostly to
simple atomic targets. We found that this approach leads to significant improvements over previous
calculations for the well documented case of the ionization of argon in equal energy sharing
geometry.

INTRODUCTION

The most complete information about the ionization process by electron impact is obtained by the (e,2e) measurements where both scattered and ejected electrons are detected in coincidence. This experimental setup has been widely used over the past thirty
years with a large variety of targets (atomic, molecular and solid) and for various kine-
matical situations (for a review see ref. [1, 2, 3]). While experimental situations using
incident electron beams with energy higher than a few hundreds eV are now considered
well understood, low energy situations still present a challenge to the theory of ionization. Significant progress has nevertheless been accomplished over the past decade with
the development of the computationally intensive Convergent Closed Coupled method
(CCC) [4, 5] and the Dynamically Screened three Coulomb (DS3C) method [6, 7]. Beyond the fact, pointed out by Malegat [8], that these two methods each describe different
regions of the configuration space and therefore should not be considered as definitive,
their application, to date, is further limited to the simplest atomic systems. The DS3C
method provides a useful ansatz for the three body Coulomb final state but is formally
restricted to hydrogenic targets while the application of the CCC method to targets with
more than two electrons is limited by current computational power. Ionization processes
involving complex atomic systems are still described within the framework of the Distorted Wave Born Approximation (DWBA) and the Plane Wave Born approximation
(PWBA) for the case of molecular systems. The DWBA is a first order formulation
method that provides a less target-restricted description of the ionization process but at
the expense of only being applicable at intermediate and high energy. The latter range
of applicability is roughly dictated by the level of refinement used in the calculation of

CP635, *Atomic Processes in Plasmas: 13th APS Topical Conference*, edited by D. R. Schultz et al.
© 2002 American Institute of Physics 0-7354-0090-3/02/$19.00

the distorted waves [9, 10].

Quantum Defect (QD) theory is widely used in modern spectroscopy to characterize Rydberg states, and in the calculation of the photoionization cross sections of various atomic and molecular species [11, 12]. In this paper, we show that QD theory used within the DWBA framework, might also prove usefull for the description of ionization processes by presenting an alternative way to account for the short range interactions (static and exchange) in the calculation of the final state continuum distorted waves. Compared to the determination of the Hartree-Fock non-local operator [13], which becomes rapidly a tedious task as the size of the target increases, our method allows for a target-independent procedure which can be readily applied to much larger atomic or molecular systems. The parameters of a functional form of the electron-ion potential are optimized in order to reproduce the QD using the canonical function method [14]. The electron-ion potentials obtained for each Rydberg series are further modified to account classically for the electron-electron interaction in the final state [15]. Calculations performed within the DWBA framework for the ionization of Argon in the equal energy sharing geometry, which is reasonably well documented both theoretically and experimentally, clearly validate our approach and show significant improvements over previous treatments.

DETERMINATION OF THE PSEUDO-POTENTIALS

The potential felt by an electron in the field of an ion is not exactly Coulombic at intermediate distances. The difference between the effective and Coulomb potentials manifests itself through the QD for bound states and phase shifts for continuum states. For a given angular momentum l, the QD, μ_l, is defined as

$$E_{n,l} = -\frac{Z^2}{(n - \mu_l)^2},$$

(1)

where $E_{n,l}$ is the binding energy of the excited electron in the n, l orbital and Z the charge of the ionic core. For a given Rydberg series, as n increases, the QD becomes constant. When the electron energy is above the continuum limit, its wavefunction is characterized by a phase shift for each spin and angular momentum. The difference between the actual and the Coulombic phase shift, δ_l, is the signature of the non-Coulombic part of the electron-ion interaction. Near the continuum limit the phase shift is related to the QD through

$$\delta_l = \pi \mu_l.$$

(2)

The relation (2) establishes QD theory and has motivated various numerical methods also referred to as effective range theory [16]. Our approach is further based on the assumption that a potential (not necessarily unique), which reproduces the information contained in the bound state spectrum of the electron-ion system, can be used directly to calculate the continuum wavefunction, in our case the relevant (e,2e) outgoing distorted wave. An extensive review of the applications of model potentials has been given by Hibbert [17] (see also ref. [11]). We choose the functional form suggested by Green *et*

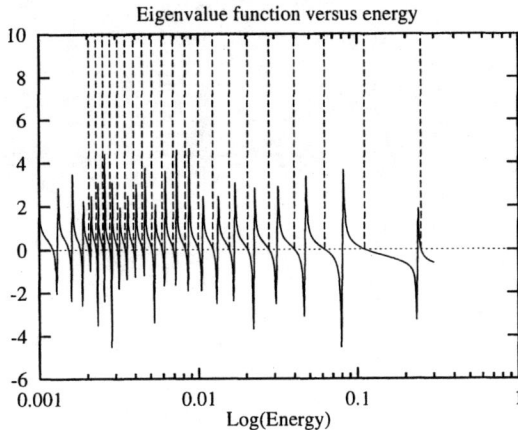

FIGURE 1. Behavior of the eigenvalue function with energy. The vertical lines indicate the eigenvalue position.

al. [18]:

$$V(r) = -(1/r)[(Z-1)\omega(r) + 1],$$

with $\qquad \omega(r) = 1/[\varepsilon_1(\exp(r/\varepsilon_2) - 1) + 1] \qquad (3)$

where ε_1 and ε_2 are parameters. The canonical function method is used to optimize the parameters ε_1 and ε_2 in order to reproduce the QD results for various Rydberg series.

The canonical function method (CFM) is a powerful means for solving the Radial Schrödinger Equation (RSE). The CFM turns the Shrödinger equation into a regular initial value problem and allows the full determination of the spectrum of the Schrödinger operator, bypassing the evaluation of the eigenfunctions. This method, (CFM), which was initially developped by Kobeissi and his coworkers [14, 19, 20] to integrate the RSE, consists in writing the general solution as a function of the radial distance r in terms of two basis functions $\alpha(r)$ and $\beta(r)$. Picking an arbitrary point r_0 at which a well defined set of initial conditions are chosen ie: $\alpha(r_0) = 1$ with $\alpha'(r_0) = 0$ and $\beta(r_0) = 0$ with $\beta'(r_0) = 1$, the RSE is solved by progressing towards the origin and towards Infinity. During the integration, the ratio of the functions is monitored until saturation signaling the stability of the eigenvalue spectrum. Figure 1 shows a typical spectrum obtained for the ^1S Rydberg series of argon.

Several advantages of the CFM are worth mentioning. First, the RSE integration belonging to the Singular Boundary Value class is avoided by transforming the problem into an initial value one with well-defined starting values for the basis functions $\alpha(r)$ and $\beta(r)$. Second, the RSE eigenvalue problem is solved, and the eigenfunctions are simultaneously determined once the roots of the eigenvalue function (EF) are found. The EF is defined as the difference of the ratio $\alpha(r)/\beta(r)$ at saturation for large and small r. The EF, depicted in Fig. 1 above, behaves typically as a tan function when the spectrum has stabilized after the saturation of the basis function ratio with r. The speed and accuracy of the method have also been tested and compared to standard

TABLE 1. Pseudo-Potential obtained for each of the first three Rydberg series of Argon.

Series	ε_1	ε_2
l=0	3.625	1.036
l=1	3.62	1.06
l=2	3.6344	1.036

integration algorithms in a variety of cases and for a wide range of potentials. In potential optimization problems, the method can be of great value due to its speed, accuracy, and ease of programming. Additional details about the procedure can be found in [21].

In the present application, we used the experimental energy levels from the Moore's tables [22] to optimize the pseudo-potentials. Table 1 summarizes the values of the ε_1 and ε_2 parameters obtained in each cases. The determination of these three pseudo-potentials enable the calculations of up to three partial waves, S, P, and D.

SCATTERING METHOD AND RESULTS

The DWBA method has been extensively used and documented over the years. Details of the formulation and implementation of the method used in the present work can be found in the following reference [9] and will not be repeated here. Instead, we focus on the specifics of the calculations performed and the modifications introduced to describe the equal energy sharing geometry of interest in the present study. Within a standard DWBA model and when a symmetric kinematic is considered (i.e. equal energy sharing between the two outgoing electrons), the exact unsymmetrized T-matrix element is approximated as

$$\langle \mathbf{k}_a \mathbf{k}_b \Phi_{J_i L_i M_i}^{ion} | T | \Phi_{JLM}^{atom} \mathbf{k}_0 \rangle \equiv \langle \chi^{(-)}(\mathbf{k}_a) \chi^{(-)}(\mathbf{k}_b) | V | \phi_{LM} \chi^{(+)}(\mathbf{k}_0) \rangle, \qquad (4)$$

and both final state distorted waves, $\chi^{(-)}(\mathbf{k}_a)$ and $\chi^{(-)}(\mathbf{k}_b)$ are chosen as eigenfunctions of the electron-ion system. In this situation, the exact collision state with final-state boundary condition, $\Psi^{(-)}(\mathbf{k}_a, \mathbf{k}_b)$, is approximated as the product of two outgoing distorted waves. Within this approximation both distorted waves are evaluated using a potential of the form (3), purely Coulombic asymptotically, and thereby neglecting any e-e correlation in the final state.

It is well known [6, 15, 23] that the exact final state wave function can be reasonably approximated as the product of two Coulomb waves provided the effective charges satisfy the condition,

$$\frac{Z_a^{eff}}{k_a} + \frac{Z_b^{eff}}{k_b} = \frac{1}{k_a} + \frac{1}{k_b} - \frac{1}{|\mathbf{k}_a - \mathbf{k}_b|}. \qquad (5)$$

In the present case, we focus on the equal energy sharing geometry $\mathbf{k}_a = -\mathbf{k}_b$ and impose on the effective charge the condition

$$Z_a^{eff} = Z_b^{eff}. \tag{6}$$

Accordingly, we modify the form of our pseudo-potentials and replace in (3) the unit charge by an effective charge Z^{eff} which satisfy (5) and (6):

$$\lim_{r \to \infty} Z^{eff} = \frac{3}{4}. \tag{7}$$

This choice of screening charge insures that each outgoing electron experiences asymptotically the exact classical force . It was further suggested that the use of such effective charges suffices to take into account the electron correlation for the ionization of rare gas atoms in equal energy sharing geometry [23].

Fig 2 shows the variation of triply differential cross section (TDCS) as function of the scattering angle for various values of the excess energy, equally shared between the two outgoing electrons. The results, obtained using the pseudo-potentials defined in the previous section for the outgoing distorted waves, are compared to calculations where the Furness McCarthy local-exchange potential [24] is used in the exit channels and to experimental measurements [25]. It should be further noted that in both models the same local exchange approximation is used in the entrance channels. In addition, since the experimental measurements are not absolute, they have been systematically normalized with our model at a scattering angle $\theta = 270$ degrees except for an incident energy $E_i = 17.8eV$ (i.e. $E_{exc} = 2eV$) where the experiment is normalized to obtain the best visual fit to the theory.

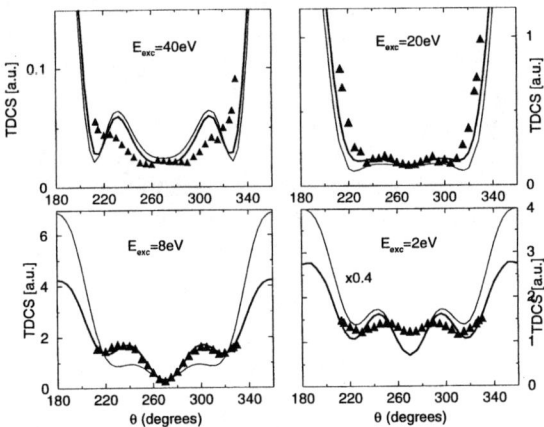

FIGURE 2. Behavior of the TDCS as a function of scattering angle in equal energy sharing geometry for various values of the excess energy as indicated on each graphs. DWBA calculations (thin line), DWBA calculations + pseudo-potentials (thick line), experiment (triangle).

For the highest incident energy considered here, $E_0 = 55.8$ eV (i.e. $E_{exc} = 40eV$), little differences in both the shape and magnitude of the TDCS are noticeable between the two theoretical models. While the use of pseudo-potentials in the calculations of the outgoing distorted waves slightly reduces the variation in amplitude of the TDCS and hence marginally improves the agreement with the experimental data, the main feature of the experimental measurements are clearly not reproduced. This suggests, at first, that the disagreement with the experimental measurements cannot be attributed to the treatment of the exchange interaction in the exit channel. Further, this result also validates our procedure for an energy interval (i.e 20 eV for each outgoing electron) where the Furness-McCarthy local exchange potential is still in its low energy limit of validity. As the incident energy decreases to $E_0 = 35.8$eV (i.e. $E_{exc} = 20$eV), the improvement resulting from the use of the pseudo-potentials calculated in the previous section is now clearly noticeable. This is particularly the case of the angular region between 230 and 320 degrees. Although improved, the TDCS is still slightly underestimated outside this particular angular range.

Interestingly, the agreement with experimental data is most improved as the incident energy is decreased to a point where each outgoing electron escapes from the ion with an energy of 4eV. The differences between the two theoretical models clearly highlight the effectiveness of the pseudo-potentials at this energy and the improvement obtained using such a description of the final state interactions. It should also be pointed out that at this particular energy value, the local exchange potential used for the description of the incoming distorted waves still stands in its lower energy region of validity ($E_0 = 23.8$eV). Finally, when the incident energy is further lowered to $E_0 = 17.8$eV, the agreement with experimental data slowly deteriorates. Figure 2 shows, at this particular kinematic, the DWBA calculations normalized with our model at a scattering angle of 270 degrees. It can readily be seen that the amplitude of the variation of the TDCS as a function of scattering angle is largely over-estimated by both theoretical models with little differences where experimental measurements are available. The most striking difference between the two calculations lies in the magnitude of the predicted TDCSs. The inclusion of the final state interactions largely reduces the overall magnitude of the TDCS and noticeably outside the angular region accessible experimentally. As the experimental measurements are not absolute it not possible to make any further conclusions regarding this result. Further, at such a low incident energy ($E_0 = 17.8$eV), the exchange approximation used in the entrance channel is most likely responsible for the remaining discrepancies in terms of the amplitude of the angular variation of the TDCS. However, while we are presently investigating alternatives to improve this situation, absolute measurements of the TDCS in this energy region would be most valuable.

CONCLUSION

In this work, we showed that QD information provides means to select and tune a class of pseudo-potentials that can be used in ionization studies, when properly modified to account for the electron correlation in the final state. The case of ionization of argon in the equal energy sharing geometry, shows that this procedure provides an efficient treatment of the final state interactions, extending beyond the energy region where the commonly used Furness-McCarthy local exchange potential is valid, and yet providing a much simpler alternative to the complete Hartree-Fock treatment. This might turn out to be particularly valuable when considering the ionization of larger atomic or molecular systems. Finally, despite the fact that the pseudo-potentials were optimized to reproduce the experimental energy spectrum of argon, the method should not be seen as semi-empirical as energy spectrum obtained from *ab-initio* calculations could be used as well.

ACKNOWLEDGMENTS

S.M. and K.F. are grateful to the "Laboratoire des Collisions Electroniques et Atomiques" at the University of Bretagne Occidentale (UBO-France) for an invitation in summer 1999 during which most of this work was performed.

REFERENCES

1. McCarthy, I. E., and Weigold, E., *Phys. Rep.* **27C**, 275 (1976).
2. McCarthy, I. E., and Weigold, E., *Rep. Prog. Phys.* **54**, 789 (1991).
3. McCarthy, I. E., and Weigold, E., *Electron–atom collisions*, Cambridge University Press, (1995).
4. Bray, I., Konovalov, D. A., and McCarthy, I. E., *Phys. Rev. A* **43**, 5878 (1991).
5. Bray, I., and Stelbovics, A., *Adv. At. Mol. Phys.* **35**, 209 (1995).
6. Berakdar, J., *Phys. Rev. A* **53**, 2314 (1996).
7. Brauner, M., Briggs, J., and Klar, H., *J. Phys. B* **22**, 2265 (1989).
8. Malegat, L., Selles, P., and Kazansky, A., *Phys. Rev. A* **60**, 3667 (1999).
9. McCarthy, I. E., *Aust. J. Phys.* **48**, 1 (1995).
10. Rasch, J., *Thesis*, University of Cambridge (1996).
11. Aymar, M., Green, C. H., and Luc-Koenig, E., *Rev. Mod. Phys.* **68**, 1015 (1996).
12. Jungen, C., *Molecular Applications of Quantum Defect Theory*, Institute of Physics Publishing, (1996).
13. Winkler, K. D., Madison, D. H., and Saha, H. P., *J. Phys. B* **32**, (1999).
14. Kobeissi, H., *J. Phys. B* **15**, 693 (1982).
15. Peterkop, *Theory of Ionization of Atoms by Electron Impact*, Colorado Associated University Press, Boulder (1977).
16. Seaton, M. J., *Rep. Prog. Phys.* **46**, 167 (1983).
17. Hibbert, A., *Adv. At. Mol. Phys.* **18**, 309 (1982).
18. Greem, A. E. S., *Phys Rev* **184**, 1 (1969).
19. Koebeissi, H., Fakhredine, K., and Kobeissi, M., *Int. J. Quantum Chem.* **40**, 11 (1991).
20. Kobeissi, H., and Fakhreddine, K., *J. Phys. II (France)* **1**, 899 (1991).
21. Tannous, C., Fahkredine, K., and Langlois, J., *J. Phys. IV (France)* **9**, 71 (1999).
22. Moore, C. E., *Atomic energy levels*, NBS Publications (1971).
23. Pan, C., and Starace, A. F., *Phys. Rev. A* **45**, 4588 (1992).
24. Furness, J. B., and McCarthy, I. E., *J. Phys. B* **6**, 2280 (1973).
25. Rouvellou, B., Rioual, S., Roeder, J., Pochat, A., Rasch, J., Whelan, C. T., Walters, H. R. J., and Allan, R. J., *Phys. Rev. A* **57**, 3621 (1998).

Near-Threshold Studies of Electron-Impact Ionization

Mark D. Baertschy

JILA, 440 UCB, Boulder, CO 80309-0440

Abstract. Much of our understanding of electron impact ionization is based upon the rich body of classical and semi-classical work performed over the past 50 years. In particular, these studies have focused on behaviors in the near-threshold energy regime where excess energies are no more than a few eV. Perhaps the most well-known result is the "Wannier law" that predicts that the total ionization cross section as a function of excess energy should obey a particular power law [1]. In this paper the results of a fully quantum mechanical calculation based on the method of Exterior Complex Scaling used to study the near-threshold behavior of electron-impact ionization of atomic hydrogen.

DIFFERENTIAL CROSS SECTIONS FOR IONIZATION

In the Exterior Complex Scaling (ECS) [2] approach the scattering boundary conditions are simplified by rotation of the electronic radial coordinates beyond some distance into the upper-half of the complex plane. This allows for the calculation of a scattering wave function, on a finite region of space, with the correct boundary conditions even with more than one outgoing electron. A scattering amplitude for ionization, $\mathcal{F}(\mathbf{k_1}, \mathbf{k_2})$, as a function of the momenta of both outgoing electrons is then extracted from this wave function [3]. The magnitudes of $\mathbf{k_1}$ and $\mathbf{k_2}$ constrained by energy conservation.

The most detailed differential cross section for electron-impact ionization (e,2e) is the triply differential cross section (TDCS) which is differential in the scattering angles, Ω_1 and Ω_2, for both outgoing electrons and the energy ε_1 of one electron. The TDCS is obtained directly from $\mathcal{F}$,

$$\frac{d\sigma_{\text{ion}}}{d\varepsilon_1 d\Omega_1 d\Omega_2}(\varepsilon_1, \Omega_1, \Omega_2) = \frac{16\pi^2}{k_i^2} |\mathcal{F}(\mathbf{k_1}, \mathbf{k_2})|^2. \qquad (1)$$

Separate cross sections are calculated for both the singlet (spin zero) and triplet (spin one) cases. The two spin components are then averaged together with weighting factors of $\frac{1}{4}$ and $\frac{3}{4}$, respectively.

Other differential cross sections typically studied are defined in terms of integrals of the TDCS. The doubly differential cross section (DDCS) is obtained by integrating the TDCS over one set of scattering angles.

$$\frac{d\sigma_{\text{ion}}}{d\varepsilon_1 d\Omega_1}(\varepsilon_1, \Omega_1) = \int_{4\pi} \frac{d\sigma_{\text{ion}}}{d\varepsilon_1 d\Omega_1 d\Omega_2}(\varepsilon_1, \Omega_1, \Omega_2) d\Omega_2 \qquad (2)$$

CP635, *Atomic Processes in Plasmas: 13th APS Topical Conference*, edited by D. R. Schultz et al.

The DDCS is a function of the energy and scattering angle for just one of the outgoing electrons. It is the (e,2e) analog of the differential cross sections for elastic and discrete inelastic scattering and can be measured without using a coincidence detector.

The singly differential cross section (SDCS) is defined by integrating the DDCS over the remaining set of scattering angles.

$$\frac{d\sigma_{ion}}{d\varepsilon_1}(\varepsilon_1) = \int_{4\pi} \frac{d\sigma_{ion}}{d\varepsilon_1 d\Omega_1}(\varepsilon_1, \Omega_1)d\Omega_1 \tag{3}$$

The SDCS gives only the energy sharing information about the (e,2e) final state. One nice feature of the SDCS from a theoretical point of view is that it is separable in the total angular momentum L. Finally, the total cross section for ionization (TICS) is the integral of the SDCS.

$$\sigma_{ion}(E) = \int_0^{E/2} \frac{d\sigma_{ion}}{d\varepsilon_1}(\varepsilon_1)d\varepsilon_1 \tag{4}$$

Indistinguishability of the two outgoing electrons requires that the SDCS be symmetric about $\varepsilon_1 = \frac{1}{2}E$. By standard convention the SDCS is integrated over only half the total energy range to give the TICS.

In this paper the threshold behavior of the TICS, the SDCS, and the DDCS are studied. For the TICS and the SDCS comparison is made with two models of electron-hydrogen scattering. In both models, all angular momenta are constrained to be zero with the differences in the two models being in the two-electron potential. The Temkin-Poet model [4] uses $\frac{1}{r_>}$ (where $r_>$ is the greater of r_1 or r_2) as the two electron potenital. This model has been used extensively in testing methods for treating (e,2e). However, because of the separability of the two-electron potential the threshold behavior in this model is unphysical. The other model considered uses $\frac{1}{r_1+r_2}$ as the two-electron potential. This model, referred to as the colinear model, is consistent with the original derivation of the Wannier threshold law [1] and gives similar threshold behavior in both the TICS and the SDCS to that of (e,2e) on hydrogen.

TOTAL IONIZATION CROSS SECTION

The most famous (e,2e) threshold law is the Wannier law for the TICS,

$$\sigma_{ion}(E) \sim E^{1.127}, \tag{5}$$

which says that that the TICS goes like the excess energy, E, to the 1.127 power near threshold. In the left side of Fig. 1 the Wannier law is compared with the TICS as calculated with the ECS theory. At low excess energy the TICS, as a function of E, behaves similarly to the Wannier law. However, from the present results it is not possible to determine the exponent to more than one or two significant figures. It is generally thought that the Wannier law is not reached until below 1 eV above threshold. Unfortunately, this represents the present limit of reliable ECS calculations.

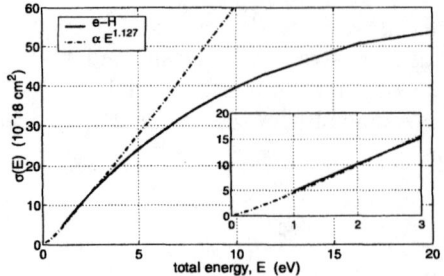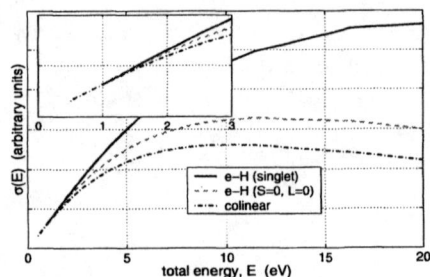

FIGURE 1. At left, total ionization cross section compared with the Wannier threshold law. At right, the total ionization cross sections for just the singlet (spin 0) case, for the $L = 0$ component of the singlet, and for the colinear model. The cross sections at right have been scaled independently so their values match at low energies.

In the right side of Fig. 1 the TICS for three different cases, all limited to the singlet spin symmetry, are compared. The upper curve is when all values of L are included, the middle curve is just the $L = 0$ component, and the lower curve is for the colinear model. Each curve has been scaled independently to make their values match near threshold. The Wannier derivation is based on the assumption that near threshold the singlet component dominates, the behavior of the total is similar to that of just the $L = 0$ component, and that the two outgoing electrons tend to move directly opposite each other.

The emerging dominance of the singlet component can be seen in the left side of Fig. 2 which shows the individual singlet and triplet components of the TICS along with the total. Indeed, the singlet component exhibits threshold behavior very near to that of the spin-averaged TICS. From the inset in the right side of Fig. 1 it can be seen that the threshold behavior of the $L = 0$ component and the complete singlet are similar below about 2 eV excess energy. Also, the colinear TICS begins to coincide with the others at somewhat less energy.

The individual L-components, as a fraction of the total, are shown in the center and right panels of Fig. 2 for singlet and triplet spin symmetries, respectively. There is no evidence that the $L = 0$ component will dominate the total at low energies. Indeed, for

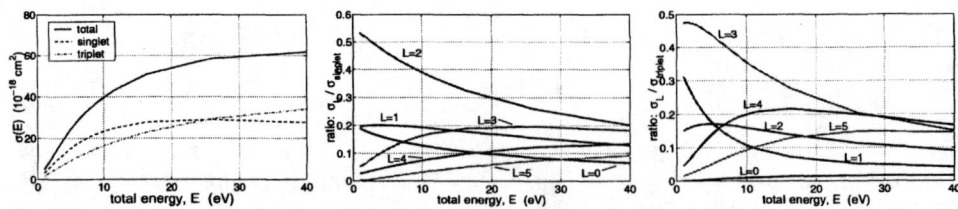

FIGURE 2. Individual singlet and triplet (spin 1) components of the TICS are shown, along with the total, at left. Individual total angular momentum (L) components of the TICS are shown for the singlet case in the center panel and for the triplet case at right. For the singlet $L = 2$ is the dominate component while $L = 3$ is for the triplet.

the triplet case, the $L = 0$ is always negligible. Clearly, the similarity in the behaviors of the total singlet and its $L = 0$ component is due more to the sum of the $L = 1$ and $L = 2$ components beginning to exhibit similar threshold behavior to $L = 0$ than it is to an emerging dominance of the $L = 0$ component.

Finally, we can infer two conclusions from the comparison of the colinear TICS to the others. The fact that the threshold behavior of the colinear model seems to tend to that of the others which, in turn, appear to agree with the Wannier law attests to the validity of the classical methods that were used in the derivation of the Wannier law. However, since the energy at which the colinear TICS coincides with the others is so low we can infer that the trend towards the threshold geometries occurs more slowly.

SINGLY DIFFERENTIAL CROSS SECTION

Some examples of the SDCS, for various excess energy E, are shown in Fig. 3. All of the curves are symmetric about the equal energy sharing point of $\varepsilon = \frac{1}{2}E$. The most striking feature of these curves is the "flattening" of the SDCS as the excess energy goes to zero. At large energies the highly asymmetric energy sharing (i.e., one "fast" and one "slow" electron) is the most favored. As the total energy decreases the SDCS becomes more isotropic until, within a few eV of threshold, all energy sharing ratios are essentially equally probable.

Similar behavior is seen when looking at the SDCS in the colinear model. At large energies the curve has a highly concave shape that flattens as the energy is lowered. Figure 4 shows three examples of the colinear SDCS near the energy where the SDCS flattens. The relative flatness of the SDCS at these energies allows us to magnify the vertical scale on the plots and see the shape of the curves in more detail. In the colinear model, the SDCS does not just flatten out as the energy decreases, it actually transforms from a

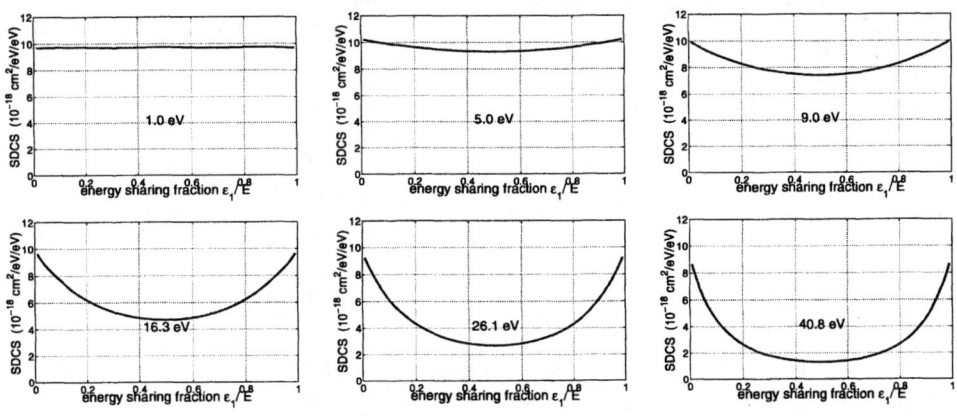

FIGURE 3. The SDCS, for various values of the excess energy E, as a function of secondary energy as a fraction of E. Each panel is labeled by the excess energy. The upper-left panel is for $E = 1$ eV, or 14.6 eV collision energy. The lower-right is for $E = 40.8$ eV, or 54.4 eV collision energy.

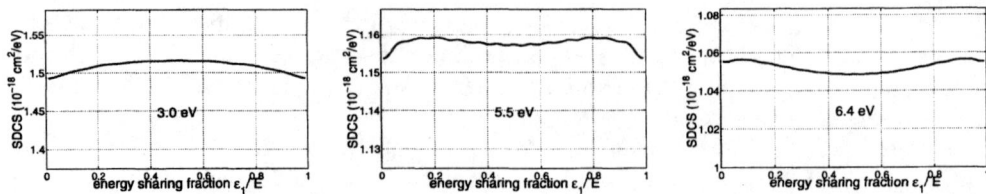

FIGURE 4. The SDCS for the colinear model at energies a few eV above threshold. At these energies the SDCS is relatively flat, note the vertical scales. The small, rapid oscillations are due to numerical error. These examples show the evolution of the SDCS curve from being convex near threshold to concave at higher energies.

slightly concave shape to a slightly convex shape. The same type of transformation from concave to convex is seen when looking at individual L-components of the SDCS for e-H scattering. However, the transitions happen at different energies for each component so the shape transformation of the complete SDCS is more complicated.

The transformation from concave to convex can be somewhat quantified by tracking the position of the maxima of the SDCS as a function of excess energy. For high energies the SDCS is peaked at the most asymmetric energies so it has a maximum at $\varepsilon = 0$ and also at the symmetric point $\varepsilon = E$. As the total energy is lowered and the SDCS begins to flatten, the maxima gradually move away from the edges of the plot into the center. This "movement" can be seen in the right and center panels of Fig. 4. When the two maxima "meet" in the middle the SDCS is convex with a single maximum at $\varepsilon = \frac{1}{2}E$.

This behavior was noticed previously by Rost [5] who studied the colinear model using semi-classical methods. He illustrated the behavior by plotting the location of the maximum (in the interval $\varepsilon \le \frac{1}{2}E$) as a function of excess energy. An analogous plot is shown in the left side of Fig. 5. In the colinear model, the maxima move away from the sides at about 7 eV excess energy and they come together at about 4 eV. For the singlet, $L = 0$ component of the e-H SDCS these same behaviors occur at energies about 4 eV

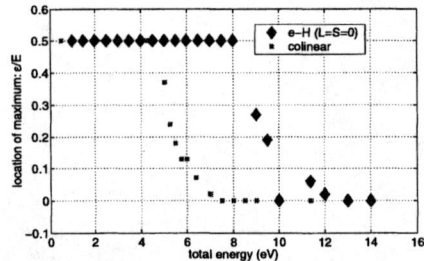
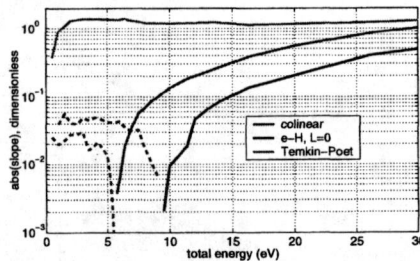

FIGURE 5. The figure at left shows the location of the maximum of the SDCS as a function of excess energy for the colinear model and the $L = 0$, $S = 0$ component of a complete e-H calculation. Near threshold the maximum is at equal-energy sharing. At several eV above threshold the maximum moves to the extreme asymmetric energy sharing. The figure at right shows the slope of the SDCS for the Temkin-Poet (top curve) and colinear (middle) models as well as the $L = 0$, $S = 0$ component of e-H (lower curve). The solid curves correspond to a concave shape and the dashed curves to a convex shape.

higher. The "movement" of the maxima in the colinear model SDCS shown in Fig. 5 is so similar to that found by Rost that one must conclude that, at least in terms of energy distributions and in the TICS, the (e,2e) process is largely classical.

One other way to analyze the relative change in shape of the SDCS as a function of excess energy is to consider the slope of a straight line approximation of half of the SDCS. This is shown in the right panel of Fig. 5. When the SDCS flattens the slope moves toward zero and when the SDCS becomes convex the slope changes sign, as indicated by the dashed lines in the figure. For the colinear model and the singlet, $L = 0$ case these slopes have very similar energy dependence. However, the SDCS in the Temkin-Poet model is decidely concave all the way down to very near threshold. For this and other reasons the Temkin-Poet model is not a reliable test of a theory's ability to correctly describe ionization cross sections at low energies.

DOUBLY DIFFERENTIAL CROSS SECTION

The final differential cross section to be considered in this paper is the DDCS. There exist very few previous studies of the DDCS with which to compare results, particularly at energies near threshold. However, there are some recent experimental measurements of the DDCS by Khakoo, *et al.* [7]. The experimental data, although still preliminary, appears to be quite good. Results for the DDCS from the ECS theory, experiment, and the Convergent Close-Coupling (CCC) theory of Bray [6] are shown in Figs. 6 and 7. For two different energies.

At 25 eV agreement among the two theories and the experiment is fairly good (Fig. 6) particularly for the more asymmetric energy sharing (upper panels). For 17.6 eV (Fig. 7) both the experiment and the calculations are more difficult, yet, the agreement is still within known limitations. Statistical error in the experiment is thought to be about 20%.

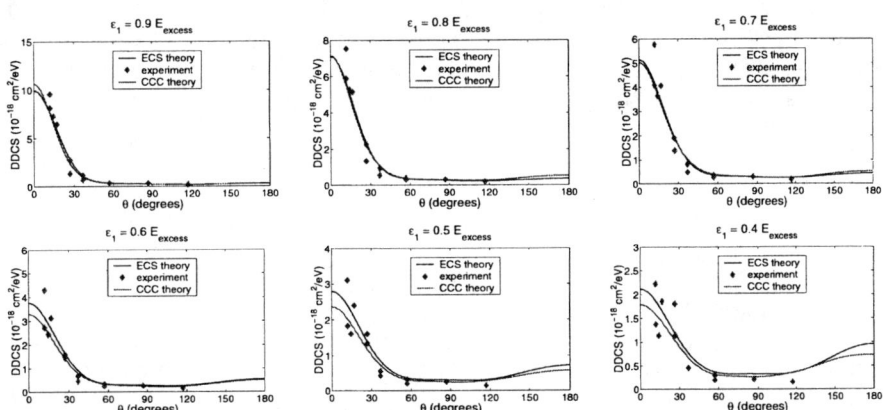

FIGURE 6. The DDCS for 25 eV collision energy. Each panel shows the angular distribution for a fixed secondary energy, ε_1 given as a fraction of the excess energy. CCC theory is due to Bray [6]. Experimental values are due Khakoo *et al.* [7].

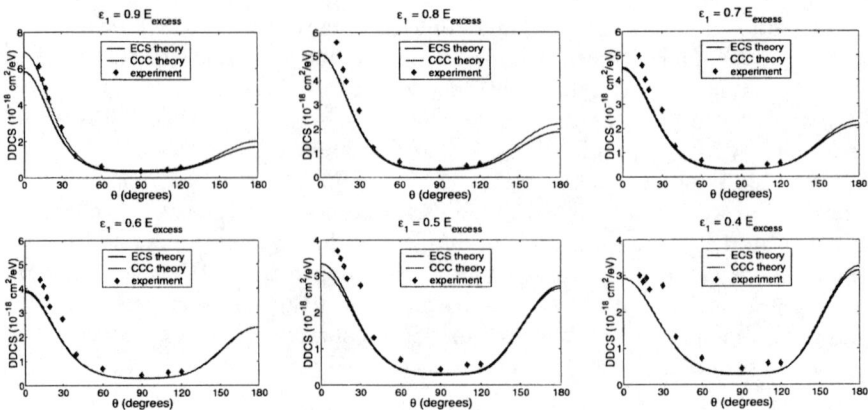

FIGURE 7. Same as in Figure 6, but for 17.6 eV collision energy.

One notable feature in the plots of the DDCS is the backward scattering peak centered at 180°. In the upper-left panel of Fig. 6 we can see that the angular distribution for "fast" electrons is highly peaked in the forward direction. As the energy of the outgoing electron is decreased we see that a backward scatering peak begins to form. The trend is more pronounced in the 17.6 eV case shown in Fig. 7. For the fast electrons (upper-left) we see a slight backward scattering peak. At equal energy sharing (lower-middle) the backward scattering peak is nearly as large as the forward peak. For "slow" electrons (lower-right) the backward peak is actually a bit larger.

This can be seen in surface plots of the entire DDCS shown in Fig. 8. Surface plots are shown for the DDCS already presented in Figs. 6 and 7 as well as for the DDCS at 15.6 eV collision energy. The backward scattering peaks are visible on the right side of each plot. For 15.6 eV and 17.6 eV the backward peak extends across the full range of energy sharing. For 25 eV, the peak is only visible for "slow" electrons. At higher energies (not shown), the backward peak is virtually nonexistent. The emergence of the backward peak at low energies is an interesting phenomenon worthy of further study. Unfortunately, the location of this peak is beyond the range of the experimental

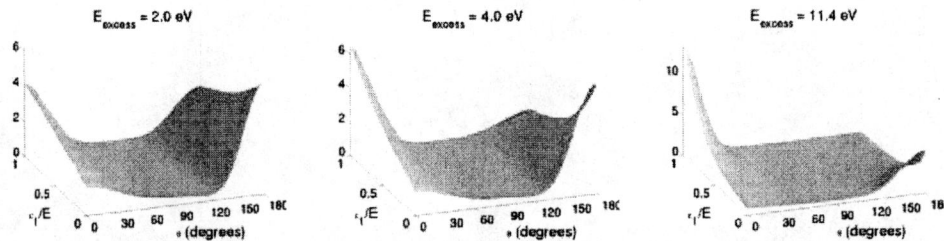

FIGURE 8. Surface plots of the DDCS for different energies, labeled by the excess energy. In each plot the front edge corresponds to "slow" electrons, the back to "fast", the left to the forward direction, and the right to the backward direction.

168

apparatus. It is hoped that modifications to the apparatus can be made that will allow for experimental verification of this peak.

ACKNOWLEDGMENTS

This work was supported by the U.S. DOE Office of Science. Computational work was carried out at the National Energy Research Scientific Computing Center at Lawrence Berkeley National Laboratory.

REFERENCES

1. Wannier, G. H., *Phys. Rev.* **90**, 817 (1953).
2. Baertschy, M., Rescigno, T. N., Isaacs, W. A., Li, X., and McCurdy, C. W., *Phys. Rev. A* **63**, 022712 (2001).
3. Baertschy, M., Rescigno, T. N., and McCurdy, C. W., *Phys. Rev. A* **64**, 022709 (2001).
4. Temkin, A., *Phys. Rev. Lett.* **49**, 365 (1982).
5. Rost, M.-M., *Phys. Rev. Lett.* **72**, 1998 (1994).
6. Bray, I., private communication (2002).
7. Khakoo, M. A., and Childers, J. G., private communication (2002).

ASTROPHYSICS

X-Ray Emission in the Solar System

T. E. Cravens

Department of Physics and Astronomy, Malott Hall, 1251 Wescoe Hall Dr., University of Kansas, Lawrence, KS 66045

Abstract. Many objects in the solar system produce x-rays, including the Sun, Venus, Earth, Mars, Jupiter, and comets. A number of emission mechanisms account for this x-ray emission including scattering and fluorescence of solar x-rays, impact excitation of atoms and molecules by energetic electrons and ions, and by charge transfer of highly charged ions with neutrals. Atomic processes play a key role in all these emission mechanisms. A brief review is provided of x-ray sources in the solar system and the emission mechanisms that are thought to be responsible for this emission.

INTRODUCTION

The solar corona is the most powerful x-ray source in the solar system, although x-ray emission has also been observed from several planets including Venus, Earth, and Jupiter, from the Moon, from a large number of comets, and from interstellar gas throughout the heliosphere. X-ray emission in astrophysics typically originates in hot collisional plasmas, such as the million degree gas residing in the solar corona, where x-rays are produced by thermal bremsstrahlung and by excitation of highly stripped ions. However, the atmospheres of planets and comets are rather cold and their x-ray emission must be powered by external sources (primarily the Sun). For example, x-rays are produced at Venus, Earth, and Jupiter by the scattering and K-shell fluorescence of solar x-rays from atoms and molecules in the atmospheres of these planets. X-rays are also produced in the terrestrial and Jovian auroral atmospheres by energetic electrons or ions, which originate in the respective magnetospheres of these planets. Cravens [1] reviewed cometary x-ray emission and also [2] reviewed x-ray emission from planets and comets in general, but much work on this topic has since been completed. In this paper, observations and models of x-ray emission throughout the solar system are reviewed.

CP635, *Atomic Processes in Plasmas: 13th APS Topical Conference*, edited by D. R. Schultz et al.

OBSERVATIONS OF X-RAY SOURCES IN THE SOLAR SYSTEM

X-rays are electromagnetic radiation with wavelength between about 0.01 and 10 nm (i.e., photons with energy between about 100 eV and 10 keV). X-rays have been observed from the following solar system objects: the Sun, Venus, Earth, the Moon, maybe Mars, Jupiter, Saturn, many comets, and from interstellar gas throughout the heliosphere. Many of the observations were made by the Roentgen satellite (or ROSAT) and, recently, observations from the Chandra X-Ray Observatory (CXO) have been reported on. Table 1 summarizes these observations and some of the characteristics of the x-ray emission. The references given are just representative in some cases, and the luminosities are only estimates based on observations reported on in rather different ways. Most of the solar system x-ray sources also vary with time, which makes it difficult to estimate a single luminosity, although the values given are "typical." Some of the individual observations are discussed later.

TABLE 1. Observed X-Ray Sources in the Solar System

Object	Some characteristics	Luminosity	Observation/references
The Sun (solar corona)	continuum and lines	10^{20} W	Numerous observations [3]
Venus	sunlit disk; C, O lines	45 MW	CXO[a] [4]
Earth - auroral	hard x-rays; variable	-	[5]
Earth – atmospheric	sunlit; N, O lines	40 MW	ROSAT[b] [6]
Earth - geocorona	variable; soft x-rays	10 MW	ROSAT [7, 8]
Mars	possible detection	3 MW	ROSAT [9]
Jupiter - disk	sunlit disk; many lines?	500 MW	ROSAT [10, 11]; CXO [12]
Jupiter - auroral	high lat., variable	1 GW	ROSAT [13]; CXO [12]
Saturn	probable – soft x-rays	400 MW	ROSAT [14]
The Moon	sunlit disk; some darkside	0.07 MW	ROSAT [15]
Comets (numerous)	soft x-rays, lines, extensive spatially, variable	1 GW	ROSAT [16], EUVE[c] [17] CXO [18, 19]
Heliosphere	part of soft x-ray background	10^{16} W	ROSAT [2, 20, 21, 22]

[a] CXO: Chandra X-Ray Observatory
[b] ROSAT: Roentgen Satellite
[c] EUVE: Extreme Ultraviolet Explorer Satellite

X-RAY EMISSION MECHANISMS

X-rays can be produced in a number of ways. Charged particle collisions with atoms or ions will produce x-rays if the particles have sufficient energy. Excitation of bound states followed by radiative de-excitation can produce line emission in the x-ray part of the spectrum. And free-free (called bremsstrahlung radiation) or bound-free transitions (which can occur as part of dielectronic recombination) can produce continuum x-ray emission. The energetic particles are often found in hot plasmas, such as the solar corona, which has a temperature of about 10^6 K. The solar corona is a strong source of both line and continuum x-ray radiation, with the line radiation coming from highly charged ions (cf. Zirin [23]).

Solar system x-ray sources other than the Sun are rather cold and the x-rays must derive their energy from external sources (cf. Cravens [2]). Collisions of fast particles can also account for auroral x-ray emission at Earth and for some of the Jovian x-ray emission. The fast electrons responsible for bremsstrahlung x-ray emission in the terrestrial aurora are accelerated to keV energies in the magnetotail from processes that ultimately are driven by the solar wind. X-rays are also produced from some objects (e.g., the atmosphere of Venus) by K-shell fluorescence associated with the absorption of solar x-rays [4, 24] or by elastic scattering of solar x-rays (e.g., at lower latitudes on Jupiter [25]). In the former mechanism, a solar x-ray ionizes an electron from the K-shell of an atom (or molecule) and the atom then de-excites by the emission of an x-ray photon or an Auger electron.

X-rays can also be produced when highly charged solar wind ions (e.g., O^{7+}, C^{6+}, Fe^{12+},...) undergo charge transfer collisions with neutral atoms and molecules they might encounter in space [26]. The ion resulting from such a collision is almost always highly excited and emits an extreme ultraviolet (EUV) or soft x-ray photon. X-ray emission from comets, from the terrestrial geocorona, and from the heliosphere has been attributed to this solar wind charge exchange (SWCX) mechanism. High charge-state ions are present in the solar wind due to the solar wind's origin in the 10^6 K solar corona.

Table 2 lists the mechanisms currently thought to be responsible for x-ray emission from different objects in the solar system. A more detailed discussion is given in the next section.

EXAMPLES OF SOLAR SYSTEM X-RAY SOURCES

Venus and Mars

X-ray emission from Venus and Mars due to scattering and fluorescence of solar x-rays was predicted [2] and K-shell emission lines from oxygen and carbon are expected to dominate the spectrum. X-ray emission was observed from Venus by CXO [4], and Figure 1 shows the first x-ray image of Venus. Note that the emission comes from the sunlit disk, as expected for scattering of solar radiation. A spectrum was also obtained by CXO and clearly shows the carbon K-shell line near .28 keV and the oxygen K-shell line at .52 keV.

TABLE 2. X-Ray Sources in the Solar System – Physical Mechanisms

Object	Dominant mechanism	References
The Sun (solar corona)	collisions: bremsstrahlung, line excitation	[23]
Venus	solar fluorescence (K-shell)	[4, 24]
Earth - auroral	collisions: bremsstrahlung	[5]
Earth – atmospheric	solar fluorescence (K-shell)	[6]
Earth - geocorona	SWCX	[7, 8, 21]
Mars	solar fluorescence, SWCX	[24, 27]
Jupiter - disk	scattering of solar x-rays	[25]
Jupiter - auroral	collisions; modified SWCX (?)	e.g., [12], [28]
Saturn	scattering of solar x-rays	[25]
The Moon	solar x-ray scattering from surface	[15]
Comets	SWCX	many refs: see review [1]
Heliosphere	SWCX	[2, 8, 20, 21]

Mars should also produce x-rays in the same way as Venus, although the expected intensities are much smaller (see Table 1). ROSAT observed several soft x-ray photons that could probably be attributed to Mars [9]. Scattering and fluorescence can account for most of this emission [24] but the solar wind interaction with exospheric neutrals of both Venus and Mars should also make a contribution [2, 27]. A simulated image of SWCX-produced x-ray emission at Mars indicates that this emission has a very different spatial morphology than the fluorescence x-ray emission, and future observations might be able to distinguish between the two mechanisms.

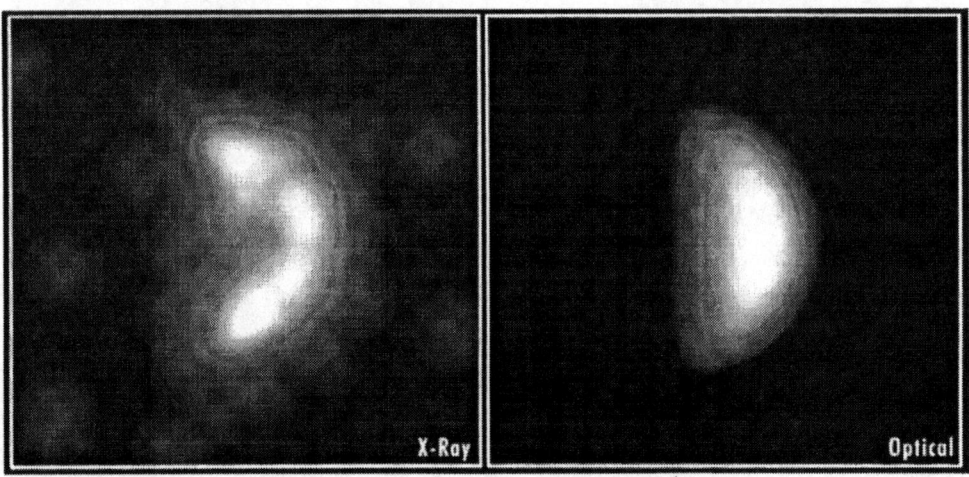

FIGURE 1. X-ray image of Venus obtained on January 13, 2001, by the ACIS-I instrument on CXO. From Dennerl et al. [4].

The Outer Planets

It has been known for many years that Jupiter emits x-rays ([29]; cf. [13]). The ROSAT images showed that the x-rays originated in two regions: low latitudes and high latitudes (i.e., auroral), but the spatial resolution was too low to define the spatial variations of the emission much beyond that simple categorization. Both bremsstrahlung associated with electron precipitation [30] and energetic ion precipitation [12, 13, 28] have been invoked to explain the auroral x-ray emission and even the low-latitude x-ray emission [10]. The ROSAT spectrum for Jupiter indicated that the photons were soft (less than a keV), but the resolution was low enough that a continuum and a multi-line spectrum could not be distinguished [13]. Energetic sulfur and oxygen ions are known to be present in Jupiter's magnetosphere, and it was suggested that fluxes of these ions could be precipitating from the middle magnetosphere into the atmosphere [31]. Heavy ions in the magnetosphere are mostly S^+, S^{++}, S^{+++}, O^+, and O^{++} but at energies of a few hundred keV/nucleon or more the ions become more highly charged when they encounter the atmosphere due to electron removal collisions [28]. X-rays are emitted by these highly charged ions due to charge transfer collisions or due to direct collisional excitation.

Some theoretical calculations suggested that scattering (about 94%) and fluorescence (due to carbon in the atmospheric methane – about 6%) of solar x-rays could account for the x-ray emission at lower latitudes at Jupiter [25] as a alternative to particle precipitation. This could explain why the Jovian emission appears to correlate with the F10.7 index of solar activity [11]. Scattering of solar x-rays should also take place at the other planets. For Saturn about 100 MW of disk emission is expected, which is roughly within a factor of 4 of the observed value [14], although the observations should also account for auroral x-ray emission.

Some exciting high spatial resolution Chandra observations of Jovian x-ray emission have recently been reported [12]. Uniform disk emission is present, confirming that scattered solar radiation is primarily responsible. And the auroral emission is located at latitudes poleward of the main auroral oval -- in the polar cap. The auroral x-rays are highly variable with a 40 minute period. Clearly, precipitation of energetic ions from the middle magnetosphere, which would produce x-rays at latitudes somewhat below the main oval, cannot be responsible for this emission [12]. Considerable discussion of alternative mechanisms has taken place (cf. [12], [32]). One suggestion is that the SWCX mechanism could be operating but with the solar wind ions penetrating down to the atmosphere in the magnetic cusp region. This will probably require considerable "extra" acceleration (hundreds of kilovolts?) beyond solar wind energies in order to get a sufficient flux of ions into the atmospheric loss cone to account for the observed x-ray luminosity.

Comets

X-ray emission from a comet was first observed by ROSAT in 1996 [16]. X-rays have subsequently been detected from almost every active comet observed [1, 33]. Following the initial discovery, several physical mechanisms were proposed, but the solar wind charge exchange (SWCX) mechanism has gradually won favor [2]. Extensive clouds of neutral gas surround active comets and the solar wind ions undergo charge transfer reactions when they encounter the neutrals. The product ions are invariably left in excited states and emit EUV and soft x-ray photons. Cometary x-ray emission has recently been reviewed [1], so only a brief summary is being provided here.

Recently, CXO has measured spectra exhibiting line emission from highly charged ions [18, 19] (see Figure 2), and these spectra confirm the viability of the SWCX mechanism. This mechanism can also explain the luminosity [26, 34], the time variations [35], and the spatial morphology [36] of the observed cometary x-ray emission.

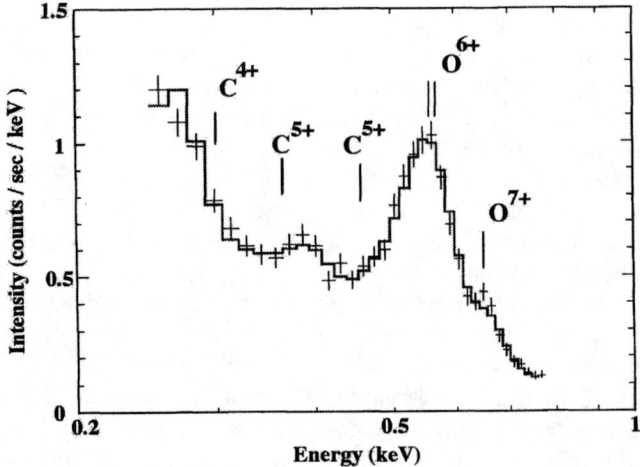

FIGURE 2. Spectrum of comet C/Linear 1999/S4 taken by the ACIS-S instrument on CXO. Reprinted with permission from *Science*, **296**, 1042-1045, copyright 2002 American Association for the Advancement of Science. The locations of a few known transition lines associated with solar wind ion species are shown.

The Heliosphere and Geocorona

A cloud of exospheric atomic hydrogen surrounds the Earth. The hydrogen geocorona provides a target for the solar wind and allows the SWCX mechanism to operate [7, 8, 21]. X-rays from the SWCX mechanism are also produced when the solar wind interacts with interstellar neutrals in the heliosphere [8, 20, 21]. Interstellar plasma is excluded from the heliosphere by the magnetized solar wind, but interstellar neutral atomic hydrogen and helium atoms are known to stream into the solar system where they have been observed. A portion of the soft x-ray background (SXRB) emission varies with time [22]. These observed temporal variations of the SXRB correlate very well with variations in the measured solar wind flux [8, 37], confirming that the SWCX mechanism must account for at least part of the SXRB. Figure 3 shows x-ray intensities as observed at the Earth from a model of the geocoronal and heliospheric emission. Variations in the solar wind flux clearly produce large variations in x-ray emission from interstellar helium and from geocoronal H, but the x-rays from interstellar H are rather steady and do not exhibit significant time variations. The flux of interstellar H into the inner solar system is severely attenuated due to photoionization by solar radiation or by charge transfer with solar wind protons so that the x-ray emission from SWCX is spread over a large volume which averages out any solar wind variations that are present.

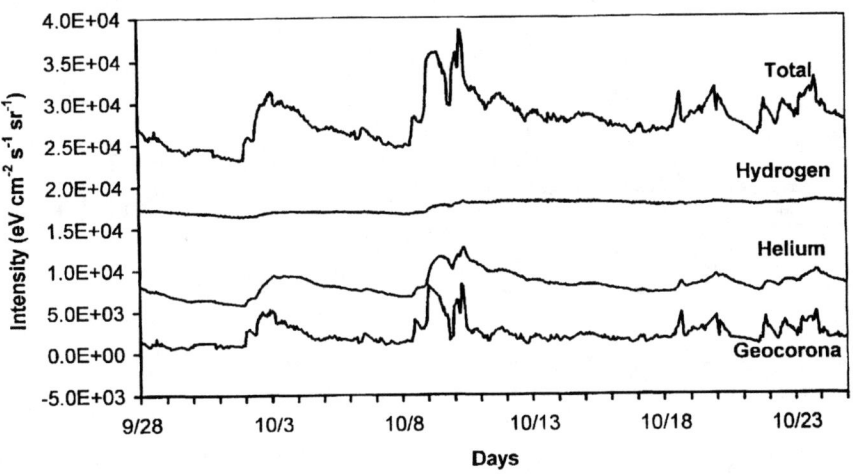

FIGURE 3. Predicted x-ray intensities from the solar wind interaction with interstellar H and He in the heliosphere and with geocoronal atomic hydrogen. From *J. Geophys. Res.*, **106**, 24883-24892, copyright 2001 American Geophysical Union.

On the other hand, attenuation of helium is much less severe and most of the observed x-ray emission comes from a smaller volume of space in the inner solar system.

SUMMARY

A review of x-ray sources in the solar system was provided in this paper. Atomic processes play a key role in all the physical mechanisms that have been invoked to explain x-ray emission throughout the solar system. With the advent of powerful tools such as the Chandra X-Ray Observatory, observations of solar system x-ray emission will increasingly be useful as a diagnostic tool for remotely probing the physical conditions in solar system objects.

ACKNOWLEDGMENTS

Support from NASA Planetary Atmospheres grant NAG5-11038 and NSF grant ATM-9815574 is acknowledged.

REFERENCES

1. Cravens, T. E., *Science* **296**, 1042-1045 (2002).
2. Cravens, T. E., *Adv. Space Res.* **26**, 1443 (2000).
3. Tobiska, W. K., et al., *J. Atm. Solar Terr. Phys.* **62**, 1233 (2000).
4. Dennerl, K., Burwitz V., Englhauser, J., Lisse, C., and Wolk, S., *Astron. Astro.* **386**, 319-330 (2002).
5. Anderson, P. C. et al., *Geophys. Res. Lett.* **25**, 4105 (1998).
6. Snowden, S. L., and Freyberg, M. J., *Astrophys. J.* **404**, 403 (1993).
7. Freyberg, M. J., pp. 113-116 in *The Local Bubble and Beyond*, eds. D. Breitschwerdt, M. J. Freyberg, and J. Trümper, Springer-Verlag, Berlin (1998).
8. Cravens, T. E., Robertson, I. P., and Snowden, S. L., *J. Geophys. Res.* **106**, 24883 (2001).
9. Liszka, C., and Holmström, M., submitted to *Icarus* (2001).
10. Waite, J. H. Jr., et al., *Science* **276**, 104 (1997).
11. Gladstone, G. R., Waite, J. H. Jr., and Lewis, W. S., *J. Geophys. Res.* **103**, 20083 (1998).
12. Gladstone, G. R., et al., in press, *Nature*, (2002).
13. Waite, J. H. Jr., et al., *J. Geophys. Res.* **99**, 14788 (1994).
14. Ness, J.-U., and Schmitt, J. H. M. M., *Astron. Astro.* **355**, 394-397 (2000).
15. Schmitt, J. H. M. M., et al., *Nature* **349**, 583 (1991).
16. Lisse, C. M. et al., *Science* **274**, 205 (1996).
17. Mumma, M. J., Krasnopolsky, V. A., and Abbott, M. J., *Astrophys. J.* **491**, L125 (1997).
18. Lisse, C. M. et al., *Science* **292**, 1343 (2001).
19. Krasnopolsky, V. A., et al., submitted to *Icarus* (2002).
20. Cravens, T. E., *Astrophys. J.* **532**, L153 (2000).
21. Cox, D. P., pp. 121-132 in *The Local Bubble and Beyond*, eds. D. Breitschwerdt, M. J. Freyberg, and J. Trümper, Springer-Verlag, Berlin (1998).

22. Snowden, S. L., et al., *Astrophys. J.* **454**, 643 (1995).
23. Zirin, H., *Astrophysics of the Sun*, Cambridge Univ. Press, Cambridge, UK (1988).
24. Cravens, T. E., and Maurellis, A. N., *Geophys. Res. Lett.* **28**, 3043-3046 (2001).
25. Maurellis, A. N., Cravens, T. E., Gladstone, G. R., Waite, J. H., and Acton, L. W., *Geophys. Res. Lett.* **27**, 1339-1343 (2000).
26. Cravens, T. E., *Geophys. Res. Lett.* **25**, 105-109 (1997).
27. Holmström, M., Barabash, S., and Kallio, E., *Geophys. Res. Lett.* **28**, 1287-1291 (2001).
28. Cravens, T. E., Howell, E., Waite, J. H., and Gladstone, G. R., J. *Geophys. Res.* **100**, 17153 (1995).
29. Metzger, A. E., et al., *J. Geophys. Res.* **88**, 7731 (1983).
30. Barbosa, D. D., *J. Geophys. Res.* **95**, 14969 (1990).
31. Gehrels, N., and Stone, E. C., *J. Geophys. Res.* **88**, 5537 (1983).
32. Cravens, T. E., abstract/presentation at *Jupiter* conference, Boulder, CO (2001).
33. Dennerl, K., Englhauser, J., and Trümper, J., *Science* **277**, 1625 (1997).
34. Schwadron, N. A., and Cravens, T. E., *Astrophys. J.* **544**, 558 (2000).
35. Neugebauer, M., et al., *J. Geophys. Res.* **105**, 20949 (2000).
36. Häberli, R., Gombosi, T. I., DeZeeuw, D. L., Combi, M. R., and Powell, K. G., *Science* **276**, 939 (1997).
37. Robertson, I. P., Cravens, T. E., Snowden, S. L., and Linde, T., *Space Sci. Rev.* **97**, 401–404 (2001).

Ionic Processes in Low Temperature Interstellar Molecular Plasmas

Nigel G. Adams, Lucia M. Babcock and Nina S. Ray

Department of Chemistry, University of Georgia, Athens, GA 30602, USA

Abstract. Astrophysical plasmas vary widely in their temperatures and compositions. Some of the most tenuous, coldest, and spatially extensive are the diffuse and dense interstellar clouds, in the latter of which stars are forming. In these regions, including circumstellar shells, ~130 molecules have been detected with ~10% being ions. The physical properties and compositions of these regions and the predominant gas phase ionic routes to the molecules are discussed. Examples of recent successes are given and some current problems highlighted.

INTRODUCTION

There are many phases in the interstellar medium, ism, varying from the hot supernovae through diffuse nebulae (HII regions), inter-cloud gas, shocked gas, cometary comae, planetary ionospheres, diffuse clouds, circumstellar shells to the cold dense and dark clouds.[1] Of these, the last three mentioned contain molecules varying in complexity from H_2 to $HC_{11}N$. The densest clouds are under gravitational collapse to form protostars, with the collapse being accelerated by cooling due to molecular emissions (in particular from CO and H_2O). During their life cycles, the stars return nuclear processed material to the ism via stellar winds and nova, and supernova explosions for the larger stars. The clouds then contain information on the nuclear processing history of the galaxy. Note that this is biased towards the more massive stars because of their shorter lifetimes (~10^6 yrs for a 15 solar mass star compared with ~10^{10} yr. for the sun). Thus, interstellar material has been processed through thousand of large star lifecycles. The ism comprises about 10% of the galactic mass and is about 1% dust grains (~0.1 μm in dia.). The conditions in the molecule-containing clouds are shown in Table 1 and differ significantly between cloud types due mainly to the differing densities. In the diffuse clouds, the densities are sufficiently low for starlight with energies below 13.6 eV (the ionization energy of the abundant H-atom) to penetrate and ionize abundant, low ionization energy atoms (e.g.,

TABLE 1. Properties of Molecule-Containing Clouds.

Cloud Type	Cloud Density (cm^{-3})	Cloud Temperature (K)	Example Clouds
Diffuse	10-100	50-150	ξ Ophiucus
Dense/Dark	10^2-10^6	10-50	Sagittarius B2, Orion Molecular Cloud
Circumstellar	10^5-10^7	200-300	IRC+10216

CP635, *Atomic Processes in Plasmas: 13th APS Topical Conference*, edited by D. R. Schultz et al.
© 2002 American Institute of Physics 0-7354-0090-3/02/$19.00

C and S).   Only small, mainly diatomic, molecules exist here because of photodissociation and the H/H_2 ratio is ~1/1. For the dense and dark clouds, the concentrations of gas and dust are sufficient to prevent significant penetration of starlight and ionization is by cosmic rays. Most of the large molecules are detected here and hydrogen is mainly as H_2. Circumstellar shells have intermediate opacities.

Molecules Detected in the Clouds

The molecules that have been detected in these regions are listed in Table 2. Note that of the 130 molecules detected, 6 identifications are in question. Indeed the identification of C_7^- as a carrier of two of the diffuse interstellar bands (DIB's) can no longer be supported in view of new laboratory spectra of C_7^- and high resolution astronomical observations of these DIBs.[4] Also, 14 positive ions have been detected, including HCO^+ (the most abundant ion detected), N_2H^+, H_3^+ and H_3O^+, which strongly supports the hypothesis that ion-molecule reactions followed by electron-ion recombination are largely responsible for the observed molecular evolution. Indeed, Herbst has concluded from gas phase ion chemical models that this predicts the observed abundances to better than an order of magnitude for ~80% of the species.[5]

TABLE 2. Molecules Observed in the Cloud Types in Table 1
Arranged by Atomicity. From McCarthy and Thaddeus[2] and Updated.[3]

2	3	4	5	6	7
H_2	H_2O	NH_3	SiH_4	CH_3OH	CH_3CHO
OH	H_2S	H_3O^+	CH_4	NH_2CHO	CH_3NH_2
SO	SO_2	H_2CO	CHO_2H	CH_3CN	CH_3C_2H
SO^+	N_2H^+	H_2CS	HC_3N	CH_3NC	CH_2CHCN
SiO	HNO	HNCO	CH_2NH	CH_3SH	HC_5N
SiS	SiH_2?	HNCS	NH_2CN	C_5H	C_6H
NO	NH_2	C_3N	H_2C_2O	HC_2CHO	c-CH_2OCH_2
NS	H_3^+	HCO_2^+	C_4H	H_2CCH_2	C_7^-?
HCl	N_2O	C_3H	c-C_3H_2	H_2C_4	
NaCl	HCO	c-C_3H	CH_2CN	HC_3NH^+	
KCl	HCO^+	C_3O	C_5	C_5N	
AlCl	OCS	C_3S	SiC_4	C_5S?	
AlF	C_2H	HCCH	H_2C_3	C_5O	
PN	HCS^+	$HCNH^+$	HC_2NC		
SiN	c-SiC_2	HCCN	HNC_3		
NH	C_2O	H_2CN	H_3CO^+		
CH	C_2S	c-SiC_3		**Total 130**	
CH^+	C_3	CH_3		**with 6 in**	
CN	MgNC	CH_2D^+?		**question**	
CO	NaCN				**10**
CS	CH_2				$(CH_3)_2CO$
C_2	MgCN		**8**	**9**	$CH_3(CC)_2CN$?
SiC	COH^+		CH_3CO_2H	C_2H_5OH	$NH_2CH_2CO_2H$
CP	HCN		HCO_2CH_3	$(CH_3)_2O$	
CO^+	HNC		CH_3C_2CN	C_2H_5CN	**11**
HF	SiCN		C_7H	$H(CC)_3CN$	$H(CC)_4CN$
SH	KCN?		H_2C_6	$H(CC)_2CH_3$	
FeO?	CO_2		CH_2OHCHO	C_8H	**13**
	AlNC				$H(CC)_5CN$

Such gas phase routes are the subject of this review. Note that some neutral radical chemistry may also occur (neutral reactions not involving radicals are known to become less efficient at low temperature due to activation energy barriers, whereas ionic reactions become more efficient) and there is the possibility of reactions on the surfaces of dust grains. However, discussion of these is beyond the scope of the present review. The observed molecules can be viewed in several ways. If the elemental composition of the molecules is examined, it is seen that only 13 of the 90 naturally occurring elements are present (excluding the uncertain detection of FeO) and these comprise a limited portion of the periodic table, as illustrated in Table 3.

TABLE 3. Restricted Periodic Table of the Elements in the Observed Molecules Showing the Number of Times Each is Detected.

H							He
86							0
Li	Be	B	C	N	O	F	Ne
0	0	0	98	46	39	2	0
Na	Mg	Al	Si	P	S	Cl	Ar
2	2	3	10	1	16	4	0
K	Ca	Ga	Ge	As	Se	Br	Kr
2	0	0	0	0	0	0	0

C-atoms (produced in the stellar triple α reaction[6]) are more common in the molecules than those containing H-atoms even though C-atoms are $\sim 10^4$ times less abundant. N and O-atoms are also prevalent, being connected to C synthesis in stars by the CNO cycle. It is interesting that Si and S are common, being produced in advanced ^{16}O burning in more massive stars, consistent with their shorter life cycles.[6] Elements produced in r and s processes (F, Cl, K, Al) in larger stars are also detected, although mainly in circumstellar shells. The molecules can also be considered from a chemical viewpoint with many inorganic compounds and organic compounds (alcohols, aldehydes, ketones, acids, hydrocarbons, amides, esters, ethers, etc.) being prevalent. Here, there are carbon chain compounds (C_n, C_nH, C_nH_2, $C_{2n}N$, C_nO, C_nS, SiC_n, $HC_{2n+1}N$, with n=1,2,3 etc.; many more chain compounds are soon likely to be detected with the laboratory identification and spectroscopic characterizations by Thaddeus[7]) and many isomers (e.g., HCO^+/COH^+, $HCO_2CH_3/CH_3CO_2H/CH_2OHCHO$) detected.

MOLECULAR EVOLUTION OF THE CLOUDS

Diffuse Clouds

In diffuse clouds, the initial ionization is by stellar photons. Abundant atoms such as the low ionization energy C and S are singly ionized and are then available to react with the abundant hydrogen. This occurs by slow radiative association

$$C^+ + H \rightarrow CH^+ + h\nu \qquad C^+ + H_2 \rightarrow CH_2^+ + h\nu \ . \qquad (1)$$

The subsequent chemistry evolves in much the same way as the dense cloud chemistry discussed below, except that the molecular size is limited by photodissociation by starlight. Diffuse cloud chemistry has been extensively reviewed previously[8] and thus only dense clouds will be emphasized here, concentrating on recent advances.

Dense and Dark Clouds

In dense and dark clouds where stellar photons cannot penetrate, ionization is by cosmic rays. This is not selective and thus the most abundant species, H_2 and He, are ionized producing singly charged H_2^+ and He^+. H_2^+ reacts with H_2 by atom abstraction producing H_3^+ (a substantial amount of kinetic data on the reactions discussed here is available in compilations[9,10]). This ion is central to gas phase molecular evolution and was recently detected by McCall et al.[11,12] in both dense and diffuse clouds (the latter detection causing some controversy; see later). It is unreactive with H_2 making it available to react with minority species. He^+ is also unreactive with H_2, and is available to react with species such as CO and N_2 (not directly detected because it is homonuclear and does not emit in the microwave or infrared, but its protonated form has been detected) by charge transfer, viz

$$He^+ + CO \rightarrow C^+ + O + He \qquad He^+ + N_2 \rightarrow N^+ + N + He, \rightarrow N_2^+ + He \ . \qquad (2)$$

C^+ can then radiatively associate with H and H_2 (equation (1)) producing CH^+ and CH_2^+ and these can H-atom abstract with H_2 producing CH_3^+

$$CH^+ \xrightarrow{H_2} CH_2^+ \xrightarrow{H_2} CH_3^+ \ . \qquad (3)$$

CH_3^+ reacts only slowly with H_2 by radiative association producing CH_5^+ and is thus available to react with minority species; it has been tentatively detected in the ism in its monodeuterated form (see Table 2) and is a central precursor for the production of many of the larger observed species.

H_3^+ being unreactive with H_2, reacts with other species predominantly by proton transfer, e.g.,

$$H_3^+ + C \rightarrow CH^+ + H_2 \qquad H_3^+ + O \rightarrow OH^+ + H_2 \ 70\%, \rightarrow H_2O^+ + H \ 30\% \ . \qquad (4)$$

The reaction with C has not been studied experimentally, but that with O has, and yields the products as indicated.[13] CH^+ can then react further with H_2 as in reaction (3) and OH^+ and H_2O^+ can react similarly producing H_3O^+. Reaction of H_3^+ with N is endothermic by 0.8 eV and thus proton transfer followed by sequential H-atom abstractions cannot occur. For a time, this reaction was believed to produce the exothermic products $NH_2^+ + H$,[14] contrary to theory,[15,16] but this has since been disproved.[17] The gas phase route to NH_4^+ is now believed to be activation of the endothermic H-atom abstraction reaction of H_2 with N^+ by kinetically excited N^+ from reaction (2)[18], viz

$$He^+ \xrightarrow{N_2} N^+ \xrightarrow{H_2} NH^+ \xrightarrow{H_2} NH_2^+ \xrightarrow{H_2} NH_3^+ \xrightarrow{H_2} NH_4^+ \ . \tag{5}$$

All the other reactions are exothermic. Ions such as, H_3^+, H_3O^+, NH_4^+ and CH_5^+ (and also HCO^+, N_2H^+, H_2CN^+, H_3CO^+) are unreactive with H_2 and are called "terminating ions" They predominantly react by proton transfer with species with higher proton affinities

$$H_3^+ \underset{H_2}{\overset{O_2}{\longleftrightarrow}} O_2H^+ \xrightarrow{N_2} N_2H^+ \xrightarrow{CH_4} CH_5^+ \xrightarrow{CO} HCO^+ \xrightarrow{H_2O} H_3O^+ \xrightarrow{HCN} HCNH^+ \xrightarrow{NH_3} NH_4^+ \tag{6}$$

or recombine with electrons, e.g.,

$$XH^+ + e \rightarrow X + H \tag{7}$$

to produce neutral species such as the observed molecules in Table 2 (when X is molecular, there can be very significant fragmentation to three products; see below). When proton transfer is endothermic, radiative associations can often occur generating much larger products in a single step. Examples of this[19] are:

$$H_3O^+ + C_2H_4 \rightarrow C_2H_5OH_2^+ + h\nu \ \text{and}$$

$$CH_3^+ + H_2O \rightarrow CH_3OH_2^+ + h\nu \qquad CH_3^+ + CH_3OH \rightarrow (CH_3)_2OH^+ + h\nu \ . \tag{8}$$

Under collisionally stabilized conditions, these reactions are know to produce the isomeric ions indicated, but the radiative associations have never been directly studied. Information for modeling radiative association in the ism is obtained from laboratory data on the collisional analog together with theoretical analyses.[20] In these examples, to produce the methanol (CH_3OH), ethanol (C_2H_5OH) and dimethyl ether (($CH_3)_2O$) observed in the ism, neutralization by electron-ion recombination or by proton transfer is required. This is the only known route to interstellar methanol.

In addition to the reactions discussed above, many other types of binary reactions are possible, many of which involve more bond breaking and making. A few examples of these are (i)

$$CH_3^+ + CH_4 \rightarrow C_2H_5^+ + H_2 \quad C_2H_2^+ + C_2H_2 \rightarrow C_4H_3^+ + H, \rightarrow C_4H_2^+ + H_2 \tag{9}$$

which build up larger hydrocarbons rapidly, (ii) H-atom abstraction from the ion or radiative association

$$C_3H_4^+ + H \rightarrow C_3H_3^+ + H_2 \ \text{and} \qquad C_3H_5^+ + H \rightarrow C_3H_6^+ + h\nu \tag{10}$$

which generate strongly bonded species and (iii) carbon or sulfur insertion reactions.

$$C^+ + C_2H_2 \rightarrow C_3H^+ + H \quad \text{and} \quad S^+ + C_2H_4 \rightarrow H_3C_2S^+ + H \ . \tag{11}$$

These are only a few representative examples of the reactions that are involved in the molecular synthesis, for which thousands of reactions are involved (obviously not all equally significant).

RECENT ADVANCES IN THE STUDY OF MOLECULAR EVOLUTION IN INTERSTELLAR CLOUDS

We have contributed in several areas (discussed below) where there have recently been advances in our understanding of the molecular evolution in interstellar clouds.

Electron Recombination of HCO$^+$

HCO$^+$ is the most abundant of the observed interstellar ions. In addition to ion-molecule reaction processes, it is lost by the recombination

$$HCO^+ + e \rightarrow H + CO \ . \tag{12}$$

This has a recombination rate coefficient which is uncertain to within a factor of 3 with the most recent experimental value being 2.2 x 10^{-7} cm^3 s^{-1},[21] and theory also disagrees on the magnitude of the rate coefficient.[22,23] What is not uncertain, however, is the excitation in the CO product. The recombination is known to produce the CO (a$^3\Pi_r$) electronically excited state in 30% of the recombination[24] with the vibrational state distribution given in Table 4, as determined from the Cameron band emissions (a$^3\Pi_r \rightarrow X^1\Sigma^+$).[25] These Cameron band emissions have also been detected from the Red Rectangle Nebula and the vibrational state distribution of the a state again determined (see Table 4)[26]. Yan et al.[27] have considered the possible sources of these interstellar emissions as reaction (12), the electron recombination of CO$_2^+$ and electron impact excitation of CO. They determined that the CO$_2^+$ column density is negligible and by comparison of the observed vibrational excitation with those in the remaining two possibilities (see Table 4), concluded that the source of the CO emissions is reaction (12). This is the first direct example of a chemical reaction beyond the solar system and adds considerable credence to the ionic production routes expounded here and elsewhere. In addition to reaction (12), Butler et al.[25] determined the N$_2$ B$^3\Pi_g$ vibrational distribution from the B$^3\Pi_g \rightarrow A^3\Sigma_u^+$ emissions in the electron recombination of N$_2$H$^+$, another observed interstellar ion. These emissions should be searched for in regions of the ism, such as the Red Rectangle, to establish the occurrence of this recombination there.

TABLE 4. Comparison of the Observed Interstellar CO a state Vibrational Distribution from the Red Rectangle Nebula with that from the Series Reactive Processes Indicated.[27]

Vibrational Level	Interstellar Observation[26]	Electron Impact Excitation of CO[28]	HCO$^+$ Recombination[25]	CO$_2{}^+$ Recombination[29]
0	1	1	1	1
1	0.5	1.3	0.3	0.5
2	0.1	1	0.2	0.3
3	0.15	0.6	0.2	0.2
4	0.2	0.4	0.2	
5	0.2	0.3	0.3	

H$_3{}^+$ in Dense and Diffuse Interstellar Clouds

A great step forward in establishing the viability of gas phase ion chemical routes to the observed interstellar molecules, was made with the detection, by infrared absorption, of H$_3{}^+$, a central ion in the ion chemical routes to the observed molecules (see above).[11,12] In dense clouds, the production and loss reactions are

$$H_2 \overset{\underset{\text{cosmic}}{}}{\underset{\text{rays}}{\rightarrow}} H_2^+ \overset{H_2}{\rightarrow} H_3^+ \overset{\underset{CO}{}}{\underset{k_{CO}}{\rightarrow}} HCO^+ \overset{\text{Further}}{\underset{\text{reactions}}{\rightarrow}} \qquad \text{where } [H_3^+] = \frac{\xi[H_2]}{k_{CO}[CO]} \qquad (13)$$

from a simple kinetic model.[11] Assuming the standard cosmic ray ionization rate, ξ, CO and H$_2$ abundances (as identified by the [ ]) and the rate coefficient, k_{CO}, for proton transfer from H$_2$ to CO, the H$_3{}^+$ density can be determined and thus also the absorbing path length using the measured H$_3{}^+$ column density. This gives the path length as ~1 parsec (a typical dense cloud dimension). The situation for diffuse clouds is not so straightforward. A similar column density to dense clouds was determined, however, the chemistry is slightly different, with the loss of H$_3{}^+$ being by the electron-ion recombination

$$H_3^+ + e \underset{\alpha_e}{\rightarrow} \text{Products} \qquad \text{and then} \qquad [H_3^+] = \frac{\xi[H_2]}{\alpha_e[e]} \qquad (14)$$

where α_e is the recombination coefficient. Here, using a value of α_e of 1.2 x 10^{-7} (300/T)$^{0.5}$ cm^3 s^{-1},[30] and reasonable values of the other parameters gives an absorbing path length of ~1 kps, a substantial fraction of the distance to the background star and this is unreasonably large. A solution to the problem was suggested by Cecchi-Pestellini and Dalgarno,[31] by postulating an absorbing path containing dense cores embedded in diffuse gas clumps. However, there is another solution, this being the result of an uncertainty in α_e of up to 2 orders of magnitude over the past 50 years. Initially values of ~2 x 10^{-7} were determined, but in 1984 experimental[36] and theoretical[37] studies found the reaction to be slow. Following this, these low or high values were obtained in a series of studies using a variety of experimental techniques.

TABLE 5. Some Determinations of Electron Recombination Coefficients, α_e, for H_3^+.

Date of Determination	α_e $(cm^3 \, s^{-1})$	Technique	Reference
1973	$2.3 \; 10^{-7}$	SA	32
1974	2.5×10^{-7}	IB	33
1977	2.1×10^{-7}	MB	34
1978	2.1×10^{-7}	MB	35
1984	$<2 \times 10^{-8}$	FALP	36
1984	Slow	Theory	37
1988,90	1.8×10^{-7}	IRAS	38, 39
1988	2×10^{-8}	MB	40
1993,94	1.15×10^{-7}	SR	30, 41
1994	$<2 \times 10^{-7}$	IRAS/LP	42
1995	1.2×10^{-7}	MB	43
2000	10^{-9} to 10^{-10}	Theory	44
2001	$\sim 3 \times 10^{-8}$	SR	45
2002	$<3 \times 10^{-9}$	SALP	46

SA, Stationary Afterglow; IB, Ion Beam; MB, Merged Beam; FA, Flowing Afterglow; LP, Langmuir Probe; SR, Storage Ring; IRAS, InfraRed Absorption Spectroscopy.

In the earlier studies obtaining high values, it was concluded that there were either impurity ions present with large α_e or residual vibrational excitation in the H_3^+ again giving high α_e. Most recently, It has been shown that there was a residual rotational energy of ~30meV in the H_3^+ in storage rings giving elevated α_e.[47] With rotational relaxation of the H_3^+, a storage ring value of $\sim 3 \times 10^{-8}$ has been obtained.[45] There is also a SALP measurement of $<3 \times 10^{-9}$,[46] with theory indicating a value between 10^{-9} and 10^{-10}.[44] If the lower value were used to determine $[H_3^+]$ from equation (14), then an absorbing path length of ~1 ps would be indicated, which is very reasonable and there would be no inconsistency in the interstellar observation.

Product Distributions for Electron-Ion Recombination

There has also been some controversy in the neutral product distribution for the recombination of H_3O^+. This is a critical ion in the ism since its recombination produces the H_2 that accelerates cloud collapse and it determines whether oxygen is tied up in H_2O or the undetectable O_2 (from the HO + O neutral reaction). Product distributions obtained for this recombination are given in Table 6.

TABLE 6. Neutral Product Distributions for the Electron Recombination of H_3O^+.

Products	Channel Exothermicity (eV)	Product Distributions (%)		
$H_2O + H$	6.4	5	33	18
$OH + H_2$	5.7	36	18	11
$OH + 2H$	1.3	29	48	67
$O + H + H_2$	1.4	30	1	4
Technique		FA^{48}	SR^{49}	SR^{50}

It can be seen that there are significant differences in the product distributions determined using different techniques and indeed using the SR technique. The reason for these differences has not yet been determined despite considerable effort.[51] The data do, however, illustrate an important point, that the channels leading to three products are very significant (a point noted by Datz[52]). This is shown more clearly in Table 7. With this degree of fragmentation varying from 20 to almost 100%, if substantiated, the production of the larger molecules by ion chemical means will be less efficient than expected and thus more effective routes to the recombining ions need to be determined. This is especially true in the case of CH_4 production by recombination of CH_5^+. Note though that there will be additional production by the three-body fragmentation of larger ions.

TABLE 7. Three-Body Channels in Electron Recombination of the Ions Indicated.

Reactant Ion	Three-Body Products	Channel Percentage	Total Percentage	Reference
H_3O^+	$OH + 2H$	29		
	$O + H + H_2$	30	59	48
NH_4^+	$NH_2 + 2H$	24	24	53
CH_5^+	$CH_3 + 2H$	65.5		
	$CH_2 + H_2 + H$	25		
	$CH + 2H_2$	5	95.5	54
H_2CN^+	$CN + 2H$	33	33	55
NH_2^+	$N + 2H$	66	66	53
CH_3^+	$CH + 2H$	16		
	$C + H_2 + H$	30	46	49
$C_2H_2^+$	$C_2 + 2H$	30	30	56

Production of Large Molecules by Radiative Association and Electron Recombination

Association reactions of the terminating ion, NH_4^+ with CH_3OH, C_2H_5OH, CH_3CHO, CH_3CO_2H, $(CH_3)_2O$ and $(CH_3)_2CO$ are rapid[57] and at interstellar cloud temperatures will have efficiencies in the range 2 to 20%. The association products are strongly bonded[58] and could be clustered species or there could be reaction of the NH_4^+ with the double bond to the O-atom (as occurs in reaction (8)) giving NH_2 and OH_2^+ groupings. For example, the association of NH_4^+ with acetic acid could proceed as

$$\text{NH}_4^+ + \text{H}_3\text{C-C=O} \;\;\rightarrow\; \text{H}_3\text{C-C-O-H}_2^+ + h\nu \tag{15}$$

with O-H substituent on the carbon and NH$_2$ below on the product.

followed by the dissociative recombination

$$\text{H}_3\text{C-C-O-H}_2^+ + e \;\;\rightarrow\; \text{H}_3\text{C-C-O-H}_2^+ + \text{H}_2 + \text{H} \tag{16}$$

with O-H above and NH$_2$ below each carbon, product labeled **Glycine**.

could yield the simplest amino acid, glycine, although note that many other channels are energetically possible and, following the precedent of Table 7, may occur. Glycine has been tentatively detected in the ism.[59,60]

ACKNOWLEDGMENTS

Financial support of NASA, Division of Planetary Sciences (Grant NAG5-8951) is gratefully acknowledged.

REFERENCES

1. Adams, N. G., and Decker, B. K., "Important Ion-Atom and Ion-Molecule Collisions in Astrophysics" in *Stellar Evolution, Stellar Explosions and Galactic Chemical Evolution*, edited by A. Mezzacappa, Bristol: Institute of Physics, 1998, pp. 89-97.
2. McCarthy, M. C., and Thaddeus, P., *Chem. Soc. Revs.* **30**, 177-185 (2001).
3. Wootten, H. A., NRAO, http://www.cv.nrao.edu/~awootten/allmols.html (2002).
4. McCall, B. J., Thorburn, J., Hobbs, L. M., Oka, T., and York, D. G., *Ap. J.* **559**, L49-53 (2001).
5. Herbst, E., "Dissociative Recombination in Interstellar Clouds" in *Dissociative Recombination: Theory, Experiment and Applications*, edited by S. L. Guberman, New York: Kluwer, 2002, in press.
6. Clayton, D. D., *Principles of Stellar Evolution and Nucleosysnthesis*, Chicago: Univ. Chicago Press, 1983.
7. Thaddeus, P., Priv. Comm. (2002).
8. van Dishoeck, E. F., "Diffuse Cloud Chemistry" in *Molecular Astrophysics*, edited by T. W. Hartquist, Cambridge: Cambridge Univ. Press, 1990, pp. 55-83.
9. Ikezoe, Y., Matsuoka, S., Takebe, M., and Viggiano, A. A., *Gas Phase Ion-Molecule Reaction Rate Constants through 1986*, Tokyo: Maruzen Co., 1987.
10. Anicich, V. G., *Ap. J. Suppl. Ser.* **84**, 215-315 (1993).
11. McCall, B. J., Geballe, T. R., Hinkle, K. H., and Oka, T., *Ap. J.* **522**, 338-348 (1999).
12. McCall, B. J., et al. *Ap. J.* **567**, 391-406 (2002).
13. Milligan, D. B., and McEwan, M. J., *Chem. Phys. Lett.* **319**, 482-485 (2000).

14. Scott, G. B. I., Fairley, D. A., Freeman, C. G., and McEwan, M. J., *Chem. Phys. Lett.* **269**, 88-92 (1997).
15. Herbst, E., DeFrees, D. J., and McLean, A. D., *Ap. J.* **321**, 898-906 (1987).
16. Bettens, R. P. A., and Collins, M. A., *J. Chem. Phys.* **109**, 9728-9736 (1998).
17. Milligan, D. B., Fairley, D. A., Freeman, C. G., and McEwan, M. J., *Int. J. Mass Spectrom.* **202**, 351-361 (2000).
18. Adams, N. G., Smith, D., and Millar, T. J., *Mon. Not. R. Astr. Soc.* **211**, 857-865 (1984).
19. Adams, N. G., and Fisher, N. D., *Adv. Gas Phase Ion Chem.* **3**, 81-123 (1998).
20. Bates, D. R., and Herbst, E., "Radiative Association" in *Rate Coefficients in Astrochemistry*, edited by T. J. Millar and D. A. Williams, Dordrecht: Kluwer, 1988, pp. 17-40.
21. Rowe, B. R., Gomet, J. C., Canosa, A., and Mitchell, J. B. A., *J. Chem. Phys.* **96**, 1105-1110 (1992).
22. Talbi, D., Hickman, A. P., Pauzat, F., Ellinger, Y., and Berthier, G., Ap. J. **339**, 231-238 (1989).
23. Kraemer, W. P., and Hazi, A. U., "Dissociative Recombination of HCO$^+$" in *Dissociative Recombination: Theory, Experiment and Applications*, edited by J. B. A. Mitchell and S. L. Guberman, Singapore: World-Scientific, 1989. pp. 61-72.
24. Johnsen, R., Skrzypkowski, M., Gougousi, T., and Golde, M. F., "Spectroscopic Emissions from the Recombination of N_2O^+, N_2OH^+/HN_2O^+, CO_2^+, CO_2H^+, HCO$^+$/COH$^+$, H_2O^+ NO$_2^+$, HNO$^+$, and LIF Measurements of the H-Atom Yield from H_3^+" in *Dissociative Recombination: Theory, Experiment and Applications IV*, edited by M. Larsson, J. B. A. Mitchell and I. F. Schneider, Singapore: World-Scientific, 2000, pp. 200-209.
25. Butler, J. M., Babcock, L. M., and Adams, N. G., *Mol Phys.* **91**, 81-90 (1997).
26. Glinski, R. J., Nuth, J. A., Reese, M. D., and Sitko, M. L., *Ap. J.* **467**, L109-112 (1996).
27. Yan, M., Dalgarno, A., Klemperer, W., and Miller, A. E. S., *Mon. Not. R. Astr. Soc.* **313**, L17-18 (2000).
28. Wicke, B. G., Field, R. W., and Klemperer, W., *J. Chem. Phys.* **56**, 5758-5770 (1972).
29. Skrzypkowski, M., Gougousi, T., Johnsen, R and Golde, M. F., *J. Chem. Phys.* **108**, 8400-8407 ((1998).
30. Sundström, G. et al., *Sci.* **263**, 785-787 (1994).
31. Cecchi-Pestellini, C., and Dalgarno, A., *Mon. Not. R. Astr. Soc.* **313**, L6-8 (2000).
32. Leu, M. T., Biondi, M. A., and Johnsen, R., *Phys. Rev. A* **8**, 413-419 (1973).
33. Peart, B., and Dolder, K. T., *J. Phys. B* **7**, 1948-1952 (1974).
34. Auerbach, D. et al., *J. Phys. B* **10**, 3797-3820 (1977).
35. McGowan, J. W., et al., *Phys. Rev. Letts.* **42**, 373-375 (1978).
36. Adams, N. G., Smith, D., and Alge, E., *J. Chem. Phys.* **81**, 1778-1784 (1984).
37. Michels, H. H., and Hobbs, R. H. *Ap. J.* **286**, L27-29 (1984).
38. Amano, T., *Ap. J.* **329**, L121-124 (1988).
39. Amano, T., *J. Chem. Phys.* **92**, 6492-6501 (1990).
40. Hus, H., Youssif, F., Sen, A., and Mitchell, J. B. A., *Phys. Rev. A* **38**, 658-663 (1988).
41. Larsson, M. et al., *Phys. Rev. Lett.* **70**, 430-433 (1993).
42. Feher, M., Rohrbacher, A., and Maier, J. P., *Chem. Phys.* **185**, 357-365 (1994).
43. Yousif, F. F., Rogelstadt, M., and Mitchell, J. B. A., in *Atomic and Molecular Physics: 4th US/Mexico Symposium*, edited by I. Alvarez, C. Cisneros, and T. J. Morgan, Singapore: World-Scientific, 1995, pp. 343-351.
44. Orel, A. E., Schneider, I. F., and Suzor-Weiner, A., *Phil. Trans. Roy. Soc. Lond. A* **358**, 2445-2456 (2000).
45. Larsson, M., "Studies of Electron-Molecular Ion Dissociative Recombination using Ion Storage Rings" in *Dissociative Recombination: Theory, Experiment and Applications*, edited by S. L. Guberman, New York: Kluwer, 2002, in press.
46. Glosik, J. et al., "Recombination of H_3^+ with Electrons: Low Limit of the Rate Coefficient" in *Dissociative Recombination: Theory, Experiment and Applications*, edited by S. L. Guberman, New York: Kluwer, 2002, in press.
47. Zajfman, D., "Three Boby Kinematic Correlation in the Dissociative Recombination of H_3^+" in *Dissociative Recombination: Theory, Experiment and Applications*, edited by S. L. Guberman, New York: Kluwer, 2002, in press.

48. Williams, T. L., Babcock, L. M, and Adams, N. G., *Mon. Not. R. Astr. Soc.* **282**, 413-420 (1996).
49. Vejby-Christensen, L. et al., *Ap. J.* **483**, 531-540 (1997).
50. Neau, A. et al., *J. Chem. Phys.* **113**, 1762-1770 (2000).
51. Adams, N. G., and Babcock, L. M., "Identity and Degree of Excitation of the Products of Dissociative Electron-Ion Recombination," *Dissociative Recombination: Theory, Experiment and Applications IV*, edited by M. Larsson, J. B. A. Mitchell and I. F. Schneider, Singapore: World-Scientific, 2000. pp. 190-199.
52. Datz, S., *J. Phys. Chem. A* **105**, 2369-2373 (2000).
53. Vikor, L. et al., *Astron. Astrophys.* **344**, 1027-1033 (1999).
54. Semaniak, J. et al., *Ap. J.* **498**, 886-895 (1998).
55. Semaniak, J. et al., *Ap. J. Suppl. Ser.* **135**, 275-283 (2001).
56. Derkatch, A. M., et al. *J. Phys. B* **32**, 3391-3398 (1999).
57. Kerns, M. S., Fisher, N. D., and Adams, N. G., "The Production of Large Molecules in Dense Interstellar Clouds" in *Stellar Evolution, Stellar Explosions and Galactic Chemical Evolution*, edited by A. Mezzacappa, Bristol: Institute of Physics, 1998, pp. 215-219
58. Adams, N. G., Babcock, L. M., Mostefaoui, T. M., and Kerns, M. S., *Int. J. Mass Spectrom.* (2002) in press.
59. Combes, F., Rieu, N.-Q., and Wlodarczak, F., *Astron. Astrophys.* **308**, 618-622 (1996).
60. Snyder, L. E., *Origin of Life and Evolution of the Biosphere* **27**, 11 (1997).

Helioseismic Tests of Radiative Opacities

Joyce A. Guzik, Corinne Neuforge-Verheecke, John J. Keady,
Norman H. Magee, Jr., and Paul A. Bradley

Los Alamos National Laboratory
Los Alamos, NM 87545-2345 USA

Abstract. During the past fifteen years, thousands of solar acoustic oscillation modes have been measured to remarkable precision, in many cases to within 0.01%. These frequencies have been used to infer the interior structure of the sun and test the physical input to solar models. Here we summarize the procedures, input physics and assumptions for calculating a standard solar evolution model. We compare the observed and calculated sound speed profile and oscillation frequencies of solar models calibrated using the new Los Alamos LEDCOP [1] and Livermore OPAL [2] Rosseland mean opacities for the same element mixture. We show that solar oscillations are extremely sensitive to opacities, with opacity differences of only a few percent producing an easily detectable effect on the sound speed and predicted frequencies [3]. The oscillation data indicate that agreement would be improved by an opacity increase of several percent below the convection zone for both the LEDCOP and OPAL opacities.

INTRODUCTION

During the past 15 years, the frequencies of millions of global solar acoustic modes of oscillation have been measured to high precision, in some cases to within a few parts in a million. These modes are stochastically excited by turbulence in the upper part of the convective envelope and are called pressure modes (p-modes) because the restoring force involved in the oscillation is due to the pressure gradient. The Sun acts as an acoustic cavity that traps these waves, and modes of different frequencies and angular degree penetrate to different depths in the solar interior. Since the frequencies depend on the interior sound speed gradient, they can be used to probe solar structure. Using solar oscillation frequencies to infer the structure of the Sun and to test the physical input used in solar models is called "helioseismology". For additional information, see the review articles in the May 31, 1996 issue of *Science* [4].

The conditions in the solar interior, where radiative diffusion transports energy, span the temperature and density range from 15.6 million K (1.3 keV) and 150 g cm^{-3} in the solar center, to 2.2 million K (0.19 keV) and 0.2 g cm^{-3} at the base of the convection zone. The radiative opacity is crucial to determine the structure of the Sun and the frequencies of the acoustic modes, since the opacity determines the temperature gradient, and hence affects the sound-speed gradient. With increasing opacity toward the solar envelope, convection becomes a more efficient means of energy transport than radiative diffusion. In the Sun's convective envelope, comprising the outer 30 percent of the radius, the temperature gradient becomes nearly adiabatic, and the sound speed is instead determined entirely by the mean molecular

CP635, *Atomic Processes in Plasmas: 13th APS Topical Conference*, edited by D. R. Schultz et al.

weight and the efficiency of convective energy transport. Near the solar surface, at temperatures of 12,000 K or less, radiative opacities, including molecular opacities, once again become important. Helioseismic frequency inversion techniques, used to determine the Sun's interior structure, have been developed to eliminate sensitivity to uncertainties in the near-surface structure.

Here we summarize evolution modeling procedures and physics of standard solar models. We then compare two solar models: one calculated with the Livermore OPAL opacities [2], and one calculated with the recent Los Alamos LEDCOP (Light Element Detailed Configuration Opacity) opacities [1], in light of the helioseismic data and inferences (oscillation frequencies, sound speed profile and location of the base of the convection zone). We discuss the reasons for the differences between the OPAL and the LEDCOP opacities. Finally, we comment on the sensitivity of solar structure to other input physics or sources of uncertainty, and the prospects for inferring solar interior opacities from helioseismic data.

SOLAR EVOLUTION MODELING

To obtain a calibrated model for the present Sun, we solve the equations of stellar structure for a sequence of time steps from the onset of the nuclear reactions to the present solar age. The equations that are solved include those for mass conservation; hydrostatic equilibrium; energy production and loss; and energy transport from the interior to the surface via radiative diffusion, electron conduction, or convection. These one-dimensional models are calculated assuming spherical symmetry, negligible mass loss or mass accretion; no rotation or magnetic fields; an initially homogeneous chemical composition; and assuming that the Sun's luminosity is generated mainly by conversion of hydrogen to helium via fusion reactions in the core. Modern solar models also include diffusive settling of helium and heavier elements relative to hydrogen. This diffusion is very slow, and impeded by the convective mixing in the envelope; however, during the Sun's 4.5 billion-year lifetime, about 10 percent of the initial helium and ~8 percent of the surface heavier elements can settle from the convection zone. This diffusion has a significant effect on the Sun's structure (see, e.g., [5], [6], [26]).

Solving the equations requires physical data, including: radiative and conductive opacities (as a function of temperature, density, and element composition); an equation of state, giving the pressure, energy, and other thermodynamic quantities as a function of temperature, density, and composition; nuclear reaction rates; element diffusion coefficients; and a treatment for convective energy transport.

Calculating a solar model is an iterative process. We typically divide the model into several hundred mass shells from the center to the surface, and the evolution into several hundred time steps. One begins by adopting an initial helium mass fraction (Y_o), initial mass fraction of elements heavier than H and He (Z_o), and an initial guess for the ratio of the mixing length to pressure scale height, α, that regulates the convective efficiency. Y_o, Z_o, and α are adjusted so that the evolved model reaches the observed luminosity $L_\odot$, radius $R_\odot$, and present surface ratio of heavy element to hydrogen mass fraction, Z/X at the present solar age. Because of diffusion, the present surface Z/X is not equal to Z_0/X_0.

The present and assumed constant mass of the Sun is $1.9891\ 10^{33}$ g [7]. The Sun does lose some mass via the solar wind, but only at the negligible rate of $2\ 10^{-14}\ M_\odot$ per year. We adopt $R_\odot = 6.9599\ 10^{10}$ cm, $L_\odot = 3.846\ 10^{33}$ erg s^{-1}, and the solar age determination of 4.52 ± 0.04 Gyr [8]. We adopt the present Z/X=0.0245 and element mixture from the Grevesse & Noels 1993 solar element abundance determination [9].

We evolve our models using an extensively updated version of the Iben code [10], described in more detail in [11] and [12]. The code includes the treatment of Burgers [13] to calculate the thermal, gravitational and chemical diffusion of the electrons and 9 additional isotopes of H, He, C, N, O, Ne and Mg (see [5] for details). We smoothly join the opacities that we use in the solar interior (either OPAL or LEDCOP) to the low-temperature tables of Alexander & Ferguson [14] used at the surface by a sinusoidal average between 7500K and 9500K. Our convection treatment is the standard mixing length theory [15]. We adopted the SIREFF analytical equation of state [11]. All charged-particle nuclear reactions are taken from Angulo et al. [16]. See [3] and [12] for additional details on the nuclear reaction rate calculations.

HELIOSEISMIC COMPARISONS

Table 1 summarizes the properties of our two evolution models that are calibrated to the present solar luminosity, radius, and Z/X, using either the OPAL or LEDCOP opacities.

TABLE 1. Properties of Calibrated Solar Evolution Models.

	OPAL Opacity Model	LEDCOP Opacity Model
Initial H mass fraction X_0	0.7100	0.7122
Initial He mass fraction Y_0	0.2703	0.2680
Initial element mass fraction Z_0	0.0197	0.0198
Mixing Length/Pressure scale height (α)	1.7738	1.7651
$T_{central}$ (10^6 K)	15.66	15.66
$\rho_{central}$ l (g cm^{-3})	152.2	150.8
$Y_{central}$	0.6375	0.6350
$Z_{central}$	0.0208	0.0209
$R_{convection\ zone\ base}$ ($R/R_\odot$)	0.7135	0.7177
$T_{convection\ zone\ base}$ (10^6 K)	2.195	2.148
$Y_{convection\ zone}$	0.2408	0.2382

Figure 1 compares the opacity differences as a function of radius for the two models. The convection zone of the LEDCOP model is shallower (see Table 1) due to the fact that the Los Alamos opacities are up to 6% lower than the OPAL opacities near the base of the convection zone, as can be seen in Figure 1. The OPAL model has a convection zone base location in better agreement with the helioseismic inference of $R/R_\odot = 0.713 \pm 0.001$ [17]. The two models have slightly different composition, density and temperature profiles due to the small adjustments in composition and the mixing-length parameter required to calibrate the model to current solar conditions, and the opacity difference at a given radius is reduced by this calibration. The absolute differences between OPAL and LEDCOP opacities for the same density, composition, and temperature profiles are only slightly higher (up to 2%) than seen in Fig. 1 for a portion of the solar interior (see Fig. 1 of [3]).

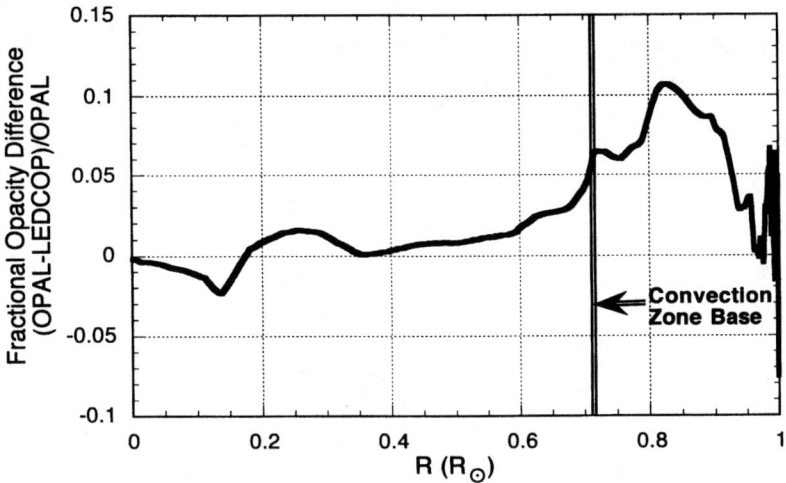

FIGURE 1. Relative opacity difference vs. fractional solar radius for two calibrated solar models using OPAL and LEDCOP opacities. The vertical lines indicate the convection zone base location.

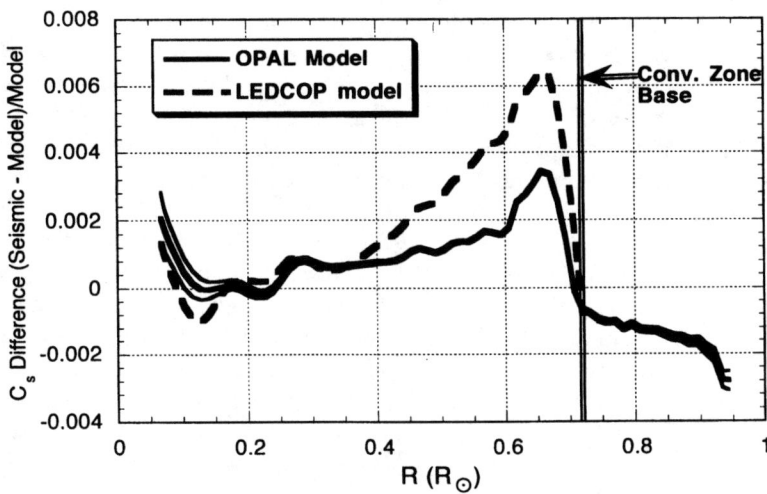

FIGURE 2. Relative sound speed differences between two solar models and the seismic inversion [18], $(c_{seismic} - c_{model})/c_{model}$, as a function of the fractional radius $R/R_\odot$. The errors on the inversion are shown for the OPAL model curve, and are approximately as large as the line width except near the solar center. The vertical lines indicate the convection zone base location.

Figure 2 compares the sound speed differences between the models and the sound speed profile inferred from helioseismic inversions by Basu et al. [18]. The error bars on the inversion are quite small compared to the differences between the model and inferred sound speed, indicating that significant improvement is necessary. However, note that the sound speed profiles of these models agree with the inferred sound speed to within a fraction of a percent. The agreement is better for the OPAL model from the base of the convection zone to $R/R_\odot = 0.12$.

We use the linear non-adiabatic pulsation code of Pesnell [19] to calculate the p-mode oscillation spectrum of our models. A solar oscillation mode can be characterized by three numbers: the radial order n, giving the number of radial nodes in the eigenfunction; the angular degree ℓ, giving the number of node lines on the surface; and the azimuthal order m, giving the number of node lines through the pole when spherical symmetry is broken, e.g. by rotation. Higher-order modes have higher frequencies and penetrate less deeply. Also, low-degree modes penetrate deeper into the Sun than do high-degree modes. For the low-degree frequency comparisons ($\ell = 0$, 1, 2, 3), we use a hybrid set of observational data described in [3, 12] chosen to maximize the number of observed low-degree modes in the set and minimize the observational uncertainties. For the intermediate-degree frequency comparisons, we use the data from [20].

Figure 4 shows observed minus calculated (O-C) nonadiabatic frequency differences (μHz) vs. calculated frequency (μHz) for some low-degree ($\ell = 0$, 1, 2 and 3) and intermediate-degree ($\ell = 5$, 10, 15 and 20) p-modes from our calibrated models. The turning points for these modes are well below the convection zone base. Note first that the agreement is excellent, with the differences no larger than a few μHz out of a few thousand, or within 0.1%. The observational uncertainties for these modes are less than 0.1 μHz, so that the trends in the frequency differences are significant. The reasons for the trends have been investigated (see, e.g., [20]). The upward trend in O-C frequency at low frequency can be removed [21] by decreasing the adopted value for the present solar radius by about 400 km, as was recently derived [22]. The downward trend at higher frequencies can be removed by a very slight adjustment to the sound speed gradient at the top of the solar convection zone, between 9,000 and 12,000 K, obtainable by improving the convection treatment and including turbulent pressure effects. An overall upward or downward shift in O-C frequency for these low and intermediate-degree modes with turning points below the convection zone base can be induced by changing the convection zone depth. The O-C frequencies of the LEDCOP model are generally higher than those of the OPAL model because the convection zone is shallower, and further from the inferred value of 0.713 $R_\odot$. The dispersion in frequency difference as a function of degree is also larger for the LEDCOP model. The modes of higher ℓ penetrate more deeply, and are therefore sensitive to different integrated regions of the Sun from the surface to the turning point. The dispersion therefore reflects the larger difference between the model and actual sound speed gradient below the convection zone evident in Fig. 2.

198

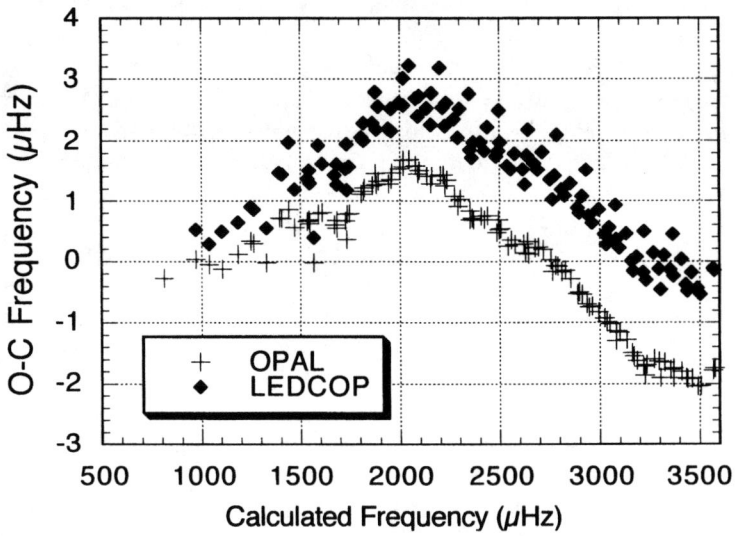

FIGURE 3. Observed minus calculated (O-C) nonadiabatic frequency differences (μHz) vs. calculated frequency (μHz) for low-degree (ℓ = 0, 1, 2 and 3) and intermediate-degree (ℓ = 5, 10, 15 and 20) p-modes of calibrated OPAL and LEDCOP models.

COMMENTS ON OPACITY DIFFERENCES

The OPAL and LEDCOP opacities differ by about 6% at the base of the convection zone. We have examined the LEDCOP opacity tables to understand the reasons for the differences, and what can be done to reduce the discrepancy in future comparisons. There appear to be three causes for the differences: the opacity models, the interpolation methods, and the resolution of the temperature grids used by the two tables.

Little information about the opacity models can be extracted from the astrophysical opacity tables, since they contain information for almost 30 elements, which has been integrated over frequency and then has been interpolated in R ($=\rho/T^3$, where T is in units of 10^6 K). Fortunately, there have been a series of opacity workshops [23], in which OPAL, LEDCOP, and many other opacity codes have been compared in detail. At the latest meeting [24], the opacities for the Grevesse and Noels 1993 element mixture were compared for solar center conditions. The OPAL results were 3% higher than LEDCOP. There are no comparisons at the physical conditions of the convection zone base, but some pure element cases, mainly iron and carbon, are relatively close to these physical conditions (T=2 10^6 K, log R= −1.5).

Concentrating principally on iron and taking into account the relative contributions to the mixture from the individual elements, we estimate that the LEDCOP opacities intrinsically are 2.5 ± 2% lower than OPAL for this (T, R) regime. We believe that this is due to differences in line transition energies, different level abundances obtained by the two equations of state, and continuum lowering models and treatment of far line wings, especially for the H-like and He-like Stark profiles.

The pure element LEDCOP opacities are calculated on a temperature and chemical potential grid, which allows the elements to be combined into mixtures. This table is then linearly interpolated in density to the final T-R astrophysical table grid. A spline interpolation is then used to obtain opacities for all X, Z, T and R. The spline interpolation has been checked and is able to reproduce the tabular values to ~ 1%. When the interpolated opacity at the convection zone base was compared with a direct opacity calculation by LEDCOP, the interpolated value was 3.5% lower than the actual calculation, with an uncertainty of 1% due to the spline interpolation. Independent comparisons for oxygen confirm that the linear interpolation routines produce values that are 4 to 5% low for oxygen in this region of the T-R table. Note that oxygen is the most important contributor to the opacity at the base of the solar convection zone [25].

A final source of discrepancy is due to the different logarithmic temperature grids used by OPAL and LEDCOP. Each table has 10 temperatures per decade, but with different spacing. The LEDCOP table does not have an opacity value near $2 \cdot 10^6$ K and log R = −1.5, whereas the OPAL table does (Fig. 4). This point is an inflection point in the opacity curve and without this point, the spline-interpolated opacity at the base of the convection zone is too low by 1.5 ± 0.5%.

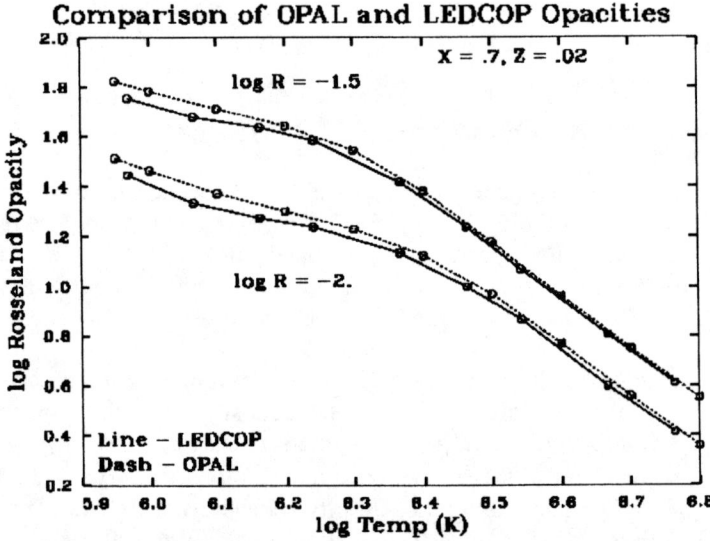

FIGURE 4. Comparison of OPAL and LEDCOP opacities for solar abundance mixture. For conditions near the solar convection zone base (log T = 6.3 and log R = −1.5), the temperature grid spacing results in a higher value for the OPAL opacity by ~1.5% compared to the interpolated LEDCOP opacity.

In summary, more than half of the opacity difference between OPAL and LEDCOP at the base of the convection zone is due to interpolation errors and the choice of the temperature grid: (3.5 ± 1%) + (1.5 ± 0.5%) = (5 ± 1.5%). These problems can be reduced or eliminated by calculating more grid points for the original elemental calculations and using this finer mesh to produce astrophysical tables with more temperatures and at least twice as many R curves. There is still a fundamental difference of 2.5 to 3% between the OPAL and LEDCOP opacities at the convection zone base. Detailed OPAL vs. LEDCOP comparisons at the center of the Sun show that the opacities can vary by a few percent due to small differences in the physics choices made when generating the opacities. Therefore, it is reasonable to admit an error bar of at least 5% on the opacity calculations due to uncertainties in physical models. The fundamental difference between the OPAL and LEDCOP opacities is well within this margin of error and cannot be fully resolved until the next opacity workshop.

COMMENTS ON OTHER SOURCES OF UNCERTAINTIES

We next discuss whether the remaining small differences between the calculated and seismically inferred solar structure are due to errors in the opacities or to some other input physics or assumptions. To what precision can solar oscillations constrain opacities? How large an opacity difference is significant? As discussed above, the treatment of the solar surface layers is the reason for some of the trends in the direct frequency comparisons, but the effects of the surface layers can be removed in the inversions for the sound speed profile. We and other researchers have investigated the sensitivity of solar models to the choice of equation of state [11], nuclear reaction rates [12], and element abundances [12]. We find that the solar model structure and predicted oscillation frequencies are affected more by a few percent opacity change, for example, by switching between the LEDCOP and OPAL opacities, than they are by switching between different modern EOS treatments, by varying the abundance mixtures within the estimated uncertainties, or by varying the nuclear reaction rates within their uncertainties. The proposed reduction in solar radius of 400 km (see, e.g., [22]) also has a small, but non-negligible, effect on inferred sound-speed profile [17]. Note that the new solar element abundance determinations of Grevesse and Sauval [28] are lower (Z/X = 0.0230) than the Grevesse and Noels [9] abundances adopted for our models (Z/X = 0.0245). The new abundances result in even lower opacities at the convection zone base, and a larger discrepancy between calculated and inferred sound speed [12].

The sound speed discrepancy between models and the helioseismically inferred profile is significant at the base of the convection zone. Some of this discrepancy may be resolved by altering the diffusion-produced composition profile at the base of the convection zone, perhaps through changes in the diffusion coefficients, or mixing due to differential rotation or gravity modes (see, e.g., [6], [29]). Several studies have shown that much of the discrepancy below the convection zone can be removed by a prescribed change in the opacity profile of 1 to ~5% [26, 27], well within the uncertainties in opacity calculations. Calculations have been done including the effects of radiative levitation of individual elements in addition to diffusive settling; these calculations show that radiative levitation effects are quite small, amounting to a change of only 0.5% in opacity due to the different resulting element distribution, and a change of 0.06% in sound speed below the convection zone [30]. With a better understanding of potential mixing and diffusion processes in the solar interior, it may be possible to use helioseismology to constrain opacities to within a few percent throughout the solar interior.

CONCLUSIONS

The observed oscillation modes that propagate through the solar interior provide an excellent test of the physics used in solar models. Opacity differences of only a few percent have a significant effect on solar structure as inferred from solar oscillations. At this level, the fineness of the opacity table grid and interpolation errors can dominate differences in results, and accurate interpolation becomes critical. The helioseismic tests performed in this paper show that the recent Los Alamos LEDCOP opacities (http://www.t4.lanl.gov) can produce solar models that agree with the current helioseismic constraints nearly as well as models produced with the Lawrence Livermore OPAL opacities. Both of these opacity sets produce much better agreement with helioseismic data than the old Los Alamos Opacity Library [31] tables, for which the opacities were too low by as much as 20% for conditions below the solar convection zone. We strongly recommend the use of the most recent LEDCOP or OPAL opacity set for astrophysical applications. The solar oscillation data indicate that agreement would be improved by further opacity increases of up to several percent below the solar convection zone. However, the solar element abundances and solar radius must be more accurately determined, and processes such as diffusive element settling and mixing below the convection zone must be better understood before we can conclude that such an opacity increase is warranted.

ACKNOWLEDGMENTS

This work was supported in part by NASA Astrophysics Theory Program grant S-30934-F. We made use of NASA's Astrophysics Data System Abstract Service and the OPAL web site http://www-phys.llnl.gov/Research/OPAL/indcx.html. We are grateful to Carlos Iglesias for providing us with results in advance of publication. This work was performed under the auspices of the U.S. Department of Energy by the University of California Los Alamos National Laboratory under contract W-7405- ENG-36.

REFERENCES

1. Magee, N.H., Jr., et al., *Astrophysical Applications of Powerful New Databases*, eds. S.J. Adelman and W.L. Wiese, San Francisco, ASP, ASP Conf. Ser. Vol. 78, 1995, p. 51; http://www.t4.1anl.gov.
2. Iglesias, C.A., and Rogers, F.J., *Astrophys. J.* **484**, 943 (1996).
3. Neuforge-Verheecke, C., Guzik, J.A., Keady, J.J., Magee, N.H., Jr., Bradley, P.A., and Noels, A., *Astrophys. J.* **561**, 450 (2001).
4. Gough, D.O., et al., *Science* **372**, 1281-1283; Gough, D.O., et al., *Science* **372**, 1296-1300; Christensen-Dalsgaard, J., et al., *Science* **372**, 1286-1290 (1996).
5. Cox, A.N., Guzik, J.A., and Kidman, R.B., *Astrophys. J.* **342**, 1187 (1989).
6. Gabriel, M., *Astron. Astrophys.* **327**, 771 (1997).

7. Cohen, E.R. and Taylor, B.N., in *Codata Bulletin 63*, Boulder, NBS, 1986, p. 1.
8. Guenther, D.B., Demarque, P., Kirn, Y.-C., and Pinsonneault, M.H., *Astrophys. J.* **387**, 372 (1992).
9. Grevesse, N., and Noels, A, in *Origin and Evolution of the Elements*, eds. N. Prantzos, E. Vangioni-Flam, and M. Cassé, Cambridge, Cambridge Univ. Press, 1993, p. 15.
10. Iben, I., Jr., *Astrophys. J.* **138**, 452 (1963); *Astrophys. J.* **141**, 993 (1965); *Astrophys. J.* **142**, 421 (1965).
I I. Guzik, J.A., and Swenson, F.J., *Astrophys. J.* **491**, 967 (1997).
12. Neuforge-Verheecke, C., Goriely, S., Guzik, J.A., Swenson, F.J., and Bradley, P.A., *Astrophys. J.* **550**, 493 (2001).
13. Burgers, J.M., *Flow Equations for Composite Gases*, New York, Academic, 1969.
14. Alexander, D.R., and Ferguson, J.W., private communication (1995).
15. Angulo, C., et al., *Nucl. Phys. A* **656**, 3 (1999).
16. Cox, J.P., and Giuli, R.T., *Principles of Stellar Structure*, New York, Gordon and Breach, 1968.
17. Basu, S., in *Sounding Solar and Stellar Interiors*, IAU Symp. 181, eds. J. Provost and F.X. Schmider, Dordrecht, Kluwer Academic Publishers, 1997, p. 137; Basu, S., *Mon. Not. Roy. Astron. Soc.* **298**, 543 (1998).
18. Basu, S., Bahcall, J.N., and Pinsonneault, M.H, *Astrophys. J.* **529**, 1084 (2000).
19. Pesnell, W.D., *Astrophys. J.* **363**, 227 (1990).
20. Schou, J., & Tomczyk, S., m2 table, http://www.hao.ucar.edu/public/research/mlso/LowL/data.html, 1996.
21. Guzik, J.A., in *Proceedings of the SOHO 6/GONG 98 Workshop, 'Structure and Dynamics of the Interior of the Sun and Sun-like Stars,'* ESA SP-418, 1998, pp. 417-425.
22. Brown, T.M., and Christensen-Dalsgaard, J., *Astrophys. J.* **500**, L195 (1998).
23. Rickert, A., *Journal of Quantitative Spectroscopy and Radiative Transfer* **54**, 325 (1995).
24. Rose, S. J., *Journal of Quantitative Spectroscopy and Radiative Transfer* **71**, 635-638 (2001).
25. Turcotte, S., and Christensen-Dalsgaard, J., in *Proceedings of the SOHO 6/GONG 98 Workshop, 'Structure and Dynamics of the Interior of the Sun and Sun-like Stars,'* ESA SP-418, 1998, p. 561.
26. Christensen-Dalsgaard, J., et al., in *Variable Stars as Essential Astrophysical Tools*, ed. C. Ibanoglu, Netherlands, Kluwer Academic, 2000, pp. 59-167.
27. Turck-Chieze, S., *Space Sci. Rev.* **85**, 125 (1998); Tripathy, S.C., Basu, S., and Christensen-Dalsgaard, J., in *Poster Volume: Sounding Solar and Stellar Interiors*, IAU Symp. 181, eds. J. Provost and F.X. Schmider, U. Nice, 1997, pp. 129-130.
28. Grevesse, N., and Sauval, A.J., *Space Sci. Rev.* **85**, 161 (1998).
29. Brun, A.S., Turck-Chieze, S., and Zahn, J.P., *Astrophys. J.* **525**, 1032 (1999).
30. Turcotte, S., Richer, J., Michaud, G., Iglesias, C.A., and Rogers, F.J., *Astrophys. J.* **504**, 539 (1998).
31. Huebner, W.F., Merts, A.L., Magee, N.H., and Argo, M.F., Los Alamos Scientific Report LA-6760-M, 1977.

High Resolution Astrophysical X-ray Spectroscopy

Frits Paerels

Columbia Astrophysics Laboratory and Department of Astronomy,
Columbia University, 550 W. 120th St., New York, NY 10027

Abstract. High resolution X-ray spectroscopy of cosmic plasmas has undergone a revolution, since the launch of the *Chandra* and *XMM-Newton* observatories two years ago. Both carry diffraction grating spectrometers, which are now providing a stream of very high quality, completely novel data, on every kind of astrophysical object. I will discuss some of the more spectacular spectroscopic surprises, and show how both familiar and new diagnostics provide detailed information of a new order on the radiation sources.

INTRODUCTION

It is not an exaggeration to say that X-ray spectroscopy of astrophysical sources has undergone a revolution over the last two years. Spectroscopy as a powerful diagnostic technique was of course pursued from the very inception of X-ray astronomy in the early 60's, but the very faintness of extrasolar X-ray sources made the implementation of standard laboratory intrumentation and techniques of sufficient sensitivity prohibitive on orbiting satellite observatories. With the exception of a small number of observations conducted with the Focal Plane Crystal Spectrometer on the *Einstein* Observatory (1979-1983; [1], [2]) and the diffraction grating spectrometers on *Einstein* and *EXOSAT* ([3]), all spectral information was limited to what could be deduced from proportional counter and solid state detectors.

A big step forward was taken with the advent of the X-ray CCD spectrometers on the Japan/US *ASCA* observatory ([4]). A resolution of 60-100 eV over the 300-8000 eV band allowed, for the first time, discrete spectroscopic detail to be discerned in a suprisingly large fraction of cosmic X-ray sources, but the diagnostic potential was still mainly limited to fairly rough measurements of the ionization balance ($\Delta E \sim 100$ eV is just about enough to discriminate between line emission from the He- and H-like charge states of the light elements, up to Fe).

NASA launched its *Chandra* X-ray observatory in July of 1999, and ESA launched the *XMM-Newton* observatory in December of the same year. Both observatories carry sensitive diffraction grating spectrometers (based on transmission gratings on *Chandra* and on reflection gratings on *XMM-Newton*). These spectrometers have up to two orders of magnitude higher resolving power, and up to an order of magnitude larger collecting area, than the CCD spectrometers on *ASCA*, and are complemented by broad-band CCD detectors with very large collecting power. Two and a half years of observation have yielded a very rich set of high resolution spectroscopic data, on basically all classes of

CP635, *Atomic Processes in Plasmas: 13th APS Topical Conference*, edited by D. R. Schultz et al.

cosmic X-ray sources, from stellar coronae to the diffuse intergalactic medium. In the context of this meeting, I thought it would be the most interesting to stress the large variety of the spectra, in which diagnostics familiar from the laboratory appear along with spectroscopic features unique to astrophysical conditions. Hence, I will discuss a series of objects whose spectrum exhibits an interesting spectroscopic feature, and I will briefly indicate how quantitative interpretation of the spectroscopy helps us constrain the physical conditions in the source. I will preface the discussion with a brief description of the instrumentation.

INSTRUMENTATION

The diffraction grating spectrometers on both *Chandra* and *XMM-Newton* are based on large area grazing incidence focusing X-ray optics, with the diffraction gratings placed directly behind the mirror assemblies, and imaging detectors at the spectroscopic focus (CCDs on *XMM-Newton*, and choice of either CCDs or microchannel multiplier arrays on *Chandra*). The mirrors, gratings and detectors are arranged in a classical Rowland circle geometry on both observatories. The spectrometers have no slits.

Chandra's optics have very high angular resolution (a few tenths of an arcsec), and the diffraction gratings (three sets, arranged on two assemblies that can be rotated into the focused X-ray beam on command, the High Energy Transmission Grating Spectrometer [HETGS] and the Low Energy Transmission Grating Spectrometer [LETGS]) produce wavelength resolutions of 0.012 (0.024) Å (HETGS) and 0.05 Å (LETGS) with relatively moderate line densities. *XMM-Newton* has three parallel sets of very high throughput grazing incidence mirrors shells (58 mirrors per assembly), of moderate angular resolution ($\sim$ 15 arcsec Half-Power Diameter). Two of these telescopes feed an array of 182 grazing incidence X-ray reflection gratings which intercept about half the light; the grating arrays are fixed to the telescopes, and high resolution spectroscopic data is collected on every source. The low angle of incidence on the gratings ensures large dispersion angles even at the moderate line density of the gratings (only 646 lines mm^{-1}), so that these Reflection Grating Spectrometers (RGS) still achieve a wavelength resolution of $\sim$ 0.05 Å. The *Chandra* HETGS covers the $1.5 - 25$ Å band, while LETGS, with the dedicated microchannel multiplier detector, covers the $6 - 160$ Å band. The *XMM-Newton* RGS covers the $5 - 35$ Å band. For comparison, the K-shell transitions of highly ionized C, N, O, Ne, Mg, and Si, as well as the L-shell transitions of Fe, occur in the $5 - 35$ Å band, while the K-shell transitions of S, Ar, Ca, and Fe are in the $1.5 - 5$ Å band. *Chandra* LETGS adds unique L-shell spectroscopy of the mid-Z elements (Mg through Ca), and $n = 2 - 2$ spectroscopy of the Fe L ions.

Figure 1 shows an artist's impression of the *Chandra* observatory. More information can be found on `http://cxc.harvard.edu`. Figure 2 shows *XMM-Newton*; more information is on `http://xmm.vilspa.esa.es`.

Both observatories provide complementary capabilities, with the instrumentation on *Chandra* most suited to the highest resolution investigations of bright sources, and the *XMM-Newton* RGS most suited to faster spectroscopy (or fainter sources) at the expense of some resolving power. Due to the much higher dispersion angles on RGS, this

FIGURE 1. Artist's impression of the *Chandra* observatory. The mirror aperture is visible at the front end. The detector assembly at the telescope focal plane (about 10 m away) is at the other end (courtesy NASA/CXC/SAO).

FIGURE 2. Artist's impression of the *XMM-Newton* observatory. The three mirror assemblies are visible at the front end. The detector platform at the telescope focal plane (about 8 m away) is at the other end (courtesy European Space Agency).

instrument also has a unique capability for high resolution spectroscopy of moderately extended objects (up to an arcminute in angular size); objects of order 15 arcsec or smaller appear essentially as point sources to *XMM-Newton*'s optics.

COLLISIONAL PLASMAS

It is appropriate to start the discussion with examples of collisional plasmas. These are widespread in astrophysics, and they are (still) the most familiar, the Solar Corona having provided much of the spectroscopic paradigm for X-ray spectroscopy in astronomy up to the present. Collisional plasmas have been studied extensively in the lab, and they therefore continue to provide an important point of reference as the field matures.

Stellar Coronae: Capella

The brightest coronal source, Capella, has provided a spectacular very high resolution spectrum with the *Chandra* HETGS ([5]). Figure 3 shows the spectrum (the $5 - 25$ Å band). Most of the emission lines are due to $n = 2 - 3$ transitions in Fe L ions, with prominent emission from O VIII (Lyα at 18.97 Å) and O VII (the $n = 1 - 2$ resonance/intercombination/forbidden transitions at 21.6, 21.8, 22.1 Å), as well as emission from H- and He-like N, Ne, Mg, and (weak) higher-Z emission. The brightest emission lines are due to Ne-like Fe (15.01, 15.26, 17.05, 17.10 Å). This was the first astrophysical high resolution spectrum of the 'New Era', and it caused considerable excitement: the spectrum is almost of laboratory quality. Apart from its intrinsic interest, it nicely shows the potential of spectroscopy with the current instrumentation for other fields of astrophysics.

Given its importance (the Fe L ions span a wide range of ionization potentials, and, through their ionization balance, provide a versatile diagnostic of the electron temperature distribution in collisional plasmas), the natural first question to ask is how well we can model the Fe L spectrum. This specific problem was addressed early on [6]), using a set of modern atomic structure calculations of the Fe L ions carried out with the HULLAC code ([7]). Figure 4 shows the Fe L band in the Capella HETGS spectrum, along with model calculations for the emission spectra of the individual Fe L ions, up to Fe^{21+}. These were added up with specific weights, in an attempt to model the total Fe L emission (the individual ionic emission spectra, when the electron temperature is close to the temperature at which the ionic abundance peaks, are essentially independent of electron temperature). A comparison of this model Fe L spectrum (marked 'Total (calculated)' in the Figure) with the data shows that, overall, the success is remarkable. Remaining discrepancies in the line intensity ratios in individual ions indicate that the rate equations and/or branching ratios need some further refinement, but overall, it appears that our ability to interpret astrophysical collisional spectra will not be limited by uncertainties in atomic physics calculations. A comparison of the measured Fe L ionization balance with model calculations for the collisional equilibrium ionization balance indicates that most of the Fe L emission arises in gas with a temperature of $kT_e \approx 600$ eV.

With the emission measure distribution determined unambiguously from the Fe L spectrum, it is now possible to interpret the intensities of emission lines from other elements in terms of their relative abundances. Interesting discrepancies with the stellar atmospheric abundance patterns have started to show up in many coronae, indicating

207

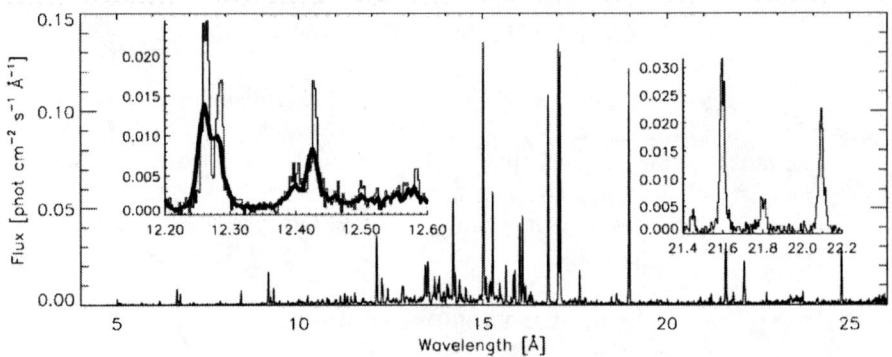

FIGURE 3. *Chandra* HETGS spectrum of Capella. The right-hand inset shows the fully resolved He-like $n = 1 - 2$ 'triplet' of O VII. Reprinted with permission from *Astrophys. J.* **539**, L41-L44, copyright 2000 American Astronomical Society.

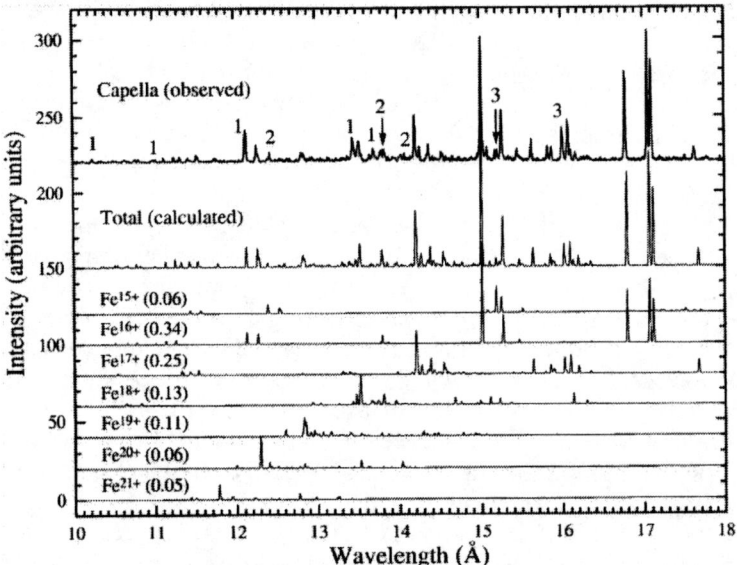

FIGURE 4. *Chandra* HETGS spectrum of Capella, with a set of model Fe L spectra, plus their weighted sum, which closely corresponds to an electron temperature of ≈ 600 eV. Reprinted with permission from *Astrophys. J.* **548**, 966-975, copyright 2001 American Astronomical Society.

a complex chain of thermal, mechanical, and electrodynamic interactions between the lower, cool stellar atmosphere, and the hot outer envelope ([8]), which are still in the process of being sorted out.

Clusters of Galaxies: Abell 1835

Clusters of galaxies are the largest gravitationally bound structures in the Universe ([9]). From an analysis of the kinematics of the Galaxies, it was concluded early on that most of the binding mass in clusters is 'dark'. In the 70's, intense diffuse X-ray emission was discovered from massive clusters, with a thermal bremsstrahlung spectrum of high temperature (kT_e characteristically of order 10 keV). A detailed study of the surface brightness distribution, beginning with the imaging X-ray telescope on the *Einstein* observatory, allowed a determination of the gravitational potential that confines this gas, and an estimate for the total mass generating the potential. The most massive clusters turned out to have total masses of order 10^{15} solar masses, of which only a fraction could be accounted for by the hot intracluster gas (and even less by the galaxies). The hot gas emits intense Fe K emission (mostly He-like Fe $n = 1 - 2$ emission), and typical Fe abundances are of order 1/3 the solar Fe abundance, suprisingly high. The origin of the intracluster gas (what fraction was stripped out of the galaxies, what fraction primordial, what fraction accreted from the intergalactic medium) is still not known with certainty.

The densities in the core of many clusters are high enough ($n_e \sim$ few times 10^{-3} cm^{-3}) that the radiative cooling time is less than the Hubble time ($\sim$ the age of the Universe), presumably comparable to the age of the clusters. Consequently, one expects gas in the core to cool radiatively, and sizeable amounts of cool gas to be present in a 'cooling flow' centered on the core of the potential well. Spectroscopic evidence for the presence of such cool gas was indeed found with the Focal Plane Crystal Spectrometer on *Einstein* in a few clusters (*e.g.* [10]), and a study of the surface brightness distribution indicated a sharp rise in the gas density in the core of many clusters, which indicated the presence of large amounts of cooling gas, probably distributed in small thermally unstable pockets in rough pressure equilibrium with the hot substrate.

The typical size of these cooling flows is of order an arcminute, and their characteristic temperatures are expected to be of order 1 keV. This makes them ideal targets for study with the RGS on *XMM-Newton*. Detailed spectroscopy of the Fe L series would allow a quantitative study of the thermodynamical history of the cooling gas, and cluster cooling flows were among the first targets studied with the RGS. The results were a total surprise.

Given the radiative loss function as a function of electron temperature, it is straight-forward to calculate the emission measure distribution for radiatively cooling gas at constant pressure from thermodynamics (the simplest possible picture of cooling in the intracluster gas ([11]). Using this emission measure distribution, one can calculate the emission spectrum for the Fe L ions; the only free parameter is the total mass flow that enters the cooling phase (basically the amplitude of the discrete spectrum). Comparison of such spectral models to the measured Fe L spectra indicated that in all cases, emission from the lower charge states of the Fe L series was significantly fainter than predicted. Figure 5 shows such a comparison for the massive cluster Abell 1835 ([12]): emission from the highest charge state of the Fe L series is readily detected, indicating incipient cooling. But the predicted accompanying intense emission from the lower charge states is not seen (especially the most intense emission line of Ne-like Fe at 15.014 Å). This pattern recurs in all cooling flow clusters studied with RGS to date ([13]). The implications for this absence of emission from the lower charge states is still being debated. We may be missing essential elements in the thermodynamical description of the cooling gas

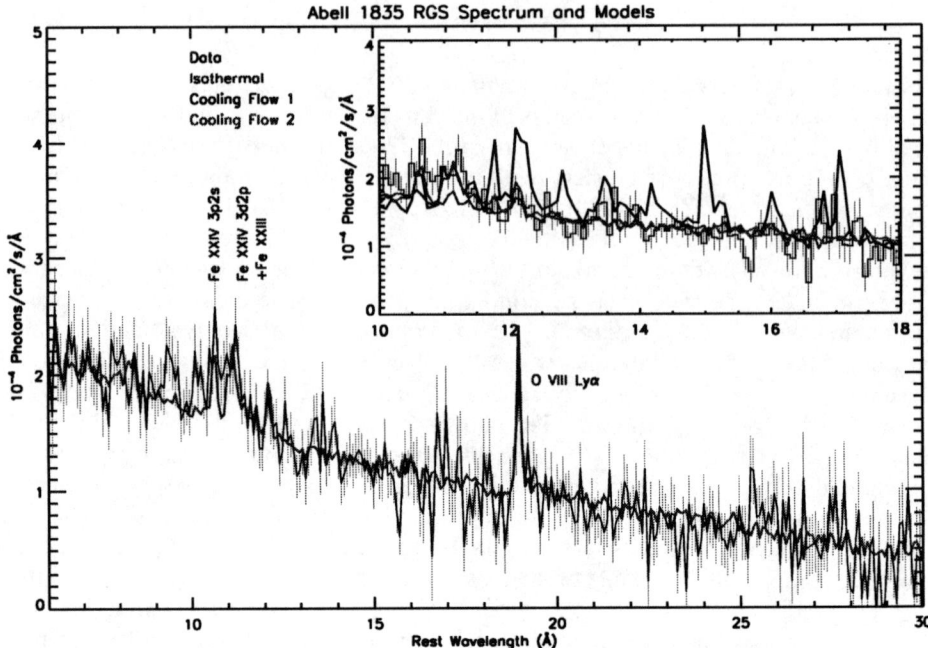

FIGURE 5. Measured spectrum of the cooling gas in cluster Abell 1835 as measured with *XMM-Newton* RGS (*jagged brown curve with error bars*), together with model prediction for isobaric radiatively cooling gas (*green and blue lines*). The inset shows the spectral region that contains most of the Fe L emission on an expanded scale. Emission from the lower Fe L charge states is missing, while emission from the highest charge state is detected, at 10.65 and 11.18 Å. All emission lines are slightly broadened due to the finite angular extent of the cooling region (the RGS has no slits). Reprinted with permission from *Astron. Astrophys.* **365**, L104–L109, copyright 2001 EDP Sciences.

(additional heating terms, neglected degrees of freedom, for instance associated with a pervasive magnetic field, etc.), but no convincing, unique explanation has emerged yet. We can be sure the problem is not with the spectroscopic models, because, as we saw earlier, the current best Fe L spectral models can account for the spectrum of stellar coronae with only minor modifications.

Hot Gas in Supernova Remnant Shocks: 1E0102.2–7219

Another prime target for the RGS are the compact (in angular size) Supernova Remnants (SNR) in the Magellanic Clouds. Here, we observe line emission from hot gas, recently shocked by the powerful shock wave sent into the Interstellar Medium by the Supernova explosion, and from a reverse shock driven into the Supernova ejecta. Again, observing small objects with the RGS allows us to reach almost the full spectral resolution of the spectrometer, producing a far more highly resolved spectrum than can be

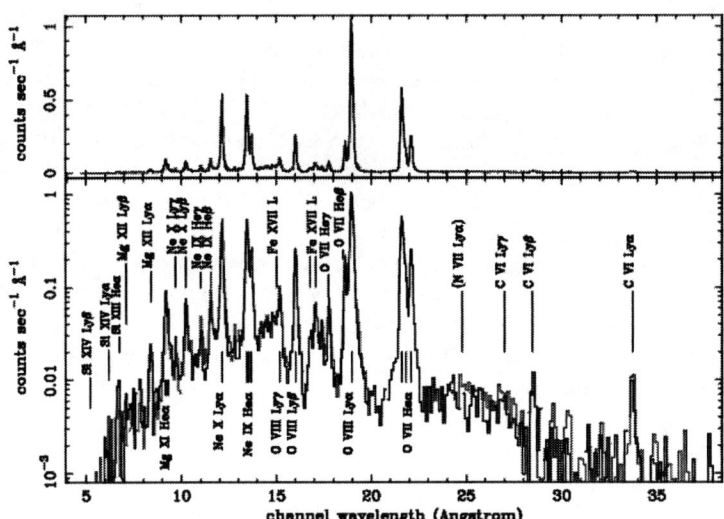

FIGURE 6. *XMM-Newton* RGS spectrum of the young Supernova remnant 1E0102.2−7219 in the Large Magellanic Cloud. Strong K-shell emission from all abundant light elements is seen, with only weak Fe L emission; this indicates that the emission is dominated by shocked stellar ejecta. A detailed study of the intensity ratios in the individual ionic spectra (*e.g.* O VIII Ly β, γ vs. Lyα) reveals deviations from the ratios expected under conditions of pure collisional ionization equilibrium. Reprinted with permission from *Astron. Astrophys.* **365**, L231-L236, copyright 2001 EDP Sciences.

obtained with either imaging CCD spectrometers, or the diffraction grating spectrometers on *Chandra*.

Figure 6 shows the spectrum of the SNR 1E0102−7219 in the Large Magellanic Cloud ([14]; this object is only one of a set of several such remnants that can be studied in this way–a complete survey is underway). In this spectrum, the lines are broadened not due to the spatial extent of the source, but due to a spread in radial velocities across the source, associated with its expansion. Prominent K-shell emission from C, N, O, Ne, Mg, and Si is seen, as well as weaker Fe L emission, indicating that the emission from this object is dominated by emission from the shocked Supernova ejecta, reflecting the abundance pattern in the exploding star. Interestingly, by complete accident the densities in the shocked media in SNR are such that for much of the early life of an SNR, the ionization balance in the shocked gas is not expected to reach collisional equilibrium. The gas is underionized with respect to the local electron temperature, and a precise measurement of the ionization balance allows a joint constraint on the density and the time since the shock passage. Such non-equilibrium conditions had been inferred from low/medium resolution spectroscopy in the past, but with fully resolved spectra we can now measure the electron temperature directly from individual ionic line spectra; in low resolution spectra, the electron temperature has to be inferred from the faint continuum, whose amplitude and shape is coupled to the measurement of the intensity of the largely unresolved emission line spectrum in such data.

This spectrum, for the first time, reveals higher-order series members ($n = 1 - 3, 4, ...$) in the He- and H-like spectra (marked in Figure 6). The intensities of these higher order series members are larger than they would be if the electron temperature were close to the value at which the abundances of the ions peak in collisional equilibrium. At high temperatures, the higher order series members brighten with repect to the $n = 1 - 2$ transition, and one explanation for the observed intensity ratios is that the gas is considerably hotter than the temperatures at which the K-shell ions of the light elements appear in equilibrium: the gas is underionized, and still evolving to higher ionization. This idea is consistent with the spatial stratification of the ions observed in the beautiful 'spectroheliogram' obtained of this object with the *Chandra* HETGS, which basically consists of a series of (overlapping) monochromatic line images ([15]). These data unambiguously show a 'reverse shock' propagating inward into the stellar ejecta, and an evolving ionization balance behind the shock wave. Calculations for the emission spectra under conditions of high temperature, and possible contributions from charge exchange reactions on neutral H atoms that drift through the shock wave, are currently underway (Ehud Behar, priv. comm.).

Ultimately, with a fully self-consistent model for the emission spectrum, we can finally unambiguously measure elemental abundances in both the Interstellar Medium and in the SNR ejecta, depending on the evolutionary state of the remnant under study, which provides important information on stellar nucleosynthesis.

X-RAY PHOTOIONIZED PLASMAS

In many astrophysical situations, relatively low density diffuse gas is exposed to an intense, externally generated hard ionizing radiation field, which drives the diffuse gas to high ionization. The most prominent cases involve diffuse gas surrounding a compact object (neutron star or black hole); accretion of gas onto the compact object generates a hard, nonthermal ionizing continuum (the details of the energy conversion and spectral formation in this accretion process are still not well understood, by the way). The resulting emission from the diffuse gas is dominated by recombination, and the discrete spectrum differs radically from the more familiar collisional spectra. In addition, we are now finding that radiative transfer effects in the strongest resonance lines is evidently important in many cases. The spectroscopy for this type of astrophysical plasma has had to be developed mostly directly from theory, but this development has met with remarkable success, as I will show below.

The Nucleus of Active Galaxy NGC1068

Most or all galaxies probably contain a massive (10^{6-9} solar masses) black hole in their center. In a small fraction of galaxies, accretion of gas onto this black hole generates a hard, ionizing continuum which affects diffuse gas in the surroundings. The resulting bright emission line spectrum marks the galaxy as an 'active galaxy', and these galaxies are collectively known as Active Galactic Nuclei (AGN). AGN come in a bewildering variety, but recently the consensus has grown that most of the variety in display is probably due to variations in a few very basic parameters, chief among which is simply orientation of the nucleus with respect to our line of sight. The black hole is presumably surrounded by an optically thick torus of gas and dust; if we happen to look down the hole in this torus, we directly observe the strong continuum source associated with the black hole, with, as we are now finding ([16],[17],[18], ands others since then), a prominent, very rich absorption spectrum arising in diffuse gas along the line of sight. If the central source is blocked from direct view, we only observe the strong emission lines arising in diffuse gas exposed to the central source, that happens to be located above the opening in the torus and is in our direct view. This dichotomy apparently persists across a huge range in intrinsic nuclear luminosity, from quasars down to the relatively low luminosity so-called Seyfert Galaxies. Objects in which the central source is blocked are called 'Type II' objects (as opposed to the ones in which the central source is visible, which are called 'Type I'). The brightest 'Seyfert-II' galaxy is the famous galaxy NGC1068, which was observed with the RGS on *XMM-Newton* early in its operational phase.

From medium-resolution spectroscopy with the CCD detectors on *ASCA* it was already known that the soft X-ray spectrum ($E < 2$ keV) of this object is completely dominated by discrete emission ([19]), but it was not possible from this data to determine the fundamental nature of the emission mechanism (collisional, in hot gas associated with vigorous star formation, or radiation driven). The RGS spectrum now unambiguously shows that all the discrete emission from the nucleus is from X-ray photoionized gas. This is not a just an academic spectroscopic exercise–if the gas had been associated with a 'starburst' instead, the emission line intensities could be used to estimate the total power release in star formation. The present finding shows that attempts to estimate the power balance in an AGN between accretion-driven power, and star-formation driven power, from a poorly resolved soft X-ray spectrum alone are bound to give unreliable results.

Figure 7 shows the RGS spectrum of NGC1068 ([20]). X-ray photoionized gas is typically cool, with characteristic electron temperatures ($kT_e \sim$ few eV) far below the electron temperatures in collisional plasmas of comparable mean ionization (which are comparable to the relevant ionization potentials). This circumstance gives rise to a powerful spectroscopic diagnostic: radiative recombination directly onto the ground state produces a characteristically narrow Radiative Recombination Continuum (RRC), whose width, moreover, is directly proportional to the local electron temperature (in collisional plasmas, the RRC's are weak, and are smeared out across a wide energy band above the ionization threshold due to the large spread in electron kinetic energies). The RRCs associated with recombination onto stripped and H-like C, N, O, Ne, and Mg are marked in Figure 7; their narrowness immediately indicates that the discrete emission must arise in photoionized gas. The RRC of C V at 31.6 Å is particularly striking. Photoionization is consistent with the appearance of the $n = 1 - 2$ 'triplets' in the He-like ions (consisting of a resonance line, two unresolved intercombination lines, and a forbidden line). In a recombining plasma, the forbidden line is the brightest, while in collisional plasmas the resonance line is the brightest. The O VII He-like triplet at 21.6, 21.8, and 22.1 Å is particularly striking in Figure 7.

Kinkhabwala et al. ([20]) have shown that the entire discrete spectrum can be quantitatively modeled. They started with a series of pure recombination spectra (one for each ion), which determines the electron temperature (from the width of the RRC) and

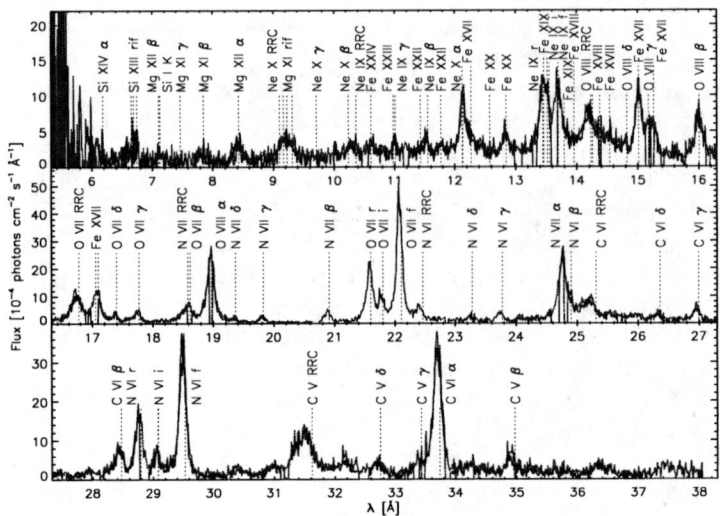

FIGURE 7. Spectrum of the nucleus of Seyfert II galaxy NGC1068, obtained with the RGS on *XMM-Newton*. Narrow RRCs are clearly visible; in the He-like 'triplets' (marked 'r', 'i', and 'f' for the resonance, intercombination, and forbidden transitions), the forbidden emission is brighter than the resonance transitions, all clear marks of recombination excitation in a photoionized plasma. The higher order series members are brighter than in pure recombination, indicating the operation of photoexcitation by the nuclear continuum source that is hidden from our direct view. From 'XMM-Newton Reflection Grating Spectrometer Observations of Discrete Soft-X-ray Emission Features from NGC 1068'; accepted for publication in *Astrophys. J.* (2002)

the total emission measure for each of the zones occupied by the idividual ions in the emitting plasma. It then appeared that all the $n = 1 - 2, 3, 4, \ldots$ resonance line intensities, in both the He- and the H-like ions, were systematically underestimated. This indicates that radiative transfer is important: continuum radiation from the hidden ionizing source is scattered into our line of sight by the strongest transitions (photoexcitation). Adding this contribution into the model produces an almost perfect fit to the measured discrete spectrum, while providing constraints on the optical depth and velocity distribution in each of the ionic zones. The measured properties of the emitting medium are consistent with it being essentially the same medium that is seen in absorption in Type I AGN, and it provides, among other things, additional constraints on a measurement of the total kinetic energy in the outflowing gas which can be compared to the total accretion power.

X-ray Binaries: Accretion onto Neutron Stars

In close, short-period binary stars containing a compact object (white dwarf, neutron star, or black hole), mass transfer can occur from the normal companion to the compact star. Gas falling into the deep potential well of the compact object gets heated to very high temperatures, and becomes a luminous X-ray continuum source. In binaries con-

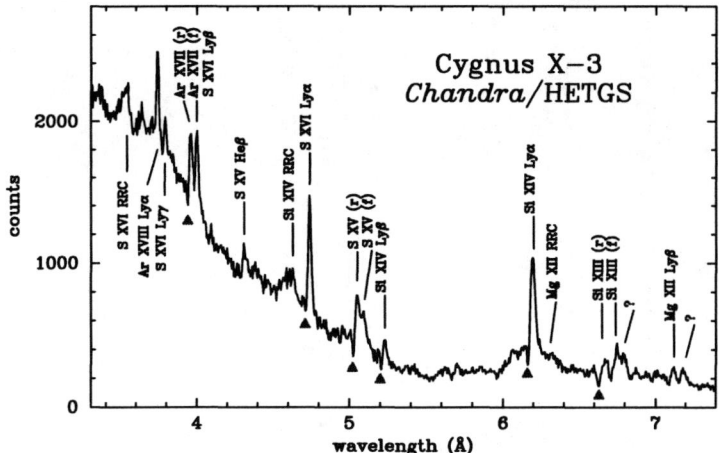

FIGURE 8. *Chandra* HETGS spectrum of Cygnus X-3. Narrow RRCs and intensity ratios in the He-like spectra indicate conditions of photoionization equilibrium. In addition, we detect narrow absorption in the strongest resonance lines. The emission lines are fully velocity-resolved (*i.e.* significantly broader than the 0.012 Å instrumental width of the spectrometer) (*courtesy Masao Sako, Caltech*).

taining a low-mass companion (*M* less than a solar mass), the mass transfer ordinarily occurs by 'overflow' of the outer envelope of the companion across the critical potential surface. In binaries containing a massive companion (several, to tens of solar masses), a (radiation driven) stellar wind of the massive companion provides gas for the accretion process. In this latter case, the diffuse stellar wind is exposed to the strong ionizing continuum emitted by gas falling onto the compact object, and the wind becomes highly photoionized, and 'lights up' in high-ionization emission lines. A detailed spectroscopic study has the potential to yield a complete empirical model for the dynamical and density structure of the stellar wind, which cannot be obtained in any other way. Study of massive stellar winds is important, because the wind mass loss can profoundly affect the evolution of massive stars, and a quantitatively correct understanding of stellar structure, evolution, and nucleosynthesis constitutes the very foundation of astrophysics. I discuss two examples, the spectra of the massive binaries Cygnus X-3 and Vela X-1.

Figure 8 shows a portion of the spectrum of Cygnus X-3, as measured with the HETGS on *Chandra*. We detect the tell-tale narrow RRCs of H-like Si and S, which indicates that the emission indeed arises in low-density X-ray photoionized gas in the stellar wind from the companion. In addition to emission lines, we see that the spectrum also exhibits weak absorption components to the strongest features. The emission line intensities constrain the emission measure in the various ionization zones in the wind, while the absorption lines measure the optical depth through the wind. All emission lines appear resolved in the HETGS spectrum, which also gives us the typical flow velocities in the various ionization zones. Combining all these constraints should allow us to derive a detailed empirical 3-dimensional model for the structure of the wind.

Figure 9 shows a portion of the *Chandra* HETGS spectrum of another massive X-ray binary, Vela X-1. The appearance of its emission line spectrum is generally very similar

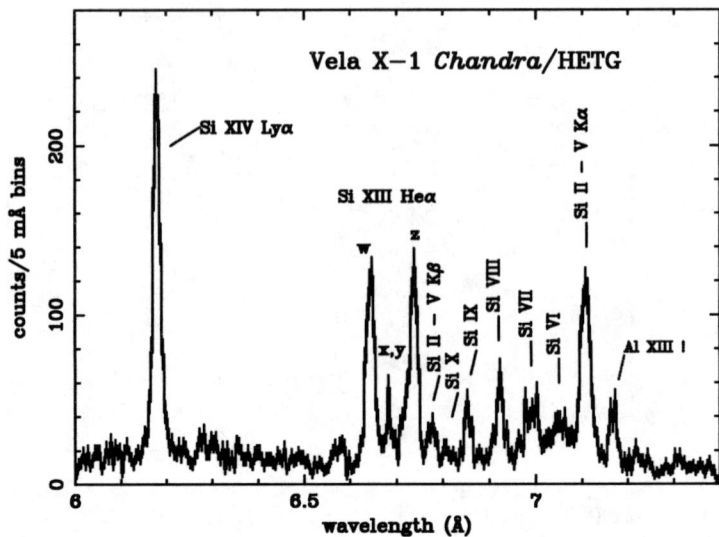

FIGURE 9. Spectrum of Vela X-1 observed with the HETGS on *Chandra* (*courtesy Masao Sako, Caltech*). This part of the spectrum reveals a full Si K shell spectrum, with fluorescent emission from most Si L ions, in addition to the He- and H-like spectra, indicating the presence of a wide range of densities in the stellar wind.

to that of Cyg X-3, but the spectrum of Vela X-1, in addition to showing emission from highly ionized material, also shows unusually strong fluorescent emission. This indicates that the wind must contain clouds of very high density (in photoionization equilibrium, the degree of ionization can be parameterized by essentially the ratio of the number density of ionizing photons, to the particle density; for a given high number density of ionizing photons, the gas has to be very dense to keep this ratio low). Figure 9 shows the Si K band, in which we detect fluorescent emission from essentially all ionization stages of Si! The distribution of intensities in the various charge states thus directly maps out a distribution of densities in a dense phase in the wind. This is interesting new information: massive stars show very strong (fully saturated) UV resonance absorption lines, due to C IV, N V, Si IV (in fact, most of the momentum transfer from the stellar UV radiation field to the wind occurs in these resonance lines), and now we see fluorescence from Si IV and adjacent ions, which will help provide additional constraints on the distribution of these important ions in the wind.

NOVELTIES: COMPTON SCATTERING AND EXAFS

Several X-ray binaries (as well as many AGN) show strong Fe Kα fluorescence, evidently excited in very dense gas exposed to the strong accretion-generated radiation field. Possible sites for the fluorescence in binaries include dense parts (or phases) of the accretion flow, and the atmosphere of the normal companion star. Figure 10 shows the part of the *Chandra* HETGS spectrum of the binary GX301 − 2 containing the $\lambda 1.87$

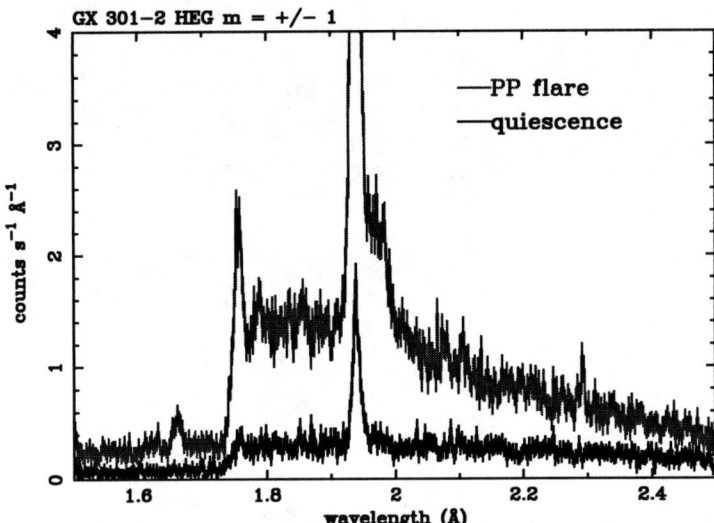

FIGURE 10. Fe Kα spectrum of GX301−2 with the HETGS on *Chandra* (*courtesy Masao Sako, Caltech*). The weak shoulder to the long wavelength side of the Kα line at 1.87 Å consists of line photons that have Compton-scattered off cold electrons (*upper histogram*). We observe a full 'Compton recoil' spectrum.

Å Fe Kα emission line. To the long wavelength side of the line is a small 'shoulder'. The significance of this feature becomes clear when we consider at what wavelength this shoulder appears to peak, before it drops off: the dropoff appears at a wavelength offset of $\Delta\lambda \approx 0.05$ Å longward of the Fe Kα wavelength. That is almost precisely twice the Compton wavelength of the electron: $\lambda_C \equiv h/m_e c = 0.0243$ Å: the 'shoulder' consists of Fe Kα photons that have been Compton-scattered off cold electrons! This feature has long been recognized as an important diagnostic (*e.g.* [21],[22]), but it has never been seen before. Recoil in a photon-electron collision results in a downshift of the photon momentum corresponding to a wavelength shift

$$\Delta\lambda = \frac{h}{m_e c}(1 - \cos\theta), \tag{1}$$

with θ the angle through which the photon is scattered with respect to the forward direction. Resolving the 'recoil spectrum' therefore directly maps out the distribution of scattering angles, which reveals information on the geometry of the source of the fluorescent photons and the scattering medium. A qualitative assessment of the shape of the feature indicates that it appears to arise from scattering through all angles (full sky filled with scattering electrons, as seen from the line photon source). In the context of the binary geometry, that probably indicates that an isotropic line photon source is embedded in the scattering medium, which is either the atmosphere of the companion star, or dense gas in the accretion flow. Careful examination of the spectrum could set limits on the relative intensity of multiply scattered photons, which gives the optical depth of the scattering medium.

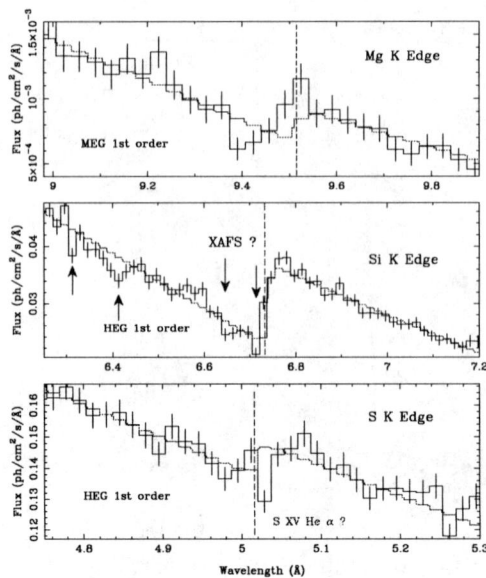

FIGURE 11. Three portions of the HETGS spectrum of GRS1915+105, near the Mg, Si, and S K edges. The spectrum at energies above the Si K edge bears the signature of the characteristic XAFS modulation of the photoelectric absorption cross section, due to absorption in solid material (dust) in the ISM. A weaker signal may be present near the Mg and S edges. Reprinted with permission from *Astrophys. J.* **567**, 1102-1111, copyright 2002 American Astronomical Society.

Finally, bright continuum sources provide the opportunity to perform absorption spectroscopy of the intervening diffuse gas, located along the line of sight. In this way, we can study the interstellar- and intergalactic medium. I will describe one spectrum that holds the potential to study the interstellar medium (ISM) in a completely new way. Much has been written recently about X-ray Absorption Fine Structure in near-edge photoelectric absorption cross sections (XAFS; Extended X-ray Absorption, etc., or EXAFS, is the old term, now in use for only certain kinds of XAFS; see for instance [23]). The effect is very simple: absorption of a photon with an energy only slightly in excess of the binding energy of an electron in an atom produces a photoelectron with a relatively large De Broglie wavelength. If the absorbing atom is embedded in an ordered structure (as opposed to being free and isolated, as in the gas phase), the photoelectron wavefunction will be scattered by neighboring charges and quantum interference results. The photoelectron wavefunction enters into the photoabsorption transition probability through the matrix element, and the photoelectric cross section will exhibit resonances in response to the order in the environment. The effect will obviously occur in solids with a crystal structure, but XAFS is seen even in diatomic molecules.

In astrophysical absorption spectra, the effect could be used to detect and measure absorption by solid particles (dust), as compared to absorption in the gas phase. Given a sufficiently bright spectrum, one could even attempt to determine the nature of the dust particles from their crystal structure.

The *Chandra* HETGS spectrum of the massive X-ray binary GRS1915+105 exhibits well-defined modulations above the Si K edge ([24],see figure 11), and possibly above the Mg and S edges. Most of this absorption arises in the Interstellar Medium of our Galaxy along the line of sight. Arrows in Figure 11 mark the positions of obvious deviations from the smooth photoabsorption transmission curve assumed in the spectral fit. It's straightforward to figure whether the energies of the resonances are on the right scale to be due to XAFS. Given a photon energy E_γ, an electron with initial binding energy χ will emerge with a De Broglie wavevector k according to

$$E_\gamma = \chi + \hbar^2 k^2 / 2m_e. \tag{2}$$

Resonance will occur whenever k satisfies a resonance condition of the form $ka \sim (n+\text{constant})\pi$, with a the lattice parameter of the solid, and n a positive integer. There is more to this condition, of course (the photoelectron wavefunction suffers phase shifts moving out of and into the Coulomb potentials of the ions, the lattice may not have a simple structure, one has to account for multiple scattering, etc.), but for the present purpose, this is good enough. Taking a of order 1 Å, we would predict to see modulations with a typical photon energy spacing (for the first two successive values of n) of order $\Delta E_\gamma \sim 40$ eV, or $\Delta\lambda \sim 0.15$ Å in photon wavelength space, at the Si K edge. Checking out Figure 11, we see that that is indeed the correct scale.

Similar modulations occur in the spectrometer efficiency curve of course (the CCD detectors contain silicon), but the CCDs have been subjected to exhaustive calibration with just these kinds of effects in mind, and Lee et al. state that the modulations in the spectrum of GRS1915+105 are significantly deeper than the modulation of the CCD efficiency curve. In addition, the effect does not seem to appear as strongly in less-absorbed sources. If correct, this would (finally) be the first detection of astrophysical XAFS.

CONCLUSION

And there is more: He-like 'triplet' spectra which clearly show the effects of collisional and photon-driven transfer between the upper levels of the forbidden and intercombination lines in hot stellar winds and X-ray binaries ([25],[26]) resonance absorption line spectra in Type I AGN ([16],[17],[18]), resonance absorption by highly ionized gas in the intergalactic medium ([27], [28]), gravitationally redshifted absorption lines in the photospheric spectrum of a neutron star ([29]), discrete emission lines in the X-ray afterglow spectra of Gamma Ray Bursts ([30],[31]), etc. So far, high resolution X-ray astrophysical spectroscopy has been spectacularly successful, and the field has approached full maturity in the astonishingly short time of about 3 years. It has been able to build on the expertise and dedication of many atomic physicists, and we look forward to continued collaboration and perhaps even more fruitful interaction between the fields in the future.

ACKNOWLEDGEMENTS

I gratefully acknowledge the assistance of all members of the RGS collaboration in the preparation of this presentation, and especially of my colleagues at Columbia Andy Rasmussen, Ali Kinkhabwala, John Peterson, Ehud Behar, Masao Sako (now at Caltech), Marc Audard, and

Steve Kahn. Funding for the development, construction, and calibration of the optics for the RGS at Columbia, and for the post-launch work on calibration, interpretation, and spectroscopy, was and is provided by NASA.

REFERENCES

1. Canizares, C. R., "High Resolution Spectroscopy and Plasma Diagnostics of Supernova Remnants," in *High Resolution X-ray Spectroscopy of Cosmic Plasmas*, edited by P. Gorenstein and M. V. Zombeck, Cambridge University Press, Cambridge, 1989, pp. 138–145.
2. Canizares, C. R., "High Resolution Spectroscopy of Thermal Plasmas," in *Imaging X-ray Astronomy*, edited by M. Elvis, Cambridge University Press, Cambridge, 1990, pp. 123–136.
3. Mewe, R., *Astr. Astrophys. Rev.* **3**, 127–168 (1991).
4. Tanaka, Y., Inoue, H., and Holt, S. S., *Publ. Astron. Soc. Japan* **46**, L37–L42 (1994).
5. Canizares, C. R. e. a., *Astrophys. J.* **539**, L41–L44 (2000).
6. Behar, E., Cottam, J., and Kahn, S. M., *Astrophys. J.* **548**, 966–975 (2001).
7. Bar-Shalom, A., Klapisch, M., and Oreg, J., *Phys. Rev. A* **38**, 1773–1784 (1998).
8. Güdel, M., *Phil. Transactions*, in press (2002).
9. Sarazin, C., *X-ray Emission from Clusters of Galaxies*, Cambridge University Press, Cambridge, 1988.
10. Canizares, C. R., Clark, G. W., Markert, T. H., Berg, C., Smedira, M., Bardas, D., Schnopper, H., and Kalata, K., *Astrophys. J.* **234**, L33–L37 (1979).
11. Johnstone, R. M., Fabian, A. C., Edge, A. C., and Thomas, P., *Mon. Not. Royal Astron. Soc.* **255**, 431–440 (1992).
12. Peterson, J. R. e. a., *Astron. Astrophys.* **365**, L104–L109 (2001).
13. Peterson, J. R. e. a., *Astrophys. J.*, in preparation (2002).
14. Rasmussen, A. P., Behar, E., Kahn, S. M., den Herder, J.-W., and van der Heyden, K., *Astron. Astrophys.* **365**, L231–L236 (2001).
15. Flanagan, K. A., Canizares, C. R., Davis, D. S., Dewey, D., Houck, J. C., and Schattenburg, M. L., *High Resolution Spectroscopy of the Supernova Remnant 1E0102−72*, American Astronomical Society Meeting nr 196, abstract 34.09, 2000.
16. Kaastra, J. S., Mewe, R., Liedahl, D. A., Komossa, S., and Brinkman, A. C., *Astron. Astrophys.* **354**, L83–L86 (2000).
17. Kaspi, S., Brandt, W. N., Netzer, H., Sambruna, R., Chartas, G., Garmire, G. P., and Nousek, J. A., *Astrophys. J.* **535**, L17–L20 (2000).
18. Sako, M. e. a., *Astron. Astrophys.* **365**, L168–L173 (2001).
19. Ueno, S., Mushotzky, R. F., Koyama, K., Iwasawa, K., Awaki, H., and Hayashi, I., *Publ. Astron. Soc. Japan* **46**, L71–L75 (1994).
20. Kinkhabwala, A. e. a., *astro-ph/0203290* (2002).
21. Hatchett, S., and Weaver, R., *Astrophys. J.* **215**, 285–290 (1977).
22. Illarionov, A., Kallman, T., McCray, R., and Ross, R., *Astrophys. J.* **228**, 279–292 (1979).
23. Lee, P. A., Citrin, P. H., Eisenberger, P., and Kincaid, B. M., *Rev. Mod. Phys.* **53**, 769–806 (1981).
24. Lee, J. C., Reynolds, C. S., Remillard, R., Schulz, N. S., Blackman, E. G., and Fabian, A. C., *Astrophys. J.* **567**, 1102–1111 (2002).
25. Kahn, S. M., Leutenegger, M., Cottam, J., Rauw, G., J., V., den Boggende, A., Mewe, R., and Güdel, M., *Astron. Astrophys.* **365** (2001).
26. Waldron, W. L., and Cassinelli, J. P., *Astrophys. J.* **548**, L45–L48 (2000).
27. Nicastro, F. e. a., *Astrophys. J.* **573**, 157–167 (2002).
28. Rasmussen, A. P., Paerels, F., Kahn, S. M., den Herder, J.-W., and de Vries, C., *Astrophys. J., in preparation* (2002).
29. Cottam, J., Paerels, F., and Mendez, M., *Astrophys. J.,* in preparation (2002).
30. Piro, L. e. a., *Science* **290**, 955–958 (2000).
31. Reeves, J. N. e. a., *Nature* **416**, 512–515 (2002).

Microphysics of Jets in Star Formation

A.E. Glassgold

New York University and University of California, Berkeley
aglassgold@astron.berkeley.edu

Abstract. Bipolar outflows are a common aspect of the accretion process that leads to the formation of low-mass stars like the Sun. During the most active accretion stages, jets are observed in the optical forbidden lines of abundant ions of oxygen, sulfur, etc.. These lines indicate that the jets have temperatures in the range from 5,000-10,000 K, hydrogen densities greater than $10^4 \, \text{cm}^{-3}$, and ionization fractions of order 10%. Significant progress has been made recently in understanding the dynamical origin of these outflows in terms of magnetocentrifugal effects. According to the X-wind model, the wind covers 180 deg in both directions and is well collimated along the poles, suggestive of a jet. An analysis is presented of the microscopic processes underlying the observed line emission from the X-wind jet that elucidates the important role of several atomic processes, espcially low-energy charge exchange.

INTRODUCTION

The gas in the Milky Way and other large spiral galaxies is extremely heterogeneous. By mass, most of it is cool ($kT \ll 1 \, \text{eV}$) and lightly ionized ($x_e \ll 1$), and in roughly circular motion close to the midplane of the galaxy. This galactic disk gas plays an important role in the evolution of galaxies because it forms stars. Star formation is a remarkable process in which cool, low-density gas is transformed into material 20 orders of magnitude more dense – and capable of long-term stability as hydrogen-burning stars. The key role of gravitational collapse in the formation of stars was recognized long ago in terms of the *nebular hypothesis*, in which stars are postulated to form from the collapse of a rotating gas cloud. Conservation of angular momentum leads to a flattened configuration now called an accretion disk. Only in the last 20-25 years have a number of key ingredients been added which transform the nebular hypothesis into a robust paradigm for the formation of low-mass stars. These new elements are well-founded in observations that probe the vicinity of young stars in the process of formation.

First, radio observations indicate that stars are formed in large gas clouds called giant molecular clouds. Their overall dimensions are of order 10-20 pc, and they have mean densities of a few thousand hydrogen molecules per cubic cm, mean temperatures of 10-20 K, and masses of order $5 \times 10^5 \, M_\odot$. These objects are inhomogeneous and contain innumerable condensations of higher than average density (in the range $10^4 - 10^7 \, \text{cm}^{-3}$) called "cloud cores". It is the collapse of these cores that initiates the dynamic phase of star formation. Second, accretion disks have been detected by a variety of new observations at infrared, radio, and optical wavelengths. A third influential discovery, made by mm radio astronomy, is that star formation is accompanied by bipolar outflows or winds, which can also be seen in appropriate optical lines as jets emanating from the immediate neighborhood of the young star. These dramatic phenomena challenge theorists to explain how powerful outflows can be generated simultaneously with the collapse that produces a new star.

Another essential ingredient in the new paradigm of star formation is the role of magnetic fields, initially to restrain the collapse of the cloud core and then to later guide the flow of accreting material to where it has to go, i.e., into the star and out in the wind. Strong magnetic fields in young stellar objects (YSOs) are observed through the Zeeman effect; they are also consistent with the ubiquitous emission of X-rays by YSOs.

The outflows accompanying accretion are usually explained in terms of magnetocentrifugal-wind models, introduced by Blandford and Payne [1] and later developed into a detailed theory of star formation by Shu and his collaborators [2]. In Shu's X-wind model, the accretion flow separates near the inner edge of the disk (also the disk co-rotation radius R_x, where the disk's Keplerian angular velocity equals that of the new star), into a wind and an accretion funnel that feeds the star. Other "disk-wind" models have been proposed in which the outflow is generated over a much larger range of disk radii [3].

Confirmation of the elements of the new nebular hypothesis requires observations capable of resolving YSOs on scales of an astronomical unit (AU) or less, especially if one wants to probe the inner workings of low-mass star formation. For nearby sources typically at a distance of 150 pc, 1 AU corresponds to an angular resolution of 7 milli-arcseconds (mas). Although this is beyond the range of current instrumentation, mm arrays and adaptive optics can now provide useful resolution at the level of ~ 100 mas, competitive with the Hubble Space Telescope. In order to understand the observable signatures of dynamical theories of star formation on these small spatial scales, one must be able to calculate the physical condition of the flows. This in turn requires detailed consideration of the underlying atomic physics of the flowing plasma. We will address these issues in the following sections using the X-wind theory of jets as the main example, as recently developed by Shang, Glassgold, Shu, and Lizano in [4], hencefoth referred to as SGSL.

MODELING THE JETS OF YOUNG STELLAR OBJECTS

The work of SGSL is based on the magnetocentrifugal-wind model of Shu et al. [5]. The MHD approximation used in that work is valid as long as the ionization fraction is not too small and the thermal pressure not too large. We then have the problem of "dressing" the model by adding appropriate physics and chemistry so that the physical conditions and then the observable intensities of the wind can be calculated. Schematically, this is a three-step process:

1. DYNAMICAL MODEL (ρ, v, B)

$\downarrow$ microscopic processes

2. PHYSICAL PROCESSES (T, x)

$\downarrow$ excitation, radiation transfer

3. INTENSITIES (I_v),

where ρ, v, and B are the density, velocity, and magnetic field; T and x are the temperature and abundances (notably the electron fraction x_e); and I_v stands for various observable (frequency-dependent) intensities. Each of the last two steps is decoupled from the previous one. I will focus here on some aspects of this methodology that emphasize atomic and plasma physics .

The jets of YSOs have been observed mainly in the forbidden lines of the neutral oxygen atom and the once-ionized sulfur ion at the wavelengths of 6302Åand 6743Å. The typical angular resolution used is of order $1''$, intermediate between the size of the source and the overall extension of the jet. These lines are excited by electronic collisions and, unless the electron density exceeds a critical density ($\sim 10^4 \mathrm{cm}^{-3}$ for S^+ and $\sim 10^5 \mathrm{cm}^{-3}$ for O) and the temperature is larger than 6,000-7,000 K, the transition is sub-thermally excited and weak. The hydrogen atom densities along the X-wind streamlines vary from $n \sim 10^{15}\,\mathrm{cm}^{-3}$ at the base down to very small values at large distances. The iso-density contours collimate towards the axis and are suggestive of a jet. Thus the X-wind should be capable of emitting the optical forbidden lines in a fashion similar to that observed if its inner collimated part can be ionized and heated to high enough levels. In order to establish whether this is the case, SGSL performed the second modeling step above by integrating the ionization and heat equation along the streamlines obtained by Shang for the steady X-wind (see Figure 2 in [2]).

The Thermal and Ionization Problem

By concentrating on the inner streamlines, SGSL were able to start with an atomic wind (predominantly H) and then check afterwards that the derived physical conditions were consistent with this assumption. The processes that determine the ionization of atomic hydrogen at temperatures of the order of 1 eV are well known to astrophysicists from considerations of the solar photosphere, galactic H II regions, and planetary nebulae. There are a number of important changes, however, that are dictated by the fact that the problem of interest involves the flow of a partially ionized gas emanating from an accreting YSO [1]. First, the individual gas parcels may not be in a local steady state because the thermal and chemical (ionization) time scales need not be small compared to the relatively short dynamical timescale s/v; s is the distance from the source and v is velocity. Second, the driver of the wind is the Lorentz force (in the presence of gravity) acting on the ions. The ions drag the mainly neutral wind by ion-neutral collisions, and in the process heat the gas (ambipolar diffusion heating). Perhaps the most important aspect of the thermal problem for the wind is that, because the density decreases along the streamlines, each mass element is cooled by expansion ("adiabatic cooling"). Adiabatic cooling is so strong for YSO winds that all conventional heating mechanisms, including ambipolar diffusion, cannot compete. Finally, the ionization fraction required

---

[1] Some but by no means all of these processes were identified in a spherically symmetric wind calculation by Ruden, Glassgold, and Shu [6], who also treated the standard collisional and radiative processes.

to explain the observed forbidden lines cannot be achieved by stellar photoionization or by collisional ionization, even if the thermal problem could be solved.

All of these features of the thermal-ionization problem for protostellar winds were realized by Ruden, Glassgold, and Shu [6]. The hope that the difficulties in heating and ionizing protostellar winds would be alleviated by using the actual 2-d X-wind solution were not realized, and SGSL had to introduce entirely new physics into the theory. The first step was to explain the high ionization level in terms of the X-rays that YSOs emit. The next step was to invoke a small level of dissipation of the mechanical or flow energy that results from shocks and turbulence associated with the fact that the X-wind is fluctuating rather than steady [7]. We shall now examine some of these new features, emphasizing the role of the atomic processes.

X-rays

Space-based X-ray observatories have established that essentially all YSOs emit X-rays. CHANDRA has gone beyond the results of earlier ROSAT and ASCA missions to find that the X-ray luminosity L_X of low-mass protostars (T Tauri stars) increases with mass. The typical X-ray luminosity for a solar-mass YSO is (E. Feigelson, private communication)

$$L_X = 5 \times 10^{30} \, \mathrm{erg \, s}^{-1}.$$

We may imagine that this is something like the X-ray luminosity of the young Sun when it was only 1 Myr old. CHANDRA has also confirmed the previous indications from ASCA that YSOs have harder X-ray spectra than were deduced from the ROSAT observations, with significant luminosity at energies above 1 keV. Another general property of YSO X-ray emission is that it is variable, fluctuating by a factor of 2 or so on time scales as short as 1 h. Larger flares (amplitude increases by more than factor of 3) occur at the rate of one every day or two.

The significance of protostellar X-rays can be illustrated by a simple rough calculation of the X-ray ionization rate (normalized as usual to one per H nucleus) at a distance r from the source:

$$\zeta_X = \frac{L_X}{4\pi r^2 kT_X} J \sigma(kT_X) \frac{kT_X}{\Delta \varepsilon},$$

where kT_X is the average energy of the incident X-rays, J is an attenuation factor, and $\Delta \varepsilon \approx 40 \, \mathrm{eV}$ is the energy to make an ion pair in a cosmic mixture (mainly H and He). Without attempting to evaluate J, which is application specific, the X-ray ionization rate at a distance of 1 AU is

$$\zeta_X \sim 6 \times 10^{-9} J \sigma(kT_X) \left(\frac{kT_X}{\Delta \varepsilon} \right) \left(\frac{\mathrm{AU}}{r} \right)^2 \mathrm{s}^{-1}.$$

for $L_X = 10^{30} \, \mathrm{erg \, s}^{-1}$ and $kT_X = 1 \, \mathrm{keV}$. This rate dominates other stellar rates close in, and it is also larger than the ionization rate for galactic cosmic rays for $r < 0.1$ pc, the dimension of a cloud core. Beyond this scale, cosmic rays and the interstellar radiation field produced by massive stars dominate the ionization of the cool disk gas.

When an X-ray ionizes a circumstellar heavy atom, it generates many secondary electrons (of order $kT_X/\Delta\varepsilon$) which can also ionize, heat, and excite the gas. Good treatments of the basic physics are available in the literature, at least for the case where $x_e < 0.1$, e.g., Dalgarno, Yan, and Liu [8]. Thermal-chemical calculations of stationary gas have also been carried out by Mahoney, Hollenbach, and Tielens [9]).

Ambipolar Diffusion Heating

We adopt a two-fluid model where all of the ions and all of the neutrals are assumed to travel with their respective velocities, whose difference is the "drift" velocity $\mathbf{w}$. It is easy to show that the heating rate due to ambipolar diffusion is (Appendix D of SGSL [4]),

$$\Gamma_{AD} = \frac{\rho_n |\mathbf{f}_L|^2}{\gamma \rho_i (\rho_n + \rho_i)^2},$$

where ρ_i and ρ_n are the total ion and neutral mass densities and γ is the ambipolar diffusion coefficient,

$$\gamma = \frac{1}{\rho_i \rho_n} \Sigma_{jk} n_j n_k m_{jk} K_{jk}.$$

The sum is over individual ions and neutrals j and k, respectively, and m_{jk} and K_{jk} are the reduced mass and rate coefficient for momentum transfer in a collision of the ion-neutral pair (j, k). Each rate coefficient is a double integral over the assumed thermal equilibrium velocity distributions of the ions and the neutrals (Draine [10]). The most abundant ion in a warm cosmic plasma is H^+, and the most abundant neutrals are H and He. The $H^+ + H$ collisions are more important than $He^+ + H$ collisions by almost two orders of magnitude, and we will only discuss H^+ collisions here. It is important to note that the rate of ambipolar diffusion heating is inversely proportional to γ.

Osterbrock [11] discussed the astrophysically relevant momentum transfer cross sections in terms of the leading $1/r^4$ potential at large distances r. In the semi-classical limit, this leads to a $1/v$ cross section and a constant Langevin rate coefficient, proportional to the square route of the neutral's polarizability. He also pointed out that, at high enough energies, shorter range interactions play a role. He and others suggested that they could be represented by a constant (equivalent geometric) cross section of the order of $10^{-15}\,cm^2$. In contrast, Draine [12] actually used theoretical calculations of high energy cross sections. There are a number of problems with these treatments when applied to $H^+ + H$ collisions in a warm astrophysical plasma: (1) The Langevin $1/v$ cross section is incompletely realized at low energies. (2) The identity of the two protons must be taken into account, as discussed by Krstić & Schultz [13]. (3) The momentum transfer must be used rather than the elastic cross section, as well as the correct thermal average, e.g., following Draine [10]. In the limit that w is small compared to the sound speed, γ is proportional to the Maxwell average of the momentum transfer cross section weighted by v^3, in contrast to the ordinary collision rate which involves the first power of v. (4) High level theoretical cross sections (that compare well with laboratory experiments) should be used.

There is of course no reason why all of these points cannot be dealt with since, after all, the H_2^+ potential energy curve is the oldest and best studied in molecular physics! Indeed, good theoretical calculations of the $H^+ + H$ cross sections have been available since 1977 (Hunter & Kuriyan [14]) for the energy range from $10^{-4} - 10\,\text{eV}$. They have been recalculated recently by Krstić and Schultz [13], along with those for $H^+ + He$ and for $H^+ + H_2$, and very close agreement is obtained. The theoretical cross sections are also in good agreement with experiments.

The $H^+ + H$ momentum transfer cross section displays a number of interesting features important in astrophysical applications. The cross sections oscillate below $10\,\text{eV}$, with such large amplitude below $1\,\text{eV}$ that it is hard to recognize the Langevin $E^{-1/2}$ behavior except in some average sense. There is a definite change in the energy dependence above $0.01\,\text{eV}$, where the cross section, averaged over small oscillations, varies weakly with energy as $E^{-1/8}$. The momentum transfer cross section is approximately $2 \times 10^{-14}\,\text{cm}^2$ near the break point at $0.01\,\text{eV}$.

To calculate the rate coefficient, SGSL fitted the average $H^+ + H$ momentum transfer cross with power laws in energy,

$$\sigma_{\mathrm{mt}} = \sigma(0.01\text{eV})\,(E/0.01\text{eV})^p,$$

with $p = -1/2$ for $E < 0.01\,\text{eV}$ and $p = -1/8$ for $E > 0.01\,\text{eV}$. This leads to values of the ambipolar diffusion coupling coefficient that are significantly larger than found in the literature. For example, for small x_e and small w, the value at $10^4\,\text{K}$ is,

$$\gamma(10^4\,\text{K}) \sim 8 \times 10^{13}\,\text{cm}^3\text{s}^{-1}\text{g}^{-1}.$$

This use of this new large coefficient means that the ambipolar diffusion heating rate, which varies inversely with γ, is reduced relative to earlier calculations. When we integrate the heat and ionization equations along the X-wind streamlines with the new γ, only a thin inner layer of the X-wind jet is heated well enough to emit the forbidden lines; the emission from the bulk of jet is too weak to observe. Thus, ambipolar diffusion cannot account for jet temperatures in the X-wind model. This is to be contrasted with the conclusions of Safier [15] and Garcia et al. [16] for self-similar disk winds. Some of the difference may be attributed to their use of γ coefficients that are too small. However, ambipolar diffusion depends on the square of the Lorentz force and thus sensitively on the magnetic field strength and configuration, so the high temperatures obtained by these authors may also reflect the differences between the magnetic field of the X-wind and self-similar disk winds.

Mechanical Heating

Since none of the conventional heating mechanisms including X-rays and ambipolar diffusion is capable of maintaining the wind at close to $10{,}000\,\text{K}$, SGSL proposed that some form of dissipation of the flow kinetic energy occurs. Observations indicate that the jets are time-dependent, e.g., in the form of pulsed jets, as discussed for example by Raga [17]. In a time-dependent jet, shock waves and turbulence are generated when fast

226

fluid elements catch up with slower ones [2]. Rather than attempting a detailed analysis of the complex physics underlying the transformation of fluctuating kinetic energy into heat (which might be studied with 3-d numerical simulations), SGSL appealed to general dimensional reasoning to express the functional form of the mechanical heating as

$$\Gamma_{mech} = \alpha \rho \frac{v^3}{s},$$

where s is the distance the fluid element has traveled along a streamline to the location of interest in the wind. The positive coefficient α is a phenomenological parameter that characterizes the magnitude of the mechanical heating.

RESULTS

The calculations done by SGSL show that the temperature of the wind flowing close to the axis has a temperature close to $10,000\,\mathrm{K}$ when α is chosen to be about 10^{-3}. The ionization fraction is about 4%. Many processes contribute, but the main thermal elements are mechanical heating, balanced roughly by adiabatic cooling. X-rays are the main ionization source. Unlike the thermal problem, where heating and cooing are in approximate balance and proceed on the dynamical timescale, X-ray ionization and recombination are not in balance except close to the source. Recombination generally proceeds on a timescale that is much longer than the dynamical timescale, and the X-ray flux is diluted by the inverse square law more rapidly than the density decrease along the axis. It is also diminished by absorption. Thus, the wind is ionized near its base and the resulting electron fraction is then frozen into the flow.

DIAGNOSTICS: THE CASE OF S^+

If we take these results on the physical properties of the flow, along with the built-in cylindrical collimation of the X-wind, we have the ingredients for an explanation of the jets observed coming from active YSOs like the low-mass T Tauri stars. The proof involves calculating appropriate emissivities, synthesizing the observable intensities, and comparing them with observations. The problem of synthesizing the images made in the forbidden lines is addressed by SGSL for the widely used forbidden optical lines of O I and S II. Synthetic images that closely resemble exiting images are obtained, and further work is in progress to deal with the specific results that can be obtained on the smallest available spatial scale (100 mas) close to the source of the jet. Comparison of theory and observations on this scale should provide the strongest tests of the X-wind theory of protostellar jets.

The spatial variation of an intensity is often of particular interest, as are particular line ratios, but the theory must also be able to predict absolute levels of the intensities.

---

[2] Hydrodynamic turbulence has often been invoked for heating interstellar clouds, e.g., Black [18].

Reasonable agreement with observation is obtained for O I, although the accuracy of absolute intensity measurements might be improved. In this case of O I, we can reliably estimate the amount of O^+ since its abundance is determined by the well known and fast (forward and backwards) charge exchange with H,

$$\frac{x(O^+)}{x(O)} \approx \frac{x(H^+)}{x(H)} = \frac{x_e}{1 - x_e}. \tag{1}$$

Thus most of the oxygen is in OI, with about 5% in OII. The case for S II is less clear and deserves special attention because it requires presently unavailable atomic data.

The problem with sulfur is obvious when we recall the large difference between the ionization potential of S (10.36 eV) and of O (13.60 eV). This means that the backward $(S^+ + H)$ and forward $(H^+ + S)$ proceed at very different rates, even at 10,000 K, so that S^+ and H^+ are less likely to be in equilibrium than are O^+ and H^+. Furthermore, the existing calculations of the rate coefficients are probably not definitive. For example, the calculations of Kimura et al. [19] leads to a rate coefficient for the forward reaction of $k(H^+ + S) \sim 10^{-10} \, cm^3 s^{-1}$ at 10,000 K, and they give a rough estimate of the backward reaction rate coefficient of $k(S^+ + H) \sim 10^{-15} \, cm^3 s^{-1}$. If the latter value is correct, then destruction of S^+ by charge exchange with H is weaker than destruction by radiative recombination, and the chemical timescale of S^+ will be long compared to the dynamical timescale. Consequently, the rate equation for S^+ has to be integrated along the streamlines, just like the equation for the total ionization. We are then led to ask what are the production and destruction processes for S^+, and how well do we know them for the temperature range of interest? The following tabulation gives some tentative answers. Question marks have been added after processes for which the underlying atomic data are either unreliable or missing.

Production of S^+ - collisional ionization of S by thermal and X-ray generated secondary electrons; charge exchange of S with H^+ (?), O^+ (?), and C^+ (?); radiative recombination of S^{2+}; charge exchange of S^{2+} with H (?).
Destruction of S^+ - radiative recombination; charge exchange with H (?), C (?), Si (?), etc. (?).

Production of S^{2+} - soft X-ray ionization of S; charge exchange of S^{3+} with H.
Destruction of S^{2+} - radiative recombination; charge exchange with H(?).

Production of S^{3+}-S^{6+} - hard X-ray ionization of S; charge exchange with H.
Destruction of S^{3+}-S^{6+} - charge exchange with H.

We have incorporated the conventional wisdom that sulfur ions with charge greater than 2 rapidly charge exchange with H to produce S^{2+} (see Stancil [20] for calculations of S^{4+} + H), but the charge exchange of S^{2+} with H is problematic (e.g., Butler & Dalgarno [21]). If charge exchange is really weak, then this ion will decay slowly, i.e., radiatively, and its rate equation will need to be integrated along streamlines. In any case, calculations of charge exchange of the sulfur ions with H at energies of 1 eV and below are badly needed, as are charge exchange cross sections of S and S^+ with the ions and

228

atoms of the abundant heavy elements.

CONCLUSIONS

We have cited two particular classes of charge transfer processes that play an important role in the physics of the jets of young stellar objects. The first is the reaction,

$$H^+ + H \rightarrow H + H^+.$$

We showed how good calculations that agree with extensive laboratory measurements allow us to make a definitive calculation of the coefficient of ambipolar diffusion heating for warm atomic regimes. On this basis we can eliminate this process for heating the X-wind and other similar protostellar jets. Perhaps another lesson is that astrophysicists need not make crude estimates of rate coefficients when good atomic cross sections are available.

In the second example, we noted the lack of good low-energy cross sections for sulfur ions,

$$S^{n+} + H \rightarrow H^+ + S^{(n-1)+}.$$

Especially urgent are the rate coefficients for the singly and doubly-ionized ions. In addition, charge exchange rates are needed for S^+ with H, Si, and other atoms with smaller ionization potentials, and for H^+, O^+, and C^+ with S. Without this information, the abundance of S II cannot be determined reliably. These reservation apply to other atoms and ions with astrophysically useful diagnostic lines, especially in problems with short dynamical time scales..

ACKNOWLEDGMENTS

This work has been supported in part by an NSF grant to New York University.

REFERENCES

1. Blandford, R.D. & Payne, D.G., Monthly Notices of the Royal Astronomical Society, 199, 883, 1982.
2. Shu, F.H., Najita, J.R., Shang, H., & Li, Z.-Y.. in *Protostars and Planets IV*, eds. V. Mannings, A.P. Boss, & Sara S. Russell, Tucson Arizona: University of Arizona, 2000, p. 789.
3. Königl and R.E. Pudritz, in *Protostars and Planets IV*, eds. V. Mannings, A.P. Boss, & Sara S. Russell, Tucson Arizona: University of Arizona, 2000, p. 759.
4. Shang, H., Glassgold, A.E., Shu, F.H., & Lizano, S., Astrophys. J. 564, 853. 2002.
5. Shu, F.H., Najita, J., Ostriker, E. C., Wilkin, F., & Ruden, S., & Lizano, S. 1994, Astrophys. J. 429, 781, 1994.
6. Ruden, S.P., Glassgold, A.E., & Shu, F.H., Astrophys. J 361, 546, 1990.

7. Shu, F.H., Shang, H., & Glassgold, A. E., & Lee,, Science, 277, 1475, 1997.

8. Dalgarno, A., Yan, M., & Liu, W.-H., Astrophys. J. Suppl. 125, 237, 1999.

9. Mahoney, P.R., Hollenbach, D.J., & Tielens, A.G.G.M., 1996, Astrophys. J. 466, 561, 1996.

10. Draine, B., Monthly Notices of the Royal Astronomical Society, 220, 133, 1986.

11. Osterbrock, D., Astrophys. J., 134, 270, 1961.

12. Draine, B., Astrophys. J. 241, 1021, 1980.

13. Krstić, P.S. & Schultz, D.R., Atomic and Plasma-Material Data for Fusion 8, 1, 1988.

14. Hunter, X. & Kuriyan, Y., Proc. Roy. Soc., A353, 575, 1977.

15. Safier, P., Astrophys. J., 408, 115, 1993.

16. Garcia, P.J.V., Ferreira, J., Cabrit, S., & Binette, L., Astron. & Astrophys. 377, 589. 2001.

17. Raga, A. C., Cantò, J., Binette, L., & Calvet, N., Astrophys. 364, 601, 1990.

18. Black, J.H., in *Interstellar Processes*, eds. D. J. Hollenbach & H. A. Thronson, Jr., Dordrecht: Reidel, 1986, p. 731.

19. Kimura, M., Gu, J.P., Hirsch, G., & Bunker, R.J., Phys.Rev. A55, 2778, 1997.

20. Stancil, P.C., Turner, A.R., Cooper, D.I., Schultz, D.R., Raković, M.J., Fritsch, W., & Zygelman, B., J. Phys. B 34, 2481, 2001.

21. Butler, S.H. & Dalgarno, A., Astron. & Astrophys. 85, 144, 1980.

X-RAY SOURCES

Interaction of Intense Laser Pulses with Noble Gas Clusters and Droplets

E. Parra, I. Alexeev, J. Fan, K.Y. Kim, S.J. McNaught, and H.M. Milchberg

Institute for Physical Science and Technology, University of Maryland, College Park, MD 20742

Abstract: A series of experiments have been performed to investigate the time dependent interaction of intense laser pulses with mass-limited noble-gas clusters and droplets. The laser-cluster interaction was explored using a variable pulse width arrangement whereas the dynamics of the droplets were investigated using a pump probe scheme. The plasma evolution was assessed by monitoring x-ray and extreme ultraviolet emission, as well as light scattering. The observed time scales for clusters and droplets are understood in terms of two time scales natural to the plasma dynamics. We also present, in the case of clusters, a 1D hydrodynamic model of the interaction in which the laser field is treated self-consistently. We find that nonuniform expansion of the heated material results in long-time resonance of the laser field at the critical density plasma layer. These simulations explain the dependence of generation efficiency on laser pulse width.

INTRODUCTION

The interaction of intense laser pulses with mass limited targets, such as clusters and droplets, has received considerable attention in the last decade.[1-7] The intense laser coupling to these targets combines absorption features of both gases and solids. Due to their near-solid internal densities, the laser heating of mass limited targets is dominated by collisional absorption, a process more typical of solid targets. This enhances the energy absorbed compared to unclustered subcritical density gases. For the same volume average density as an unclustered gas, the gas of clusters / droplets can thus be heated to higher plasma temperatures. These laser-produced plasmas (LPP) efficiently generate incoherent short wavelength light for applications such as extreme ultraviolet (EUV) lithography,[6] EUV / x-ray microscopy[8] and x-ray tomography.[9] One of the most efficient LPP sources[6,7] is produced at Sandia National Laboratory by the irradiation of a spray of xenon droplets with intense pulses from a 1500 W, Q-switched (~10 ns), 300-mJ, kHz-repetition rate laser system.[6] This interaction produces EUV radiation with conversion efficiencies comparable to those of solid targets (approximately 0.2% conversion of laser energy into EUV radiation into 2π steradians and 2% bandwidth).[6]

A goal in many EUV or x-ray source applications of intense laser-target interactions is to maximize the short wavelength light emission for a given laser pulse

CP635, *Atomic Processes in Plasmas: 13th APS Topical Conference*, edited by D. R. Schultz et al.
© 2002 American Institute of Physics 0-7354-0090-3/02/$19.00

energy. In the case of clusters / droplets, an intrinsic time scale for laser coupling efficiency is the critical density lifetime τ_{crit}, which is the time for a laser-heated target to expand to a local electron density below critical. For times t > τ_{crit}, laser light absorption will decrease significantly. Energy delivered by laser pulses at times longer than τ_{crit} should be absorbed no better than in a gas of the same average density. This initially suggests that the EUV/x-ray emissions from laser-heated mass limited targets should be a strong function of pulse width.

In this paper, investigations of optimal laser coupling time scales of clusters and droplets are presented separately in two sections. In the cluster section, we present experimental and numerical studies regarding the pulse width dependence of the laser – cluster interaction. These experiments examined the efficiency of EUV and x-ray production of argon clusters as a function of the laser pulse width. The results can be understood physically in terms of two time scales natural to the plasma: a short time scale for optimal resonant absorption at the critical density layer in the expanding plasma, and a longer time scale for the plasma to drop below critical density. Numerical simulations were carried out showing that the plasma hydrodynamic expansion is in general nonuniform with the dominant absorption mechanism being a long-lived resonant absorption at the critical density layer. Numerical simulations of the variable pulse width interaction display the optimal resonant coupling behavior consistent with the experiment.

In the droplet section, the droplet plasma dynamics is studied by using two 200-mJ, 100-ps, 1064-nm laser pulses in a variable-delay pump-probe experiment. The first laser pulse heats the argon or krypton droplets, forming a plasma, and the second pulse serves as a heater probe for measuring time scales for efficient production of EUV and x-ray emission. The time scales are observed to vary from a few hundred picoseconds to several nanoseconds, depending on the gas species, average droplet size, and type of emission measured. The disassembly of the laser-heated droplets is visualized *in situ* by imaging the scatter from a variably delayed low-intensity 532-nm probe pulse.

CRYOGENIC CLUSTER / DROPLET SOURCE

The formation of clusters and droplets occurs during the rapid flow of either gas or liquid through a nozzle into vacuum. In particular, *nanometer-sized* clusters are van der Waals bonded assemblies of $\sim10^2$-10^7 atoms, which form during the rapid cooling of the reservoir *gas* as it flows into vacuum. This is a *condensation* process. On the other hand, if the reservoir gas is maintained at high pressures and low temperatures, it undergoes a phase transition, and a liquid jet, rather than gas, is ejected when the valve opens.[10,11] Contrary to cluster formation, *micron-sized* droplets are produced through the *fragmentation* of the liquid jet.[10] A gas nozzle can be operated in either regime by careful control of the reservoir gas thermodynamic conditions.

Our cryogenic cluster / droplet source is shown in Fig. 1. A standard General Valve Corporation series 9 stainless steel solenoid valve with copper gasket seals and a Kel-F poppet is mated to a nozzle with a 0.5 mm orifice and 5° expansion half angle.

The valve is controlled using a high-speed pulser with a valve opening time of 450 μs and pulsed synchronously with the laser at 10 Hz. A close-fitting copper cooling jacket clamps a stainless steel liquid nitrogen coolant tube and two solid-state cartridge heaters to the valve body. The combination of these heating and cooling elements allows continuous control of the valve temperature between 100 and 300 K. At any given set point, the temperature remains stable to within ± 0.25 K (see Fig. 1).

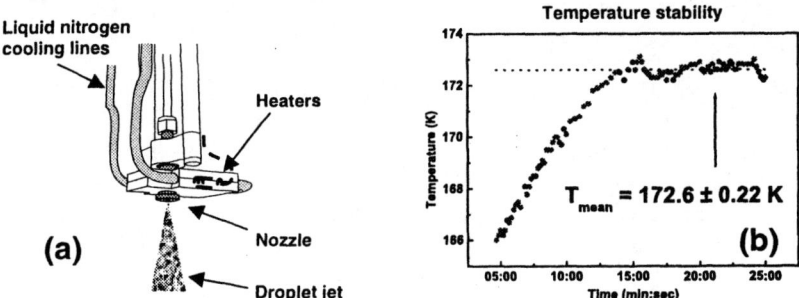

Figure 1. (a) Diagram of the cluster / droplet source showing valve, nozzle and cooling/heating jacket. (b) Transient stabilization to new set point of 172.5 K.

VARIABLE LASER PULSEWIDTH CLUSTER EXPERIMENTS

Here we measure x-ray (> 1.5 keV) and EUV ($\lambda = 2 - 44$ nm) emissions from a laser heated supersonic argon gas jet at room temperature ($T_0 = 300$K) and cryogenic temperature ($T_0 = 173$ K) using laser pulses of constant energy (50 mJ) and variable pulse width in the range 100 fs to 10 ns.

Experimental setup

For all temperature conditions, the pulsed valve was operated at a backing pressure of $P_0 = 500$ psi with a valve opening time of ~ 450 μs. The gas jet was irradiated with laser pulses from a Ti: Sapphire chirped pulse amplification laser ($\lambda_0 = 800$ nm, 10Hz repetition rate) that were synchronized with the pulsed valve. The laser pulse energy was held constant at 50 mJ and the pulse width varied in the range 100 fs to 10 ns. Control over the pulse width was achieved by either detuning the grating compressor or inserting an adjustable mask in the pulse stretcher. The laser beam was focused ~ 1 mm below the nozzle orifice, 1.3 cm downstream from the jet throat, using a f=285mm MgF$_2$ lens. The FWHM focal spot was measured to be 16.6 μm, which corresponded to a peak vacuum intensity of 1.3×10^{17} W/cm^2 for 100 fs pulses. The EUV spectra (2-44 nm) were built up by slowly scanning the exit slit of a 1.5-meter grazing incidence vacuum monochromator equipped with a windowless electron multiplier. The X-rays (>1.5 keV) were measured using a silicon photodiode (IRD AXUV-100) with 150 nm aluminum and 25 μm beryllium filters, which strongly attenuated X-rays below ~1.5 keV. Both instruments viewed the jet perpendicular to the laser propagation axis.

An important parameter for interpreting the experimental results is the cluster average size. Direct measurement of the cluster size distribution is difficult, although rough estimates are possible.[1,12] For large rare gas clusters, the final mean cluster size (atoms/cluster) has been shown to be reasonably well predicted by the expression $<N> = 33\left(\Gamma^*/1000\right)^{2.35}$ where Γ^* is the dimensionless 'Hagena parameter',[13] $\Gamma^* = k(d/\tan\alpha)^{0.85} p_0 T_0^{-2.29}$, where d is the nozzle diameter (in μm), α the expansion half angle, p_0 the backing pressure (in mbar), T_0 the pre-expansion temperature (in Kelvin) and k is a gas-specific constant ($k = 1650$ for Ar). Experiments have shown that for $\Gamma^* > 2000$ the condensed fraction in the cluster beam is above 90%.[13] Hagena-based size estimates are in reasonable agreement with other methods such as Rayleigh scattering.[14] Under our conditions, at $T_0 = 300K$ we obtain $\Gamma^*_{Ar} \sim 2\times10^5$, giving for cluster size and radius $<N>_{Ar} \sim 7\times10^6$ ($R_{Ar} \sim 40$ nm). For $T_0 = 173$ K, $\Gamma^*_{Ar} = 7\times10^5$, giving $<N>_{Argon} = 1\times10^8$ ($R_{Ar} \sim 100$ nm).

Results and discussion

Figure 2 shows the time integrated EUV spectra ($\lambda = 2 - 44$ nm) for the full range of laser pulse widths ($\tau = 100$ fs – 10 ns) for uncooled ($T_0 = 300$ K) and cooled ($T_0 = 173$ K) valve operation. The spectra have been offset for clarity. In general, EUV emission appears to be maximized for pulse widths less than ~1ps.

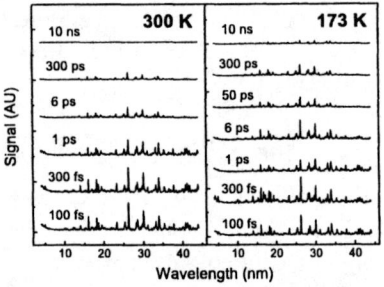

Figure 2. EUV spectra for argon clusters at T = 300 K and T = 173 K for the full range of laser pulse widths ($\tau = 100$ fs – 10 ns). The various spectra have been offset for clarity.

A more detailed view of the pulse width dependence is obtained by plotting the relative yields of lines from high ion stages. Figure 3 shows a plot of yield vs. laser pulse width for several unblended L shell ($2s2p^6 \rightarrow 2s^2 2p^5$) Ar^{9+} transitions ($\lambda = 16.6$ and 17.1 nm).[15] Under our conditions, emission from Ar^{9+} occurs primarily during electron collisional excitation, when the plasma temperature and density are high. This is deduced from the fact that ionization to Ar^{10+} from the ground state of Ar^{9+} requires 479 eV,[15] and no Ar^{10+} emission lines were observed. Therefore, we associate the yield of Ar^{9+} emission with the efficiency of laser-cluster coupling. Peak Ar^{9+} emission occurs at both valve temperatures for pulses in the range 100-300fs, with a prominent peak at T_0=173K centered at ~300fs. X-ray emission above 1.5 KeV, for the same gas jet conditions, is shown in Figure 4. In this photon energy range, K-shell Ar ion transitions, as well as bremsstrahlung, will contribute. Here it is seen that peak emission occurs at ~300fs for both valve temperatures.

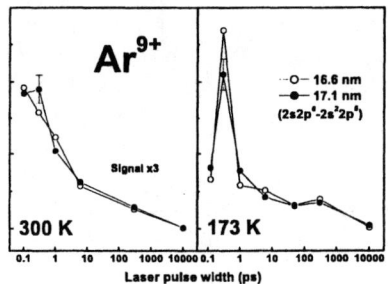

Figure 3. Time integrated argon cluster EUV emission from selected L-shell Ar^{9+} lines at T = 300K and 173 K as a function of laser pulse width. The signal scale is linear. Error bars represent maximum shot-to-shot fluctuations.

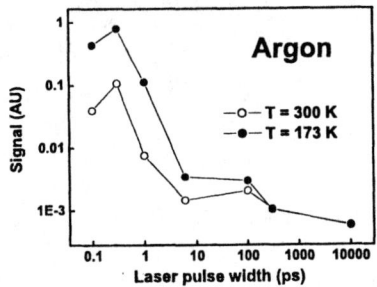

Figure 4. Argon cluster x-ray (> 1.5 keV) signal at T = 300K and 173K as a function of laser pulse width.

Our results appear to be consistent with two timescales expected from the cluster plasma hydrodynamic evolution. The first is a longer time scale τ_{crit} during which the plasma from an individual cluster relaxes to below critical density. This time scale depends on the cluster size and is given by

$$\tau_{crit} \sim \frac{R}{c_s}\left(\frac{N_{e0}}{N_{crit}}\right)^{1/3} \qquad (1)$$

where R is the cluster radius, c_s is the plasma sound speed, N_{e0} is the pre-expansion electron density of the near-solid density cluster plasma, and $N_{crit} = m\omega^2/4\pi e^2$ is the critical density. For λ = 800 nm, $N_{e0} \sim 2\times10^{23}$ cm^{-3}, $R_{Ar} \sim 40$ nm (from discussion above), and estimating $c_s \sim 2\times10^7$ cm/s for plasma temperatures of at least several hundred eV, we get $\tau_{crit}^{Ar} \sim 1.0$ ps, consistent with the maximum driving laser pulse widths for efficient emission of both EUV from Ar^{9+} and x-rays. This timescale represents maximization of the bulk thermal heating of the cluster.

A second, shorter, time scale relates to the finite time for establishing an optimal critical density layer on the periphery of the expanding cluster plasma, in which resonance absorption is maximized.[16,17] Too short a pulse does not allow a sufficient critical density layer $\delta r_{crit} = (v/\omega)L_{cr}$ to develop, where $L_{cr} = N_e|\nabla N_e|^{-1}$ is the electron density scale length in the critical density region. L_{cr} increases with time as the plasma expands. On the other hand, too long a pulse increases the density scale length so much that the penetration of the electric field to the critical density surface is reduced. Therefore, there exists an intermediate electron density scale length for optimal resonant absorption, which implies the existence of an optimal laser pulse width. For linear polarization, the optimal scale length for resonant absorption in the cluster can be estimated using the plane target result for p-polarized electromagnetic waves,[17] $L_{cr} = L_{opt} \sim 3\sqrt{3}/(8k\cos^3\theta)$, where k is the laser vacuum wave number, θ is the angle between the laser polarization and the local surface normal, and L_{opt} represents a balance between the field's penetration to the critical density surface and its projection along the density gradient. While this expression is more appropriate for the case of long scale lengths and larger targets (i.e. for micron–sized droplets)[18,19] ($kR\gg1$) rather than for the near-field case of clusters ($kR\ll1$), it gives reasonable results in the near field regime for $k=\omega_p/c$ and $\theta=0$.[16] If we set $L_{opt}\sim c_s\tau_{res}$, and use $c_s\sim2\times10^7$cm/s, $\lambda=800$nm, and $\theta\sim0$ (since for short scale lengths, resonance absorption is maximized at near-oblique incidence,[20] here near the poles of the plasma sphere), we get $\tau_{res}\sim400$fs. This is in reasonable agreement with the increase in laser pulsewidth over which the EUV Ar^{9+} and x-ray emission yields rise to their peak values. As the plasma expands beyond L_{opt} (for $t>\tau_{res}$), the local heating drops as less laser field penetrates to the critical density surface. For $t>\tau_{crit}$, as discussed earlier, the overall cluster volume heating drops, consistent with the tails in emission yield dropping off beyond ~1 ps.

Even though for the experimental conditions reported here the two timescales are similar, τ_{crit} scales with the target size whereas τ_{res}, being a more local effect, does not. This characteristic has been explored in other *variable pulsewidth experiments*, where the short time scale emission yields of krypton clusters and micron-sized krypton droplets have been compared.[18] In these variable pulsewidth studies (not shown in this paper), we find that for droplets, the timescales for peak EUV emission can increase up to several hundred picoseconds whereas the x-ray emission has a peak centered at ~300fs for *all* cases. These results strongly suggest that for short pulses, the optimal resonant absorption scenario is operable for both clusters and droplets.

Numerical modeling of the laser – cluster interaction

Previously, the laser-cluster interaction has been modeled assuming that the cluster plasma expands as a uniform density sphere (the 'uniform density' model).[1] This model has been successful in qualitatively reproducing several experimental features, such as the high degree of ionization observed through spectroscopy,[4] the observation of fast electrons and ions,[2] and resonant behavior in the laser-plasma coupling.[1,2] The assumption that the radial plasma density profile remains uniform during the expansion leads to an absorption and scattering resonance when $N_e \sim 3N_{crit}$, which last only during a very narrow time interval given by, $\delta t_{res} \approx \dfrac{2}{3}\dfrac{v}{\omega}(\dfrac{N_{e0}}{3N_{crit}})^{1/3}\dfrac{R}{c_s}$, where R is the initial cluster radius, N_{e0} is the initial electron density, c_s is the plasma sound speed, and v/ω is the normalized collision frequency at $N_e=3N_{crit}$.[16] For a typical case of $R=100$ Å, $c_s=2\times10^7$ cm/sec, $N_{e0}=2\times10^{23}$ cm^{-3}, $N_{crit}=1.8\times10^{21}$ cm^{-3} (corresponding to $\lambda_{laser} \sim 800$nm), and an estimate of $v/\omega \sim 0.03$ (assuming a scaling of $v/\omega \sim 3N_{crit}/N_{e0}$ $(v/\omega)_{solid}$, where $(v/\omega)_{solid} \sim 1$),[21] we get $\delta t_{res} \sim 3$ fs. These resonance durations are much shorter than suggested by previous pump-probe and variable pulse width measurements,[22] where resonant enhancements of absorption and scattering appear to last at least several picoseconds. Even allowing for initial cluster size distributions in the uniform density model,[22] a large discrepancy remains between this model theory and the experimental results.

In order to allow further physical interpretation of our experiments, and to improve upon the uniform density model (see above), a detailed numerical model of the laser – cluster interaction was constructed. In this model, the complete details of which are presented elsewhere,[16] the laser field is coupled to the nonequilibrium time-dependent hydrodynamics of the expanding cluster plasma. More specifically, the equation for the electric near field, $\nabla \cdot (\varepsilon E) = 0$, is coupled to a 1D radial Lagrangian hydrocode. The near field treatment, appropriate where $kR_{max} \ll 1$, is a good approximation for initial cluster sizes much smaller than a laser wavelength and for times not too late in the cluster expansion. Starting with a solid density neutral cluster, the electric field is solved at each time step using the neutral, ion and electron density profiles and the temperature profile of the previous time step. The resulting electric field ionizes and heats the plasma, driving the cluster dynamics. The dielectric response of the laser-heated cluster plasma is taken to be of the Drude type, which is appropriate for hot near-solid density plasmas with no electronic band structure. The plasma pressure and temperature are related by the ideal gas equation of state. The calculation includes field and collisional ionization, thermal conduction, the effect of ponderomotive forces on the plasma, and a time-dependent collisional-radiative model for the ionization dynamics.

As an example of conditions in our experiments, we model the interaction of a linearly polarized, 300 fs FWHM Gaussian, 800nm laser pulse of peak intensity 10^{15} W/cm^2 with a 60 nm radius argon cluster of initial atomic density 1.8×10^{22}cm^{-3}. The intensity is lower than in our experiments in order to reduce the computation time, with the goal being to provide a qualitative understanding of the interaction. Higher intensities increase the ionization rates and the plasma temperature, requiring dramatic reductions in the numerical time step demanded by numerical stability considerations. In Figure 5, we plot electron density profiles and the corresponding electric field profiles for a sequence of times in the pulse envelope. The peak of the pulse occurs at $t=0$. The electric field is scaled relative to its magnitude in vacuum at each time.

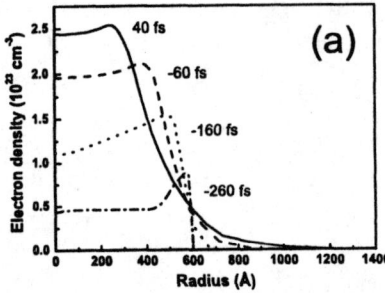 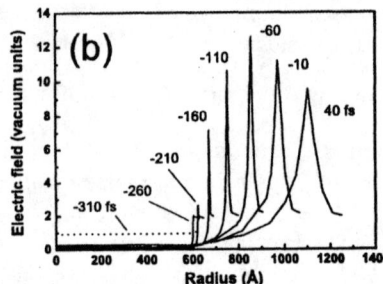

Figure 5. (a) Electron density profiles vs. time for argon clusters of initial radius 60 nm, and laser with peak intensity 1015 W/cm2, $\lambda = 800$ nm, 300 fs FWHM Gaussian pulse, and initial argon cluster atomic density 1.8x1022 cm-3. (b) Electric field profiles (normalized to vacuum value) corresponding to the electron density profiles of (a).

There are two important physical effects illustrated by this simulation. First, the electron density profiles (Fig. 5(a)) are non-uniform. This nonuniformity is a natural consequence of the expansion of a hot fluid sphere. Near the plasma-vacuum interface, where the pressure gradient is largest, the fluid velocity is highest. In the interior, the pressure gradient is weaker and the radial fluid velocity is reduced. The bump in electron density near the cluster edge at -260 fs is a result of the mass compression driven by the ablative pressure of the expanding plasma. For these conditions, a weak compression wave propagates to the cluster center as time advances.

Secondly, the electric field (Fig. 5(b)) is significantly enhanced with respect to its vacuum value in the region near $N_e \sim N_{crit}$. The resonance at N_{crit} lasts for the full duration of the pulse, arising at early times as a critical density layer expands from the cluster. At very early times ($t<-310$fs) before a plasma is formed it is seen that the relative field amplitude is just less than unity, as expected from dielectric shielding in the neutral material where $\varepsilon > 1$. Once plasma develops ($t>-300$fs), the cluster center is shielded from the electric field, so that the field amplitude there drops significantly below its vacuum value. The resonance is then maintained throughout the pulse duration as long as the cluster plasma does not expand below critical density.

The non-uniform plasma expansion and the presence of a long-lived resonance at N_{crit} are natural consequences of the laser plasma interaction. This phenomenology stands in contrast with the spherical uniform density model described earlier, which predicts a short lived resonance at $3N_{crit}$ throughout the plasma volume. We find no experimental or numerical evidence for the short resonant timescales predicted by the uniform density model. Once the plasma begins to expand, even with a very thin layer outside of a uniform density core, the $3N_{crit}$ resonance disappears. This resonance would also disappear owing to aspheric plasma geometry, as is likely the case because the near field laser-cluster interaction is really a two dimensional problem.[16] Both of these considerations apply even for very small clusters. On the other hand, the N_{crit} resonance arises naturally from the nonuniform hydrodynamics, is independent of plasma geometry (due to its locality) or size, and explains the long-lived resonant behavior seen experimentally here and in other work.[22]

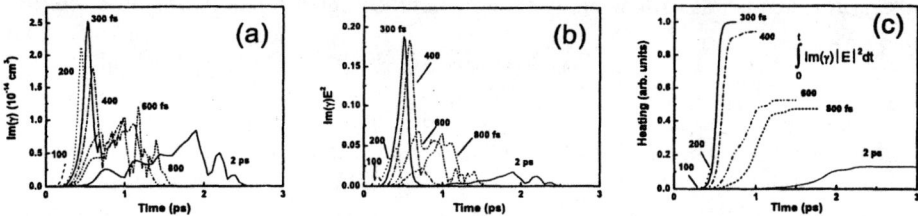

Figure 6. (a) Imaginary part of argon cluster polarizability vs. time for pulse widths 100fs, 200fs, 300fs, 400fs, 600fs, 800fs, and 2 ps at fixed energy. Cluster initial radius 60 nm. Intensity range 6×10^{15} W/cm^2 (100fs pulses) through 3×10^{14} W/cm^2 (2ps pulses). (b) Instantaneous & (c) Cumulative heating for same laser and cluster conditions.

Using this model, we can numerically examine the laser coupling efficiency to clusters by varying the width of the irradiating laser pulse at fixed laser energy. Specifically, we examine the cluster energy absorption by calculating the imaginary part of the complex cluster polarizability γ (defined as $\mathbf{P} = \gamma \mathbf{E}$, where $\mathbf{P}$ is the cluster complex dipole moment and $\mathbf{E}$ is the external electric field). We simulate our variable pulsewidth experiment using a 60 nm radius argon cluster of initial atomic density 1.8×10^{22} cm^{-3} irradiated with variable width 800 nm pulses in the range 100fs to 2 ps, with peak intensity varying from 6×10^{15} W/cm^2 (10 fs pulse) to 3×10^{14} W/cm^2 (2 ps pulse). The peak intensity is below that in the experiment in order to reduce the computation time, with the goal being a qualitative understanding of the pulse width dependence of cluster heating. For the pulse widths used, Figure 6(a) shows curves of Im(γ) vs. time, Figure 6(b) shows the instantaneous heating rates ($\propto$ Im(γ)$|\mathbf{E}|^2$), and

Figure 6(c) shows the cumulative heating ($\propto \int_0^t$ Im(γ)$|\mathbf{E}|^2\, dt$). The Im(γ) curves

represent the time-dependent absorption capability of the different pulsewidth-prepared clusters. Inspection of these curves as well as the instantaneous (Fig. 6(b)) and time-integrated (Fig. 6(c)) heating curves for the various pulses show that the ~300 fs pulses are most efficient. Examination of the $\mathbf{E}$ and N_e profiles (not shown here) show that the proposed physical picture of the absorption maximum is correct: optimum coupling via resonance absorption occurs when the density scale length is not too short or not too long. For our conditions, that occurs at several hundred femtoseconds.

LASER PUMP – PROBE DROPLET EXPERIMENTS

In this section, the dynamics of the droplet plasma formed by the irradiation of micron-sized argon and krypton droplets with intense 100-ps laser pulses was investigated using a pump-probe scheme. The plasma evolution was assessed by monitoring delay-dependent x-ray and EUV emission, and by imaging frequency-doubled probe light scattered from the interaction region. These results are explained in terms of the plasma expansion, excitation and recombination emission time scales.

Experimental setup

In these experiments, shown schematically in Figure 7, 100-ps pulses generated by a mode-locked, 10-Hz Nd:YAG laser system were split equally into pump and probe beam lines using a waveplate and polarizer setup. The energy of both pulses was set to 200 mJ. The probe pulse was sent through a delay line, allowing it to follow the pump pulse by a variable delay, known to within 50ps, of 0 – 14 ns. The two pulses were then recombined using a polarizer and focused collinearly inside a vacuum chamber with a 150-mm lens (f/3 optics) into a low-density droplet jet, forming a plasma. The chamber was evacuated by a 350-cfm Roots pump, and the ambient pressure remained at ~10-20 millitorr during experiments. The interaction region was viewed under high magnification (10 – 30X) at both 0° and 90° to the laser propagation direction using two imaging setups and CCD cameras. The 0° (focal spot) and 90° (side scatter) cameras were used to ensure precise transverse and axial alignment, respectively, of the two foci to each other and to the droplet jet. The focal spot of a single beam was measured to be 16.2 μm FWHM, corresponding to 1×10^{15} W/cm^2 peak vacuum intensity.

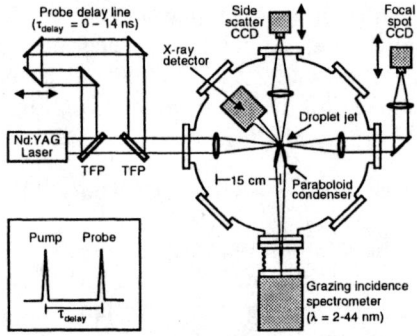

Figure 7. Schematic of the pump probe droplet experimental setup.

From droplet imaging,[10] we estimated there to be ~10-100 droplets in the interaction region, defined as the cylindrical volume whose length and width are the laser's confocal parameter and focal spot diameter, respectively. Dark-field imaging with 355-nm laser light showed that the average diameter of argon droplets (for reservoir conditions of 600 psi and 138 K) was ~5 μm,[10] while it was ~7 μm for Kr at 500 psi and 173 K (Ref. 3) and ~10 μm for Kr at 200 psi and 143 K.[10] The droplet jet was narrowed and collimated by a 500-μm diameter skimmer located 1 cm below the valve orifice and 4.5 cm above the laser-droplet interaction region. The goal was to have the laser interact with as few droplets as possible, while still allowing detection of acceptable EUV and x-ray signals. Light from a CW He-Ne laser was scattered from the droplet flow and collected by a photomultiplier tube. The transient scattered signal from the leading edge of the flow indicated an average droplet flow speed of ~200 m/s.[10] Under these conditions, we estimate that the droplets likely have become ice by the time they traverse the 5.5 cm distance from the orifice to the laser focus.[10]

Approximately 5% of the EUV radiation emitted by the plasma was collected by a rhodium-coated paraboloidal condenser at 90° from the laser axis, and directed to the entrance slit of a grazing incidence vacuum spectrometer. This spectrometer, whose entrance slit was 50 cm from the laser focus, utilized a windowless electron multiplier detector and measured emission in the wavelength range of 2-44 nm (see Fig. 7). EUV rays from the laser plasma reflect from the polished inner surface of the condenser at ~10-15° grazing angle and are directed toward the spectrometer's entrance slit. If the EUV source is positioned precisely at the condenser's focus ("collimated alignment"), the rays will be collimated by the condenser and result in an annular intensity profile (with a minimum on center) in the plane containing the spectrometer slit. However, if the condenser is positioned several hundred microns closer to the EUV source ("convergent alignment"), the rays will converge and produce an image of the EUV source at the spectrometer's entrance slit. Given the 50 cm source-to-detector distance, these calculations estimated that the condenser would increase the signal level measured by the spectrometer by a factor of 500 for collimated alignment, and by 2000 for convergent alignment, as compared to using no condenser at all. In the laboratory, the condenser was aligned to produce the sharpest possible image of a test optical source on the spectrometer slit. With this setup, the EUV signal at $\lambda = 26.0$ nm was measured to increase by a factor of 1600, proving the value of the condenser.

Lastly, an x-ray detector consisting of an amplified, 1-cm^2 silicon photodiode, with 25-μm Be and 150-nm Al filters, was situated at 135° from the laser propagation direction and 12.5 cm from the interaction region. This detector is sensitive to x-rays greater than ~1.5 keV.

As the delay between pump and probe pulses was varied, the EUV or x-ray signals produced by the pump pulse alone were subtracted from the total signals produced by the pump and probe to yield the time scales for radiation production by the probe. At each delay, the signals from 250 shots were averaged. Over the course of an experimental run with fixed droplet source conditions, the EUV and x-ray emission associated with the pump pulse varied little, indicating that the laser and droplet source remained stable over time (see Fig. 8a).

Results and discussion

With increasing delay between pump and probe pulses, the EUV and x-ray emission produced by the probe pulse decays, implying reduced coupling of the probe to the expanding droplet plasma over time. Figure 8a shows the EUV signals measured from the interaction of the pump and probe pulses with argon droplets. The

EUV emission associated with the probe pulse at 26.0 nm ($2p^64f \rightarrow 2p^63d$ transition in Ar^{7+} ions)[15] falls to half its maximum value after 2.1 ± 0.3 ns (circles), but the Ar^{9+} emission at 16.6 nm ($2s2p^6 \rightarrow 2s^22p^5$)[15] decays in 190 ± 60 ps (filled squares). The quoted uncertainties originate from the least squares fit parameters, where the fitting function was a second-order exponential with a constant offset. The data labeled with open squares is the 16.6-nm EUV signal from the pump pulse alone, illustrating the stability of the laser and droplet source over time. Under our conditions, emission from Ar^{9+} occurs primarily during electron collisional excitation from the ground state, when the plasma temperature and density are high (ionization to Ar^{10+} from the ground state of Ar^{9+} requires 479 eV,[15] and no Ar^{10+} emission lines were observed), while emission from Ar^{7+} can occur during recombination from Ar^{8+} (ionization to Ne-like Ar^{8+} requires 143 eV,[15] and Ar^{8+} lines *were* observed), well after the laser pulse, and at sub-critical density. Therefore, the Ar^{9+} emission is a more sensitive indicator of laser heating. Figure 8b shows the EUV signals measured from the interaction of the probe pulses with krypton droplets. The EUV spectrum is dominated by Kr^{9+} line emission at 10.0 nm ($3p^63d^84p \rightarrow 3p^63d^9$).[15] There is some evidence of emission from Kr^{10+}, but these lines are blended with others from lower ionization stages. We observed a long decay time for the Kr^{9+} EUV signal of 1.7 ± 0.2 ns for 7-μm droplets and 2.5 ± 0.4 ns for 10-μm droplets, which is consistent with this emission being associated with recombination from Fe-like Kr^{10+} ions (ionization to Kr^{10+} requires 275 eV).[15]

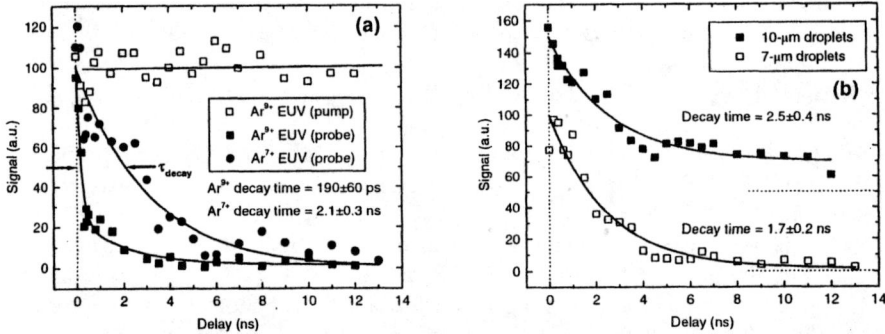

Figure 8. (a) Argon droplet (5-μm diameter) EUV plasma emission versus time delay for Ar^{9+} emission at 16.6 nm from the pump pulse (open squares), Ar^{9+} emission at 16.6 nm from the probe pulse (filled squares), and Ar^{7+} emission at 26.0 nm from the probe pulse (circles). (b) Krypton droplet EUV plasma emission produced by the probe pulse for Kr^{9+} ions at 10.0 nm for 10-μm (filled squares) and 7-μm (open squares) diameter droplets. The curves are offset for clarity, and the dotted line below each curve is the zero signal baseline.

For both argon and krypton droplets, the >1.5 keV x-ray signals shown in Fig. 9 rapidly fall off with delay. For argon droplets, the x-ray emission decay time of 160 ± 30 ps is much shorter than the Ar^{7+} EUV decay and is similar to the Ar^{9+} EUV decay. For krypton, the x-ray decay times of 210 ± 20 ps (7-μm droplets) and 295 ± 50 ps (10-μm droplets) are much shorter than the corresponding Kr^{9+} EUV decay times.

The x-ray and Ar^{9+} EUV decay times are consistent with direct laser plasma excitation by the probe at early times, when a significant portion of the droplet is at or above N_{crit}.[16]

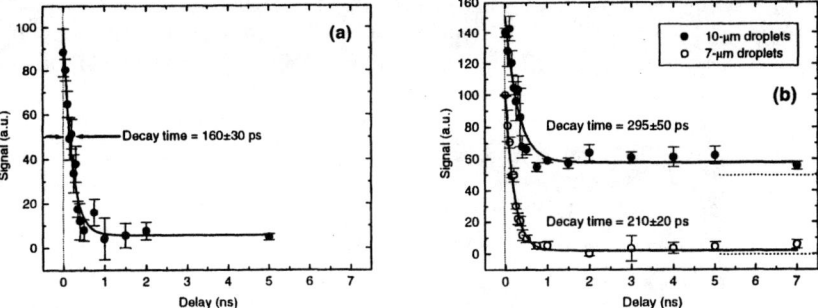

Figure 9. (a) Argon droplet (5-μm diameter) x-ray plasma emission (>1.5 keV) versus time delay from the probe pulse. (b) Krypton droplet x-ray plasma emission produced by the probe pulse for 10-μm (filled circles) and 7-μm (open circles) diameter droplets. The curves are offset for clarity, and the dotted line below each curve is the zero signal baseline.

The 'droplet disassembly time' τ_{crit} is defined earlier to be the time it takes the plasma to expand to below the critical density (here ~10^{21} cm^{-3} for 1064 nm laser pulses), whereupon laser-plasma coupling drops significantly. Figure 10 shows the theoretical droplet disassembly time (Eqn. 1) as a function of droplet diameter using the typical values $N_{e0} = 2 \times 10^{23}$ cm^{-3}, $N_{cr} = 10^{21}$ cm^{-3}, and $c_s = 10^7$ cm/s (for a plasma temperature of several hundred eV). Included in the plot are our measured 5-μm argon droplet x-ray and Ar^{9+} EUV decay times (160 and 190 ps, respectively), and 7-μm and 10-μm krypton droplet x-ray decay times (210 ps and 295 ps, respectively). According to Eqn. 1, τ_{crit} ~ 140 ps for 5-μm diameter argon droplets, ~200 ps for 7-μm Kr droplets, and ~280 ps for 10-μm Kr droplets. The measured and calculated values are in reasonable agreement. We therefore associate these time scales with the expansion of the droplet plasma to below critical density.

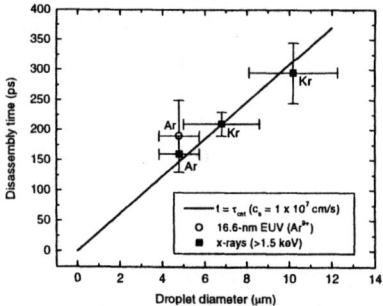

Figure 10. Laser droplet plasma disassembly time versus droplet diameter. Shown are the calculated disassembly time (solid curve), measured x-ray decay times (squares) and measured 16.6-nm Ar^{9+} EUV decay time (circle).

The significantly longer decay times for probe pulse generation of Ar^{7+} and Kr^{9+} EUV emission (see Fig. 8) are representative of the recombination process, which lasts longer than the disassembly time: the duration of the Ar^{7+} and Kr^{9+} signals from the pump pulse alone was ~ 4 ns. This time scale was estimated by deconvolution of these EUV signals with the EUV signal observed from the rapidly-decaying Ar^{9+} excitation emission, which was assumed to follow the impulse response of the electron multiplier.

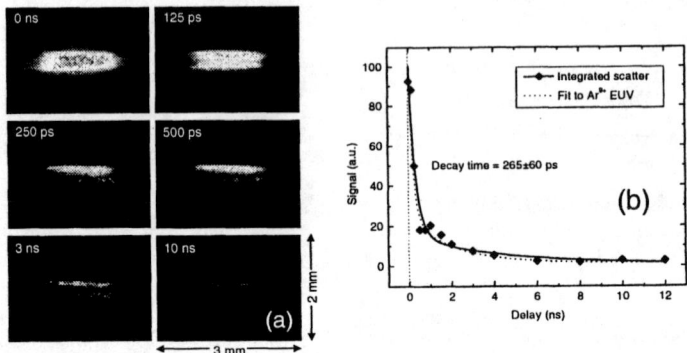

Figure 11. (a) Scattered light images, using 532 nm probe light, of the interaction region for 5.0-µm argon droplets at various time delays relative to the pump pulse. The channel in which droplets have expanded to below critical density begins to form at 125 ps. (b) Integrated scatter signal versus probe delay. The decay time associated with the green probe scatter (solid curve) is in good agreement with that for the 16.6-nm radiation from Ar^{9+} ions (dotted curve).

Our explanation above for the observed timescales for probe pulse EUV and x-ray generation requires the droplet plasma to drop below critical density (or 'disassemble') within a few hundred picoseconds. We check this by scattering a very weak, variably delayed probe pulse from the plasma generated by the pump in the droplet stream. It is expected that the scattering is greatest when the plasma of an individual droplet is at or above critical density. The original 1064nm probe beam was frequency-doubled to 532 nm and its energy reduced to 1 mJ. The droplet plasma was formed by a 200-mJ pump pulse and the non-perturbing green probe light was focused collinearly with the pump beam and scattered from the droplets in the focus. Because of the differing wavelengths, the axial positions of the infrared and green foci differed slightly, and the green probe beam overfilled the pump interaction region. Images of the scattering region were relayed to the 90° CCD camera and recorded as a function of probe delay. An interference filter placed between the focus and the camera ensured that only the 532-nm scattered light was imaged. Figure 11 shows 20-frame averages of these images for 5-µm argon droplets at several time delays relative to the pump pulse. The speckle in the images originates from scattering from the individual droplets. An elongated region of reduced scattering, or channel, was seen to form within a few hundred picoseconds where the focused infrared pump light was most intense, indicating that the droplets in this region had mostly disassembled by this time. To quantify the amount of scattering as a function of time, the CCD signal was integrated inside a box which best represented the channel region. These integrated values are plotted on the right side of Fig. 11. A second-order exponential fit was made to these data, as with the EUV and x-ray data. The decay time is 265±60 ps, in agreement with the 190-ps decay time given by the Ar^{9+} EUV emission, and also in reasonable agreement with Ar droplet x-ray emission decay time.

Our pump-probe investigation of droplet plasmas has shown that the measured decay in the probe pulse yields of EUV and x-ray emission is at least as sensitive to the type of emission monitored as it is to the droplet species and mean size. We therefore infer that the optimum *single pulse duration* for best radiation generation efficiency is therefore determined by the nature of the emission as well as the droplet characteristics. The most efficient generation of x-ray or EUV *excitation* emission requires pulses with $\tau \leq \tau_{crit}$ (for our droplets, $\tau_{crit} \sim 100\text{-}300$ ps). On the other hand, the several nanosecond decay in the probe pulse yield of EUV *recombination* emission from droplet plasmas suggests that for a few nanosecond Q-switched laser pulse of a few hundred millijoules, reasonable efficiency may be achieved even though the droplet plasma may have fallen below critical density early in the pulse. Even though pulses with $\tau \leq \tau_{crit}$ will always be absorbed most efficiently, laser-heated maintenance of a reservoir of ions that feeds recombination (for example, Ar^{8+}) may not require the above-critical density and high temperature demanded by EUV excitation emission or x-ray emission. The most efficient pulse for generation of EUV recombination emission is therefore one whose duration is several effective recombination times.

ACKNOWLEDGEMENTS

The authors acknowledge useful discussions with C. Tarrio, who designed the condenser, and G. D. Kubiak. This work is supported by the National Science Foundation and the EUV-LLC.

REFERENCES

1. Ditmire, T., et al., *Phys. Rev. A* **53**, 3379 – 3402 (1996).
2. Tish, J.W.G., "Clusters in intense laser fields," in *Atoms, Solids, and Plasmas in Super-Intense Laser Fields*, edited by D. Batani et al., New York: Kluwer Academic / Plenum Publishers, 2001, pp. 99 – 118.
3. Ditmire, T., et al., *Phys. Rev. A* **57**, 369 – 382 (1998).
4. McPherson, A., et al., *Nature (London)* **370**, 631 – 634 (1994).
5. Lezius, M., et al., *Phys. Rev. Lett.* **80**, 261 – 264 (1998).
6. Tichenor, D.A., et al., "System integration and performance of the EUV engineering test stand," in *Emerging Lithographic Technologies V*, Proc. SPIE 4343, 19-37 (2001).
7. Fiedorowicz, H., et al., *Opt. Commun.* **184**, 161-7 (2000); Hertz, H.M., et al., "Liquid-target laser-plasma sources for EUV and X-ray lithography," in *EUV, X-ray, and Neutron Optics and Sources*, Proc. SPIE **3767**, 2-9 (1999); Richardson, M., et al., *Opt. Commun.* **145**, 109-12

(1998); Schriever, G., et al., *J. Vac. Sci. Technol. B* **17**, 2058-60 (1999); Shimoura, A., et al., *Appl. Phys. Lett.* **75**, 2026-28 (1999).

8. Kirz, J., et al., *Q. Rev. Biophys.* **28**, 33 (1995).
9. Levine, Z.H., et al., *Appl. Phys. Lett.* **74**, 150 – 152 (1999).
10. Parra, E., et al., *Rev. Sci. Instrum.* **73**, 468-75 (2002).
11. Smith, R.A., et al., *Rev. Sci. Instrum.* **69**, 3798-804 (1998).
12. Abraham, O., et al., *J. Chem. Phys.* **75**, 402 – 411 (1981).
13. Hagena, O.F., and Obert, W., *J. Chem. Phys.* **56**, 1793 – 1802 (1972).
14. Zweiback, J.S., Ph.D. thesis, University of California, 1999 (UVRL Report No. LR-134008).
15. Kelly, R.L., *J. Phys. Chem. Ref. Data Suppl.* **16**, 1 (1987); Shirai, T., et al., *J. Phys. Chem. Ref. Data* **24**, 1577 (1995).
16. Milchberg, H.M., et al., *Phys. Rev E* **64**, 056402 (2001).
17. Ginzburg, V.L., *Propagation of Electromagnetic Waves in Plasmas* New York: Pergamon, 1970.
18. Parra, E., et al., *Phys. Rev. E* **62**, R5931 – R5934 (2000).
19. McNaught, S.J., et al., *Appl. Phys. Lett.* **79**, 4100 – 4102 (2001).
20. Milchberg, H.M., and Freeman, R.R., *J. Opt. Soc. Am. B* **6**, 1351 – 1355 (1989).
21. Milchberg, H.M., et al., *Phys. Rev. Lett.* **61**, 2364 – 67 (1988).
22. Zweiback, J., et al., *Phys. Rev. A* **59**, R3166 (1999); Zweiback, J., et al., *Opt. Express* **6**, 236 – 242 (2000).

MAGNETIC CONFINEMENT FUSION

Understanding of Neutral Gas Transport in the Alcator C-Mod Tokamak Divertor

D. P. Stotler[*], C. S. Pitcher[†], C. J. Boswell[†], B. LaBombard[†], J. L. Terry[†], J. D. Elder[**] and S. Lisgo[**]

[*]Princeton Plasma Physics Laboratory, Princeton University, P. O. Box 451, Princeton, NJ 08543-0451, USA
[†]MIT Plasma Science and Fusion Center, NW17, Cambridge, MA 02139, USA
[**]University of Toronto Institute for Aerospace Studies, 4925 Dufferin St., Toronto M3H 5T6, Canada

Abstract. A series of experiments on the effect of divertor baffling on the Alcator C-Mod tokamak provides stringent tests on models of neutral gas transport in and around the divertor region. One attractive feature of these experiments is that a trial description of the background plasma can be constructed from experimental measurements using a simple model, allowing the neutral gas transport to be studied with a stand-alone code. The neutral-ion and neutral-neutral elastic scattering processes recently added to the DEGAS 2 Monte Carlo neutral transport code permit the neutral gas flow rates between the divertor and main chamber to be simulated more realistically than before. Nonetheless, the simulated neutral pressures are too low and the deuterium Balmer-α emission profiles differ qualitatively from those measured, indicating an incomplete understanding of the physical processes involved in the experiment. Some potential explanations are examined and opportunities for future exploration are highlighted. Improvements to atomic and surface physics data and models will play a role in the latter.

INTRODUCTION

The development of models for plasma transport and plasma-material interactions in the periphery of magnetically confined fusion devices aids in our understanding of the underlying physical processes and should lead to tools facilitating the design of future experiments [1]. An integral component of these models is a description of the behavior of neutral atoms and molecules generated by plasma-material interactions and volume recombination processes. Mean free paths for such atoms and molecules can be large relative to the plasma and device scale lengths, requiring a kinetic description of their transport. Furthermore, the physics and geometric detail needed for the desired solution accuracy vastly exceed that achievable by analytic means. Instead, the Monte Carlo techniques developed for simulating neutron transport have been adapted and extended [2] to fill this need. Monte Carlo codes have the advantage of permitting the incorporation of arbitrarily complicated geometry [3, 4] and physics into the simulations. The limit on the realism of these simulations usually hinges on the correctness of the physics involved.

The physics input to these codes can be broken down into plasma profiles, sources of neutrals, and atomic and surface physics models. The first two items, which include the electron and ion density, temperature, and flow velocity, can be obtained from analytic

CP635, *Atomic Processes in Plasmas: 13th APS Topical Conference*, edited by D. R. Schultz et al.
© 2002 American Institute of Physics 0-7354-0090-3/02/$19.00

expressions, edge plasma transport codes, or models based on direct experimental measurements. We focus on the last approach since it seems to be the least uncertain given the difficulty of solving the plasma transport equations in realistic situations. Issues of diagnostic interpretation do exist [5], but lie beyond the scope of this paper.

We will describe simulations of specific experiments on the Alcator C-Mod tokamak that test the atomic and surface physics models and data used in the Monte Carlo neutral transport code DEGAS 2 [6] and illustrate how those physical processes directly impact the observable quantities. The results of the simulations differ significantly from the corresponding experimental values. We will examine potential explanations for the discrepancies by determining the sensitivity of the simulated values to changes in the assumptions underlying the simulations.

We conclude that no single satisfactory explanation can be found within the conventional model of scrape-off layer plasma transport. More detailed and realistic descriptions of the atomic and surface physics processes occurring in magnetic fusion devices may lead to improved agreement with the experimental measurements or at least eliminate some of the existing uncertainties. However, a true resolution will likely require advances in plasma transport models.

ALCATOR C-MOD EXPERIMENTS

A series of experiments on the Alcator C-Mod tokamak has directly addressed the impact of neutral gas flows between the divertor region and the main chamber (Fig. 1) on the plasma behavior. Prior to these experiments, the C-Mod device had been considered to effectively baffle neutral hydrogen and impurity species generated in the divertor so that they cannot easily make their way into the main chamber, penetrate into the confined plasma and lead to a deterioration of its pressure through radiative cooling or through effects on plasma microinstabilities.

A set of bypass valves has been installed on C-Mod to permit the neutral conductance between the divertor and main chamber to be essentially doubled in as little as 20 ms. With the bypass valves closed, diagnostic openings provide intrinsic pathways for neutrals to reach the main chamber. A principal result of these experiments is that opening the bypass reduces the divertor neutral pressure by a factor of two [7]. The current flowing through the bypass from the divertor to the main chamber is thus inferred to remain constant. Even more surprisingly, the plasma parameters, Balmer-α emissions, and several global characteristics of the plasma do not change substantially either.

The conclusion drawn is significant and astonishing: the Alcator C-Mod divertor operates as if it were unbaffled [7]. Neutral transport simulations with DEGAS 2 undertaken to aid in the understanding of these experiments [4] supported a hypothesized explanation based on a one-dimensional model [8]. Subsequent experiments and analysis [9] suggest that we still do not have a complete understanding of these results.

For the purposes of this paper, the key point of [4] is that the absolute values of the simulated neutral pressure are an order of magnitude too low. The particular deuterium discharge upon which the simulations are based had a line average electron density $\overline{n}_e = 1.46 \times 10^{20}$ m^{-3}. With the bypass closed (open), the pressure in the divertor plenum

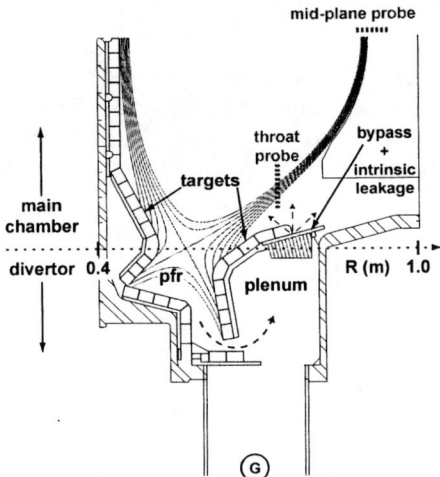

FIGURE 1. Poloidal cross section of lower half of Alcator C-Mod tokamak; R is the major radius. Plasma flows radially from closed flux surfaces across the separatrix and onto the open field lines. The "X-point" of the separatrix serves as a dividing line between the "main chamber" and "divertor" portions of the vacuum vessel. The divertor "targets" are the material surfaces first intersected by these open field lines. The flux surfaces below the X-point are referred to as the "private flux region" (pfr). Neutral atoms and molecules generated by plasma-material interactions are predominantly ionized by the plasma. Some find their way through gaps (dashed curves) in the vacuum vessel into the divertor "plenum" and ducts extending downward. The neutral pressure is measured by gauge "G". Intrinsic leakage pathways plus the bypass valves permit some neutrals to reach the main chamber from the plenum. Plasma parameters are measured by two upstream reciprocating probes and by Langmuir probes embedded in the targets.

measured by an absolute capacitance gauge (Fig. 1) was 30 (15) mTorr. The other key observable is the deuterium Balmer-α (D_α) emission pattern in the divertor region viewed by a radial array of downward looking detectors.

The extensive diagnostic set present on Alcator C-Mod permits the plasma parameters to be specified almost entirely by experimental measurements. The plasma conditions across the open flux surfaces adjacent to the core plasma ("upstream") are obtained from fast-scanning Langmuir-Mach probes at midplane and above the entrance to the divertor ("throat"). Fixed Langmuir probes in the target provide the plasma densities and temperatures there as well as the ion fluxes striking the target. Because the discharge density is well below the detachment density limit for the outer scrape-off layer, a simple "two point" model [4, 10] is expected to suffice for specifying the spatial variation of the plasma parameters between the probe locations. An ad hoc prescription for the plasma parameter variation through the private flux region was made [4] for the initial simulations. A more refined approach incorporating additional diagnostic data [11] will be discussed later in the paper.

The original simulations postulated neutral sources in the main chamber and exponential spatial decays for the plasma parameters away from the contiguous flux surfaces [4]. However, these details are not critical to this paper and have not been incorporated into the simulations described here.

DESCRIPTION OF SIMULATION AND BASELINE RESULTS

The geometry used in DEGAS 2 is built up from a simple outline of the tokamak vacuum vessel and a magnetic equilibrium computed for the C-Mod shot and time of interest [4]. The geometry and simulations are toroidally symmetric. While this is a reasonable, and typical, approximation for the plasma behavior, the vacuum vessel structure through which the neutral gas flows possesses significant non-axisymmetric features [5]. The implications of our approximation will be discussed at the end of the paper.

Two principal sources of neutrals enter into the simulations. The first is recycling of plasma fluxes striking the divertor target plates. Physically, incoming plasma ions are accelerated toward the targets through a narrow potential that naturally forms to slow the more rapid flow of electrons towards the surface, resulting in no net current to the surface. The ions are neutralized by the electron cloud near the surface prior to coming in contact with surface atoms. In DEGAS 2, an ion is sampled from a Maxwellian distribution at the local ion temperature and drift velocity. A value for the sheath potential is computed using one of a few different, simple models. The result is an increment, typically a few times the local electron temperature, to the ion's energy representing its acceleration through the sheath.

Specific models describing the interactions of neutral atoms, such as those arising from incident ions, and molecules with material surfaces are incorporated into DEGAS 2. In general, the output of these models is a characterization of one or more product velocity distributions and values for the probability of each. In particular, the binary collision code TRIM [12] provides probabilities for reflection of deuterium atoms on molybdenum (all surfaces in these simulations) as a function of incident energy and angle. The TRIM data also contain probability distributions characterizing the outgoing atom's energy and direction relative to the incident direction. Typical reflection coefficients are between 0.5 and 0.6; reflected atoms usually leave the surface with a significant fraction of their incident energy.

The magnitude of tokamak plasma fluxes is large enough that the near-surface layers of the targets are quickly saturated. Typically, the plasma density reaches a steady state without additional fueling [7]. The simulation assumes that the number of incident ions and atoms per second not reflected (i.e., are absorbed) is balanced by an equal rate of molecular desorption. These deuterium molecules have an energy characteristic of the wall temperature, taken to be 300 K. Their outgoing angular distribution is preferentially normal to the surface, but isotropic in azimuthal angle. All incident *molecules* are assumed to be absorbed, but likewise balanced by an equal rate of molecular desorption (i.e., no reflection).

The second neutral source in the simulations is volume recombination of deuterium ions and electrons. The recombination rate is obtained from a collisional-radiative model [13, 14] that also provides the rate for multi-step electron impact ionization of deuterium atoms. The collision cross-sections used in the model have been taken from [15]. The collisional-radiative model assumes that the divertor plasma is optically thin; we will address this issue later in the paper. The rates and kinetic treatment of molecular dissociation and ionization are described in [16].

Scattering of deuterium atoms and molecules off deuterium ions is effected with differential cross-sections calculated using state-of-the-art quantum mechanical techniques

254

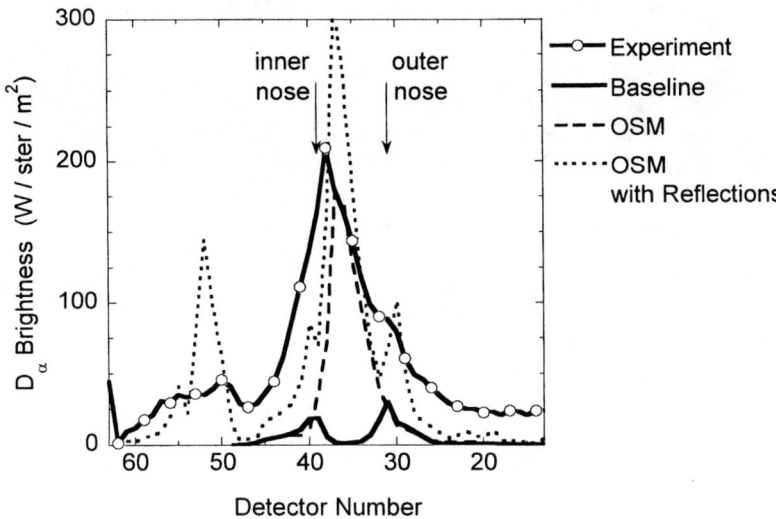

FIGURE 2. Comparison of experimental data from the array of downward looking D_α detectors with several simulations.

[17]. The atom - ion interaction incorporates both classically identifiable charge exchange and elastic scattering channels. For computational efficiency, a minimum scattering angle is enforced with a constraint that the momentum transport cross-section be unaltered [18]. The differential scattering is handled using cumulative probability tables for the cosine of the scattering angle.

A simple, iterative BGK treatment of neutral-neutral elastic scattering is used [18, 19]. For the observed pressures, the ratio of the neutral-neutral mean free path relative to a typical scale length can be as low as ~ 0.01 for molecules and > 1 for atoms. The transitional conditions occurring in the problem demand a nonlinear kinetic treatment similar to the one being used here.

The baseline simulations corresponding to the bypass open and closed states yield molecular pressures near the location of the gauge of 0.74 and 0.97 mTorr, respectively. Both results are more than an order of magnitude smaller than the measured values of 15 and 30 mTorr. Likewise, the baseline simulated D_α signal is everywhere well below the experimental curve (Fig. 2). Furthermore, the radial variation of the two curves is qualitatively different.

The trend of the neutral pressure and flow rate through the bypass as the bypass conductance is varied [4] are consistent with Pitcher's analytic model [8]. However, the magnitude of the pressure and D_α disagreement detracts from the credibility of the simulations as a comprehensive picture of the experimental behavior. Furthermore, they indicate a serious shortcoming or omission in some aspect of the neutral transport model (including, possibly, the input experimental data). Such under-predictions of the neutral pressure are not new [5, 20, 21]. We will now examine possible explanations for the current discrepancies.

ASSESSMENT OF MODEL COMPONENTS

Both the neutral pressure and the overall magnitude of the D_α emission profile scale directly with the strength of the neutral sources [4]. The D_α signal is generally indicative of a nearby source of neutral atoms. The light can be the result of a cascade down from the continuum (recombination) or due to excitation of atoms.

The two principal characteristics of the recycling source are the total current of particles reaching the divertor targets and the reflected fraction. For a given recycling current, the neutral pressure and D_α signals are sensitive to the plasma model used to interpolate between the Langmuir probe data. For example, a somewhat ad hoc lowering of the plasma density and temperature immediately in front of the target effectively increases the recycling source by reducing the fraction of recycled atoms that are ionized just as they come off the surface [4].

The rest of the connection between the recycling source and the neutral pressure measured some distance, perhaps meters, away is governed by the elastic interactions between the plasma ions, recycled atoms and molecules. Recycled ions reflected as atoms will have energies comparable to the local electron temperature (a few eV or larger). Atoms resulting from dissociation of molecules have Frank-Condon energies of 2 - 3 eV. In the plasma region, the atoms are also tightly coupled to the local ion population through elastic scattering, i.e., charge exchange. The result is an atom temperature in the divertor of roughly 2 eV. On the other hand, molecules start off at the wall temperature, equivalent to an energy of only 0.03 eV.

Elastic scattering transfers momentum from the plasma ions and warm atoms to the molecules, raising their pressure [8]. Turning off the elastic scattering processes between the molecules and other species in one sensitivity test resulted in more than a factor of two decrease in the total neutral pressure [4]. Their addition represents a significant improvement in the realism of the code's atomic physics [22]. The data underlying these processes and their implementation in DEGAS 2 have now been thoroughly validated [14, 18]. No further enhancements or modifications capable of explaining the neutral pressure discrepancy are foreseen.

The surface reflection coefficient determines the fraction of incident ions that are reflected and, hence, the balance between relatively fast atoms and slow molecules in the stream of recycled neutral particles. This balance in turn has an influence on the neutral pressure through the momentum exchanges mediated by the elastic scattering processes. Pitcher's semi-analytic model of the divertor bypass experiments produced molecular pressures of > 10 mTorr [8]. The biggest contributor to the apparent discrepancy with the complete DEGAS 2 runs is the semi-analytic model's simplified treatment of the interactions of the neutral species with material surfaces. We can emulate that treatment in DEGAS 2 by specularly reflecting all ions / atoms incident on the divertor targets. All other surfaces are taken to yield only desorbed molecules. The result is an increase of the neutral pressure from 0.97 mTorr to 1.72 mTorr. The D_α emission profile for this run does not differ significantly from the baseline curve in Fig. 2.

This sensitivity suggests taking a closer examination of the surface physics models used in DEGAS 2. However, because the material conditions found inside an operating tokamak are difficult to duplicate in the laboratory and because diagnosing tokamak plasma-material interaction in situ is even more problematic, improvements to the idealized descriptions currently used in the simulations will be slow in coming.

The source rate for recombination is determined by the plasma parameters input to DEGAS 2 via the collisional-radiative model for hydrogen. In the baseline simulation, the total source of recombined atoms is 1% of the ion flux to the targets. The resulting D_α emission profile peaks near the divertor targets, whereas the experimental signal possesses a single broad peak centered about the private flux region (Fig. 2). This discrepancy suggests a much stronger recombination source in the private flux region than is provided in the baseline model.

Tomographic reconstruction of the D_γ emission [23] also indicates much more recombination than is predicted with the simple plasma model used to set up the input to DEGAS 2. The D_γ emission peaks in the private flux region between the inner target and the X-point. Ad hoc manipulations of the private flux region plasma intended to replicate the peak D_γ emission result in a centrally peaked D_α emission pattern and an increase in the neutral pressure to 1.88 mTorr [4].

The Onion Skin Method (OSM) code [24] provides a more physically accurate approach to this task, as well as to estimating the plasma parameters between the probe locations in the main scrape-off layer [11]. The OSM model solves the plasma conservation equations in one dimension along field lines using the available experimental data as constraints. In the main scrape-off layer, the plasma sources due to neutral species are self-consistently computed with the EIRENE Monte Carlo neutral transport code. In the private flux region, the plasma description contains a number of free parameters that can be varied to yield an optimal match between the simulated and experimental D_α and D_γ profiles. The effect of Lyman-α radiation trapping on the recombination rate is estimated using the EIRENE-computed neutral density to determine the local photon mean free path. The resulting plasma densities and temperatures are then cross-checked with Stark broadening measurements. The result is again a complete, two-dimensional description of the plasma density and temperature built up from the available diagnostic data with a minimal number of assumptions.

Using this plasma as input to DEGAS 2 yields a neutral pressure of 2.02 mTorr and an improvement in the comparison with the experimental D_α emission pattern (Fig. 2). Despite the increased sophistication of the plasma model, the neutral pressure is again too low and the D_α profile still differs significantly from the one observed experimentally. Note that the DEGAS 2 results match those produced by EIRENE as part of the OSM solution procedure. The independent simulations with DEGAS 2 not only provide confirmation of the EIRENE results, but serve as useful comparisons with the other DEGAS 2 work described in this paper.

Two physical effects missing from this DEGAS 2 calculation are radiation trapping and light reflections off of the metal surfaces inside Alcator C-Mod. For typical Alcator C-Mod divertor parameters, Lyman-α and β are trapped [25]; Balmer-α is optically thin. Incorporating these effects into the code would lead to changes in recombination (similar to those in OSM-EIRENE) and ionization rates as well as to the ionization balance. In the DEGAS 2 simulations, the most significant impact of radiation trapping would be an increase in the ionization rate, leading to a *reduction* in the neutral pressure below that of the baseline simulation. Radiation trapping should be self-consistently incorporated into these simulations, although doing so may not improve the comparison with experimental data. An escape factor formalism [25] is feasible, but represents only a rough approximation. Some attempts at more sophisticated treatments have been made [25, 26]. A comprehensive, but practical, model for including radiation transport into divertor plasma and neutral transport is needed.

The second missing physical effect is the reflection of the D_α light off of the shiny molybdenum tiles in the Alcator C-Mod divertor. Such reflections complicate interpretation of the divertor-viewing chords. The reflectivity of molybdenum is roughly 50% at the 6560 Å D_α wavelength. To assess the potential importance of this effect, we extend the integration chords used in DEGAS 2 to simulate the divertor-viewing diagnostic to include several 50% specular reflections. The simulation based on the OSM plasma was then run again. As shown in Fig. 2, the reflections broaden the central D_α emission peak, but not enough to explain the discrepancy with the experimental data. Of course, this change to the simulated diagnostic signal has no effect on the neutral pressure. More realistic treatments of these reflections, including its incorporation into the OSM-EIRENE analysis, will be reported elsewhere.

DISCUSSION

Ro-vibrational excitation of molecules alters the rates and, hence, the relative importance of molecular and molecular ion reactions [27, 28]. Furthermore, via the "molecular assisted recombination" (MAR) mechanism [27], it can provide an additional source of recombination for deuterium ions. Heretofore, such effects have been largely ignored in divertor simulations. Collisional-radiative models involving vibrationally excited molecules have been assembled [28, 29] and compared with experimental data [27, 30, 31]. Additional and higher quality cross-sections are needed to complete these models [28]. Furthermore, the virtually non-existent data on rotationally excited states must be measured or computed and incorporated into them [27]. Because ro-vibrationally excited states are metastable on typical neutral transport time scales, they must be explicitly transported in codes such as DEGAS 2 and EIRENE. The consequent addition of 17 or more species to the code significantly increases the complexity and computational requirements of these simulations [30].

The importance of MAR to tokamak divertor plasmas remains contentious [27, 30]. Using the rates provided in [28], we estimate the neutral source due to MAR to be 2% of the total neutral source in the baseline DEGAS 2 simulation and 7% for the OSM plasma case. This incremental source is too small to explain the discrepancies in neutral pressure and D_α emission.

The structures through which neutral gas escape the divertor to reach the pressure gauges and the main chamber are not toroidally symmetric [5]. In particular, the gaps between the private flux region and the gas box, as well as the divertor bypass and intrinsic leakage pathways, all vary toroidally. Because a net neutral flow is maintained along the neutral pathway, pressure drops exist across these constrictions. The magnitude of the pressure drop is affected by the size of the opening and by the mean free path of the neutral species. The simulated gaps in DEGAS 2, however, are axisymmetric. Without performing three-dimensional simulations, determining the resulting impact on the neutral pressure is difficult. While the error could be significant, we suspect that the neutral pressure discrepancy is due to a vastly less subtle effect.

The rapidly evolving picture of scrape-off layer plasma transport may provide just such an effect. Recent observations of filamentary structures [32] and broad main chamber scrape-off layer profiles [33, 34] indicate that the "standard" characterization of the scrape-off layer as having diffusive transport and exponentially decaying plasma profiles is qualitatively incomplete. Furthermore, new data and associated analyses indicate that the bulk of the neutrals being recycled into the main chamber are arising from the main chamber walls, not from the divertor surfaces [33, 34], as had hitherto been supposed. The revised description of scrape-off layer behavior as intermittent and non-diffusive [35] invalidates some of the assumptions behind the models used to generate the plasma input to DEGAS 2. Furthermore, the strong ($\sim 100\%$) fluctuations reported in these observations imply correspondingly large fluctuations in the physical quantities measured by the diagnostic signals (e.g., the D_α emission profile). The absence of tens-of-kilohertz time resolution in these signals complicates the comparisons with the DEGAS 2 simulations. The origins of the discrepancies reported in this paper may only become clear once our global picture of scrape-off layer transport has been appropriately revised and issues such as diagnostic interpretation have been addressed.

SUMMARY

We have described neutral transport simulations of Alcator C-Mod experiments. Consistently low neutral pressures and D_α emission profiles qualitatively different from those observed indicate deficiencies in the physics model underlying the code. Aspects of the atomic and surface physics data have been examined as potential areas for future improvement. The surface physics model appears particularly deserving of more careful validation and / or increased realism. The effects of radiation trapping and ro-vibrationally excited molecules should be incorporated into the simulations; comprehensive and practical descriptions of these processes are needed. The greatest deficiencies in these simulations appear to lie in the models used to tie the experimental data together into the complete two-dimensional plasma profiles input to DEGAS 2. Experimental and theoretical efforts to better understand the plasma transport embodied in these models are currently underway.

ACKNOWLEDGMENTS

This work was supported by US DOE Contract Nos. DE-AC02-76-CHO-3073 (PPPL) and DE-FC02-99-ER5-4512 (MIT). The authors would also like to acknowledge support from IPP-Jülich, the Alcator C-Mod team, and the University of Toronto.

REFERENCES

1. ITER Physics Expert Group on Divertor, ITER Physics Expert Group on Divertor Modelling and Database, and ITER Physics Basis Editors, *Nucl. Fusion*, **39**, 2391–2469 (1999).

2. Tendler, M., and Heifetz, D., *Fusion Tech.*, **11**, 289–310 (1987).
3. Feng, Y., Sardei, F., and Kisslinger, J., *J. Nucl. Mater.*, **266-269**, 812–818 (1999).
4. Stotler, D. P., Pitcher, C. S., Boswell, C. J., Chung, T. K., LaBombard, B., Lipschultz, B., Terry, J. L., and Kanzleiter, R. J., *J. Nucl. Mater.*, **290–293**, 967–971 (2001).
5. Niemczewski, A., Hutchinson, I. H., LaBombard, B., Lipschultz, B., and McCracken, G. M., *Nucl. Fusion*, **37**, 151–163 (1997).
6. Stotler, D. P., and Karney, C. F. F., *Contrib. Plasma Phys.*, **34**, 392–397 (1994).
7. Pitcher, C. S., Boswell, C. J., Goetz, J. A., LaBombard, B., Lipschultz, B., Rice, J. E., and Terry, J. L., *Phys. Plasmas*, **7**, 1894–1903 (2000).
8. Pitcher, C. S., Boswell, C. J., Chung, T., Goetz, J. A., LaBombard, B., Lipschultz, B., Rice, J. E., Stotler, D. P., and Terry, J. L., *J. Nucl. Mater.*, **290–293**, 812–819 (2001).
9. Lipschultz, B., LaBombard, B., Pitcher, C. S., and Boivin, R., *Plasma Phys. Controlled Fusion* (2002).
10. Pitcher, C. S., and Stangeby, P. C., *Plasma Phys. Controlled Fusion*, **39**, 779 (1997).
11. Elder, J. D., Lisgo, S., Stangeby, P. C., LaBombard, B., Lipschultz, B., Terry, J. L., Pitcher, C. S., Boswell, C., Porter, G., and Reiter, D., *Bull. Am. Phys. Soc.*, **45**, 320 (2000).
12. Biersack, J. P., and Eckstein, W., *Appl. Phys.*, **34**, 73 (1984).
13. Weisheit, J. C., *J. Phys. B*, **8**, 2556–2564 (1975).
14. Degas 2 user's manual (2002), `http://w3.pppl.gov/degas2/Doc/degas2_all.pdf`.
15. Janev, R. K., and Smith, J. J., *At. Plasma-Mater. Interaction Data Fus.*, **4**, 1–180 (1993), supplement to the journal Nucl. Fus.
16. Stotler, D. P., Skinner, C. H., Budny, R. V., Ramsey, A. T., Ruzic, D. N., and Turkot, R. B., Jr., *Phys. Plasmas*, **3**, 4084–4094 (1996).
17. Krstic, P. S., and Schultz, D. R., *At. Plasma-Mater. Data Fus.*, **8**, 1 (1998), supplement to the journal Nucl. Fus.
18. Kanzleiter, R. J., Stotler, D. P., Karney, C. F. F., and Steiner, D., *Phys. Plasmas*, **8**, 5064–5069 (2000).
19. Reiter, D., et al., *J. Nucl. Mater.*, **241–243**, 342 (1997).
20. Haas, G., et al., *J. Nucl. Mater.*, **196–198**, 481–484 (1992).
21. Allen, S. L., Rensink, M. E., Hill, D. N., Perkins, D. E., Jackson, G. L., Mahdavi, M. A., and the DIII-D Research Team, *J. Nucl. Mater.*, **162–164**, 80–92 (1989).
22. Stotler, D. P., Pigarov, A. Y., Karney, C. F. F., Krasheninnikov, S. I., LaBombard, B., Lipschultz, B., McCracken, G. M., Niemczewski, A., Snipes, J. A., Terry, J. L., and Vesey, R. A., "DEGAS 2 neutral transport modeling of high density, low temperature plasmas," in *Proceedings of the Sixteenth International Conference on Fusion Energy*, International Atomic Energy Agency, IAEA, Vienna, 1997, vol. 2, pp. 633–639.
23. Boswell, C. J., Terry, J. L., LaBombard, B., Lipschultz, B., and Goetz, J. A., *J. Nucl. Mater.*, **290–293**, 556–560 (2001).
24. Fundamenski, W., Stangeby, P. C., and Elder, J. D., *J. Nucl. Mater.*, **266–269**, 1045–1050 (1999).
25. Terry, J. L., Lipschultz, B., Bonnin, X., Boswell, C., Krasheninnikov, S. I., Pigarov, A. Y., LaBombard, B., Pappas, D. A., and Scott, H. A., *J. Nucl. Mater.*, **266–269**, 30–36 (1999).
26. Scott, H. A., Wan, A. S., Post, D. E., Rensink, M. E., and Rognlien, T. D., *J. Nucl. Mater.*, **266–269**, 1247–1251 (1999).
27. Krasheninnikov, S. I., *Physica Scripta*, **T96**, 7–15 (2002).
28. Pigarov, A. Y., *Physica Scripta*, **T96**, 16–31 (2002).
29. Greenland, P. T., *J. Nucl. Mater.*, **290–293**, 615–618 (2001).
30. Fantz, U., Reiter, D., Heger, B., and Coster, D., *J. Nucl. Mater.*, **290–293**, 367–373 (2001).
31. Ohno, N., Nishijima, D., Takamura, S., Uesugi, Y., Motoyama, M., Hattori, N., Arakawa, H., Ezumi, N., Krasheninnikov, S., Pigarov, A., and Wenzel, U., *Nucl. Fusion*, **41**, 1055–1065 (2001).
32. Maqueda, R. J., Wurden, G. A., Zweben, S., Roquemore, L., Kugel, H., Johnson, D., Kaye, S., Sabbagh, S., and Maingi, R., *Rev. Sci. Instrum.*, **72**, 931–934 (2001).
33. Umansky, M. V., Krasheninnikov, S. I., LaBombard, B., and Terry, J. L., *Phys. Plasmas*, **5**, 3373–3376 (1998).
34. LaBombard, B., Boivin, R. L., Hughes, J., Lipschultz, B., Mossessian, D., Pitcher, C. S., Terry, J. L., Zweben, S. J., and Alcator Group, *Phys. Plasmas*, **8**, 2107–2117 (2001).
35. D'Ippolito, D. A., Myra, J. R., and Krasheninnikov, S. I., *Phys. Plasmas*, **9**, 222–233 (2002).

Atomic Line Shapes in the Presence of an External Magnetic Field

M.L. Adams[*], R.W. Lee, H.A. Scott, H.K. Chung[†] and L. Klein[**]

[*] *Department of Nuclear Engineering, Massachusetts Institute of Technology, NW16-230, 167 Albany St., Cambridge, MA 02139, USA*
[†] *University of California, Lawrence Livermore National Laboratory, P.O. Box 808, Livermore, CA 94551, USA*
[**] *Department of Physics and Astronomy, Howard University, Washington, DC 20059, USA*

Abstract. Both the theoretical basis and computational approach for extending the capabilities of a spectral line broadening code are presented. Following standard line broadening theory, the effects of an external magnetic field are incorporated into the atomic Hamiltonian and plasma average. In the presence of an external magnetic field the atomic Hamiltonian angular properties are altered — atomic energy levels are perturbed and the spectral emission line is polarized. The magnetic field introduces a preferential axis that changes the plasma average. These extensions have been incorporated into a new spectral line broadening code that is applied to several problems of importance to the understanding of tokamak edge plasmas. Applications fall into two broad categories: 1) determination of local plasma properties from distinct line shape features; and 2) consideration of global plasma phenomenon, such as radiation transport. Observable features of the Zeeman effect make H_α a good magnetic field diagnostic. H_β does not make a good electron density diagnostic since the Zeeman effect is comparable to the Stark effect for a majority of tokamak edge plasma conditions. When optically thick lines exist the details of the spectral line shapes are shown to significantly influence the transport of radiation throughout the system.

1. INTRODUCTION

Plasma properties in the tokamak edge typically span several orders of magnitude at any given time during a typical experiment [1, 2]. The system properties span the following range: $10^{12} < n_e(\text{cm}^{-3}) < 10^{16}$, $0.1 < T_e(\text{eV}) < 500.0$ and $0.1 < B(\text{T}) < 10.0$; where n_e is the electron density, T_e is the electron temperature and B is the magnetic field strength (assuming the permeability is one). While this large variation in plasma conditions poses many interesting theoretical and computational challenges, for the present spectral line shape study the high-density low-temperature (HDLT) edge plasma regions are of particular importance [3].

HDLT plasmas are found in detached divertor experiments near the divertor target plate and in MARFEs (an axisymmetric radiating phenomenon that is poloidally localized in the plasma edge). In Alcator C-Mod, HDLT plasma properties are of the order: $n_e \sim 10^{15}\text{cm}^{-3}$, $T_e \sim 1.0\text{eV}$, $B \sim 6\text{T}$ and $L \sim 5\text{cm}$; where L is a characteristic spatial extent of the region. What makes these plasmas especially interesting is that they are partially ionized (there exists a significant fraction of neutral particles in the edge plasma) and they interact strongly with line radiation (optical depths are much greater than unity). Thus, in HDLT plasmas the details of spectral line shapes will be important

CP635, *Atomic Processes in Plasmas: 13th APS Topical Conference*, edited by D. R. Schultz et al.
© 2002 American Institute of Physics 0-7354-0090-3/02/$19.00

to the spatial and frequency variation of the radiation field, which in turn influences the spatial variation in level populations and ultimately the transport of (non-radiative) energy. It is this connection between energy transport and the details of spectral line shapes that motivates this research.

This paper proceeds as follows. *Section 2* begins by reviewing standard line broadening theoretical and computational methods [4]. The effects of an external magnetic field are then systematically incorporated into both the atomic Hamiltonian and the plasma average. The resulting code is named TOTALB. *Section 3* applies TOTALB several problems of importance to the understanding of tokamak edge plasmas. *Section 4* provides concluding remarks and proposes potential future research directions.

2. LINE SHAPE FORMALISM

In this section the pertinent elements of the current embodiment of atomic spectral line shape calculations are discussed. Then effects of an external magnetic field on the plasma average and atomic Hamiltonian are considered. We note that while the atomic Hamiltonian modification is complete, future work is needed to rigorously incorporate magnetic effects into the plasma average.

2.1. General spectral line shape

Following Baranger [5] and Kolb and Griem [6], it is customary to begin with the quantum electrodynamical formula for the power radiated by an atom in an electric dipole transition from an initial state (i) to a final state (f):

$$P(\omega) = \frac{4\omega^4}{3c^3}\phi(\omega) \tag{1}$$

$$\phi(\omega) = \frac{1}{2\pi}\int_{-\infty}^{\infty}\Phi(t)e^{i\omega t}dt \tag{2}$$

$$\Phi(t) = \sum_{if}\rho_i\langle\psi_i(0)|\mathbf{d}|\psi_f(0)\rangle\langle\psi_f(t)|\mathbf{d}|\psi_i(t)\rangle \tag{3}$$

where $\phi(\omega)$ is the spectral line shape (normalized to unity $\int\phi(\omega)d\omega = 1$) and $\Phi(t)$ is the time correlation function. $\sum_{if}$ represents a sum over all final states and an average over all initial states that contribute to the line shape, ρ_i is the initial density matrix of the system, $|\psi(t)\rangle$ represents the state (in the Schrödinger picture) of the system and $\mathbf{d}$ is the time-independent electric dipole moment of the system. The quantity $\frac{4\omega^4}{3c^3}$ can be added after the calculation of $\phi(\omega)$ since the frequency variation of ω^4 is small compared to $\phi(\omega)$; this transfers the focus from the power radiated to the spectral line shape. To connect with dynamical theories the initial focus is further transferred to $\Phi(t)$.

Following Fano [7, 8], the correlation function can be simplified by first using general operator techniques and then separating an emitting atom (a) from its surrounding plasma (p). This allows the general correlation function to be written in the form:

$$\Phi(t) = \text{Tr}_a \left[\mathbf{d} \exp \left\{ -\frac{i}{\hbar}(L^a - i\Gamma)t \right\} \rho_i^a \mathbf{d} \right]_{Av} \tag{4}$$

where L^a is the Liouville operator pertaining to the atomic Hamiltonian, Γ is the relaxation theory collisional operator, ρ^a is the initial atomic density matrix, the trace is performed over all atomic states and the average (Av) is over plasma states.

Following Griem [9], the spectral line shape calculation considers those electric dipole transitions between bound states that contribute to a well defined frequency range. The associated atomic states are grouped into an upper manifold (with states $|\alpha\rangle$) and a lower manifold (with states $|\beta\rangle$). Dipole transitions between states of the same manifold and collisional transitions between different manifolds are not considered. The plasma average is handled by simultaneously introducing an electron model and an ion microfield model into the correlation function:

$$\Phi(t) = \int P(E) \langle\langle \alpha\beta | \mathbf{d} \exp \left\{ -\frac{i}{\hbar}(L_E^a - i\Gamma)t \right\} \rho_i^a \mathbf{d} | \alpha'\beta' \rangle\rangle dE \tag{5}$$

where $P(E)$ is the quasi-static electric ion microfield distribution function, L_E^a is the Liouville operator pertaining to the atomic Hamiltonian for a particular ionic field (E), Γ is now the electron collision operator and $|\alpha\beta\rangle\rangle$ represents the atomic state in Liouville (or line) space. Regardless of the specific models used in the plasma average there exists a general computational approach for solving this problem.

Following Calisti *et al.* [4], the spectral line shape computational approach centers around solving the complex eigenvalue problem for a fixed ionic field and then averaging over possible ionic fields. Since the correlation function must be real, which requires that $\phi(-\omega) = \phi^*(\omega)$, the line shape equation can be written:

$$\phi(\omega) = \frac{1}{\pi}\text{Re} \int_0^\infty \Phi(t) e^{i\omega t} dt \tag{6}$$

$$= \int P(E) \frac{1}{\pi}\text{Re} \int_0^\infty \langle\langle \alpha\beta | \mathbf{d} \exp \left\{ -\frac{i}{\hbar}(L_E^a - i\Gamma)t \right\} \rho_i^a \mathbf{d} | \alpha'\beta' \rangle\rangle e^{i\omega t} dt dE \tag{7}$$

$$= \sum_{j=1}^{n_E} W_j \sum_k \frac{c_{1k}[\omega - x_k(E_j)] + c_{2k} y_k(E_j)}{[\omega - x_k(E_j)]^2 + y_k(E_j)^2} \tag{8}$$

where the integration over E has been replaced by a sum that represents an average over discrete ionic fields with weights given by W_j. Let $|k\rangle\rangle$ be the basis in which $L_{E_j}^a - i\Gamma$ is diagonal, obtained through use of a transformation matrix T_{E_j}, with complex eigenvalues $z_k(E_j) = x_k(E_j) + i y_k(E_j)$. The coefficients c_{k1} and c_{k2} depend on $\mathbf{d}$ and T_{E_j}. In order to implement this formula atomic structure information is needed.

An optimal set of atomic data is obtained by using the fact that atomic structure information only enters in the selection of states that contribute to radiation in a specific frequency range and the evaluation of the Liouville operator matrix elements. In Liouville space, or line space, the Liouville operator matrix elements are:

$$\langle\langle \alpha\beta | L_E | \alpha\beta \rangle\rangle = \hbar\omega_{\alpha\beta} \tag{9}$$

$$\langle\langle \alpha\beta | L_E | \alpha\beta' \rangle\rangle = \langle \beta | \mathbf{d} | \beta' \rangle E \tag{10}$$

$$\langle\langle \alpha\beta | L_E | \alpha'\beta \rangle\rangle = -\langle \alpha | \mathbf{d} | \alpha' \rangle E \tag{11}$$

where $\hbar\omega_{\alpha\beta}$ is the energy difference between states $|\alpha\rangle$ and $|\beta\rangle$, $\langle\beta|\mathbf{d}|\alpha\rangle$ is the electric dipole matrix element in Hilbert space and E is the ionic field strength. Thus, numerically speaking, the atomic data information must include: 1) an index to label each state; 2) quantum information ascribable to each state; and 3) electric dipole matrix elements between states.

From the diagonal element of the Liouville operator matrix elements, it is clear that energy level information is needed to describe each transition; this information is also needed for the selection of states mentioned above. Regarding the electric dipole matrix elements, in their current form they require complete specification of the relevant quantum numbers of the states involved. Due to the spherical symmetry properties of atoms with complex atomic structure the matrix elements can be simplified by using the Wigner-Eckart theorem [10] as follows:

$$\langle\gamma JM|\mathbf{d}_q^{(1)}|\gamma'J'M'\rangle = (-1)^{J-M} \begin{pmatrix} J & 1 & J' \\ -M & q & M' \end{pmatrix} \langle\gamma J\|\mathbf{d}^{(1)}\|\gamma'J'\rangle \tag{12}$$

where tensor-operator notation has been explicitly used, the second term on the right-hand-side (RHS) is a Wigner 3-j symbol and the square of the last reduced matrix element on the RHS is known as the electric dipole line strength. The electric dipole line strength includes radial matrix information and is independent of angular momentum coupling schemes — a detail that allows atomic structure codes based on either LS or JJ angular momentum coupling schemes to generate data for the line shape code. Now the only atomic data necessary to calculate the electric dipole matrix elements are the $2J+1$ value of each state, which specifies the possible orientations of the 3-j symbol, and the reduced matrix elements between states.

2.2. Plasma average modifications

The plasma average consists of two parts: the electron collision operator, which is calculated within the framework of binary collision relaxation theory, and the quasi-static electric ion microfield. An external magnetic field introduces a preferential axis into the system that destroys the arbitrary orientation of the electric dipole operator and can alter the dynamical properties of the plasma.

The quasi-static ion microfield model assumes that the emitting atom is surrounded by a spherically symmetric plasma. The electric dipole moment can be chosen to point in an arbitrary direction and the ion microfield distribution must be a function only of ionic field strength. With the introduction of a preferential axis, the spherical symmetry is broken and the integration over the microfield distribution becomes anisotropic:

$$P(E)dE \rightarrow P'(E_\perp, E_\parallel)dE_\perp dE_\parallel \tag{13}$$

where the subscripts refer to the direction of the magnetic field. This modification amounts to an average over electric field directions relative to the magnetic axis. The modification increases computational time by introducing a double integration into the plasma average and tripling the number of non-zero electric dipole matrix elements that enter the complex eigenvalue problem. Note that while this modification to the microfield integration has been made, the microfield itself is assumed unchanged by the introduction of a magnetic field. The following discussion considers the limitations of this model.

Ion dynamics might be important when an external magnetic field is introduced into the spectral line shape formalism. The quasi-static ion microfield model is valid provided the half-width at half-maximum (HWHM) of the normalized line profile ($\Delta\omega_{1/2}$) is much greater than the ion plasma frequency (ω_{pi}). In the presence of a magnetic field the Lorentz force ($q\mathbf{v} \times \mathbf{B}$) increases the plasma frequency and decreases the region of validity of the quasi-static ion microfield model [11]. Thus, a more general criterion for neglecting ion dynamics, which accounts for the asymmetry of the ion plasma frequency, is:

$$\Delta\omega_{1/2} \gg \omega_{LH} \geq \omega_{pi} \qquad (14)$$

where the lower hybrid resonant frequency (ω_{LH}) is the ion plasma frequency perpendicular to the magnetic field. If ion dynamics is important then anisotropy should be considered further.

The electron collision operator is valid provided $\Delta\omega_{1/2}$ is much less than the electron plasma frequency (ω_{pe}). In the presence of a magnetic field the Lorentz force increases the plasma frequency and increases the region of validity of the electron collision operator. The general condition for the validity of the electron collision operator can be expressed as:

$$\Delta\omega_{1/2} \ll \omega_{pe} \leq \omega_{UH} \qquad (15)$$

where the hybrid resonant frequency (ω_{UH}) is the electron plasma frequency perpendicular to the magnetic field. While the external magnetic field strengthens the condition it also introduces anisotropy.

The consideration of magnetic effects on the motion of charged particles results in a constraint on the binary collision relaxation model straight-line path (SLP) approximation. The condition for validity of the SLP approximation is that the Larmor radius must be greater than the Debye length ($r_L > \lambda_D$); or similarly the Larmor frequency must be greater than the electron plasma frequency ($\omega_L = \frac{qB}{2m_e} < \omega_{pe}$). Explicitly writing the plasma property dependence involved in this condition places an upper limit on the magnetic field strength: $B(T) < 4.5\sqrt{n_{e14}}$, where n_{e14} is in units of 10^{14}cm^{-3}. Thus, for most HDLT edge plasma phenomenon the SLP approximation is valid.

2.3. Atomic Hamiltonian modification

In the presence of an external magnetic field the atomic Hamiltonian angular properties are altered — atomic energy levels are perturbed and the line shape is polarized. We use the non-relativistic one-electron atom Hamiltonian to illustrate this modification and at the same time introduce the approximations employed. Generalization to atoms with complex structure is then accomplished through the use of Racah algebra.

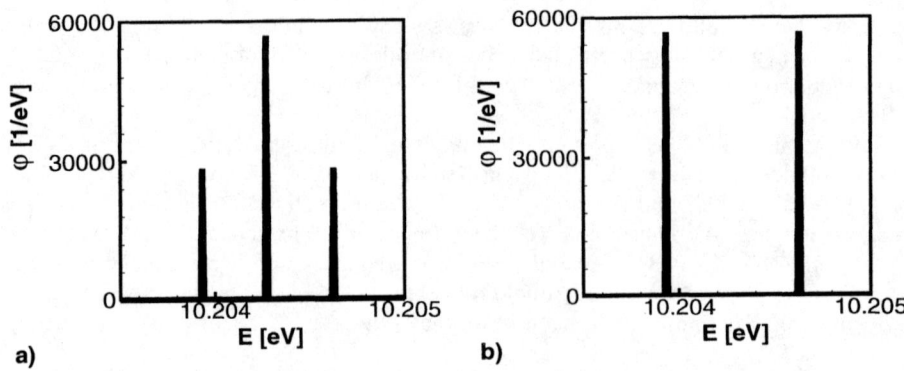

FIGURE 1. Zeeman components of Hydrogen Ly_α in a $B = 6T$ magnetic field. The observation direction is a) perpendicular to the magnetic axis ($\beta = \pi/2$) and b) along the magnetic axis ($\beta = 0$).

The atomic Hamiltonian for a one-electron atom in a uniform external magnetic field, with vector potential $\mathbf{A} = -\frac{1}{2}\mathbf{R} \times \mathbf{B}$, is given by:

$$H = \frac{1}{2m_e}[\mathbf{P} - q\mathbf{A}(\mathbf{R})]^2 + V(\mathbf{R}) \tag{16}$$

$$= \left\{\frac{P^2}{2m_e} + V(\mathbf{R})\right\} - \frac{\mu_B}{\hbar}\mathbf{L}\cdot\mathbf{B} + \frac{q^2B^2}{8m_e}R_\perp^2 \tag{17}$$

where $\mu_B = \frac{q\hbar}{2m_e}$ is the Bohr magneton, $\mathbf{L} = \mathbf{R} \times \mathbf{B}$ and $R_\perp^2 = R^2 - \frac{(\mathbf{R}\cdot\mathbf{B})^2}{B^2}$ is the projection of $\mathbf{R}$ onto the plane perpendicular to $\mathbf{B}$ [12]. The term proportional to B is called the paramagnetic term; the term proportional to B^2 is called the diamagnetic term. Assuming B is small ($B < 10^3$T) the following energy ordering applies:

$$\frac{\Delta E_1}{\Delta E_0} \sim \frac{\Delta E_2}{\Delta E_1} \ll 1 \tag{18}$$

where the subscript denotes order with respect to B. Such an energy ordering is valid for magnetic confinement fusion, where $B < 10$T, and can be written in terms of plasma properties as:

$$\mu_B B = \hbar\omega_L \ll E_I = \frac{1}{2}\alpha^2 m_e c^2 \tag{19}$$

where α is the fine structure constant. Thus, the diamagnetic term is negligible compared to the paramagnetic term and will not be considered.

The change in frequency and polarization of the emitted radiation due to the paramagnetic term is collectively called the Zeeman effect. These effects are illustrated in Fig. 1. The line shapes depicted in Fig. 1 were calculated with TOTALB using non-relativistic atomic data and neglecting both plasma broadening and Doppler broadening. The transition energies and relative magnitudes agree with first order perturbation theory [12].

For atoms with complex atomic structure, the paramagnetic term only affects angular properties of the atom. The general paramagnetic term can be written as:

$$\mathbf{L} \cdot \mathbf{B} \rightarrow [\mathbf{J} + (g_s - 1)\mathbf{S}] \cdot \mathbf{B} \tag{20}$$

where $g_s \cong 2.00232$ is the anomalous gyromagnetic ratio for electron spin. Now the complete atomic Hamiltonian can be written as:

$$H = H_0 + H_S + H_Z \tag{21}$$

where H_0 represents the full complex atomic Hamiltonian (that may include fine structure effects); H_S (Stark) and H_Z (Zeeman) are given by:

$$H_S = -\mathbf{E} \cdot \mathbf{d}^{(1)} \tag{22}$$
$$H_Z = -\mathbf{B} \cdot \boldsymbol{\mu}^{(1)} = \mu_B \mathbf{B} \cdot [\mathbf{J} + (g_s - 1)\mathbf{S}] \tag{23}$$

To evaluate the paramagnetic matrix element let $\mathbf{B} = B\hat{z}$ and $g_s = 2$. The matrix element takes the form:

$$\frac{1}{\mu_B B} \langle \gamma J M | H_Z | \gamma' J' M' \rangle = \langle \gamma J M | J_0^{(1)} + S_0^{(1)} | \gamma' J' M' \rangle \tag{24}$$

In a manner similar to the electric dipole matrix element this equation can be simplified by application of the Wigner-Eckart theorem:

$$\langle \gamma J M | J_0^{(1)} + S_0^{(1)} | \gamma' J' M' \rangle = (-1)^{J-M} \begin{pmatrix} J & 1 & J' \\ -M & 0 & M' \end{pmatrix} \langle \gamma J || J^{(1)} + S^{(1)} || \gamma' J' \rangle \tag{25}$$

Analogous to the electric dipole line strength the square of the reduced matrix element on the RHS is called the magnetic dipole line strength. Note that the magnetic dipole line strength also includes radial matrix information and is not dependent on an angular momentum coupling scheme.

3. TOKAMAK APPLICATIONS

Applications of TOTALB fall into two categories: 1) determination of local plasma properties from distinct line shape features; and 2) consideration of global plasma phenomenon, such as particle and energy transport. This section considers both types of applications. In the numerical calculations that follow we use atomic data from the relativistic complex atomic structure code HULLAC (Hebrew University Lawrence Livermore Atomic Code) [13].

3.1. Hydrogen resonance line

In HDLT plasmas the resonance line of Hydrogen (Ly_α) has a significant optical depth ($\tau \sim \chi L > 1$, where χ is the line radiation absorption coefficient and L is a length characterizing the extent of the plasma region). Since details of the spectral line shape will affect the plasma opacity, it is interesting to quantify magnetic field effects on τ.

To begin, the line radiation absorption coefficient for a plasma in local thermodynamic equilibrium (LTE) is [14]:

$$\chi(\nu) = n_i B_{ij} \frac{h\nu_{ij}}{4\pi} \phi_{ij}(\nu) \left[1 - \exp\left\{-\frac{h\nu_{ij}}{T}\right\}\right] \sim n_i \left\{\frac{e^2}{4\pi\varepsilon_0} \frac{\pi}{mc} \phi_{ij}(\nu)\right\} f_{ij} \quad (26)$$

$$\chi_0 \sim n_{i14} \left\{\frac{0.011}{\Delta E_{eV}}\right\} f_{ij} \left[\text{cm}^{-1}\right] \quad (27)$$

where i (j) refers the lower (upper) level of a bound-bound transition, B is the Einstein B coefficient, $h\nu_{ij}$ is the line center energy, ϕ is the line profile, T is the thermodynamic temperature, f is the line absorption oscillator strength, χ_0 is the line center absorption coefficient and ΔE is the full-width at half-maximum (FWHM) of the line shape in units of electron-Volts (eV). In Eqs. 26 and 27 the terms in curly braces represent an effective cross section for photon absorption. The latter equation clearly shows that ΔE and χ_0 are inversely proportional; as ΔE increases the cross section decreases and hence χ_0 decreases. That is, as a line shape becomes broader the line absorption oscillator strength is distributed over a greater frequency range. With χ_0 given by Eq. 27 the optical depth can be estimated with a suitable choice of L.

Now the effect of an external magnetic field on opacity can be quantified by using TO-TALB. As an example, we consider the Hydrogen Ly_α line shape for an HDLT plasma ($n_e = 10^{15}\text{cm}^{-3}$ and $T_e = 1\text{eV}$) with ($B = 6\text{T}$) and without ($B = 0\text{T}$) an external magnetic field. As the magnetic field increased from $B = 0\text{T}$ to 6T the FWHM increased from $\Delta E = 0.0009\text{eV}$ to 0.0011eV. Setting $L = 1\text{cm}$ and choosing a ground state density of $n_1 = 10^{14}\text{cm}^{-3}$ yields values without (with) magnetic effects: an absorption coefficient of 5.1cm^{-1} (4.2cm^{-1}); effective photon mean free path of 0.20cm (0.24cm); an optical depth of 5.1 (4.2). Thus, magnetic line broadening reduced the optical depth by approximately 18%.

3.2. Magnetic field diagnostic

In Eq. 21 there are three terms on the RHS: H_0 pertains to the energy of a state; H_S leads to the Stark effect that is proportional to the plasma density as well as the principle quantum number (which scales approximately as n^2) [10]; and, H_Z leads to the Zeeman effect and is proportional to **B**. Thus, the shift and intensity of a spectral line shape component are dependent on the relative contributions of both the Stark effect and the Zeeman effect. Furthermore, the intensity of a component will vary with angle of observation relative to the magnetic axis due to the polarization properties of the Zeeman effect. From these line shape features, if the Zeeman effect dominates the Stark effect then both the magnetic field strength and angle of the magnetic field relative to the direction of observation can be accurately determined using TOTALB.

For HDLT plasmas in the Alcator C-Mod tokamak H_α is a prime magnetic field diagnostic candidate. High-resolution H_α measurements are made routinely in Alcator C-Mod [15] and the data clearly exhibit the Zeeman effect signature [3]. While these measurements were intended to study the Hydrogen to Deuterium ratio and plasma fluctuations, they are ideally suited for magnetic field diagnostic purposes.

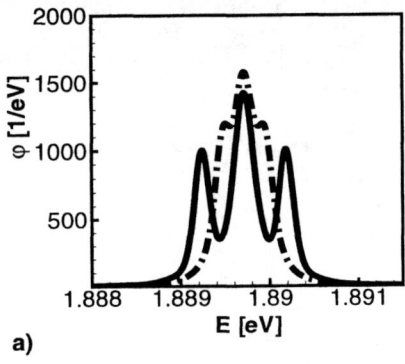
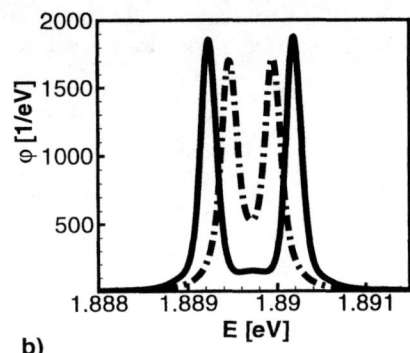

FIGURE 2. a) Magnetic field strength variation from $B = 4T$ (dash-dot line) to $B = 8T$ (solid line). Plasma conditions are: $n_e = 10^{15} cm^{-3}$ and $T_e = 1eV$. The angle of observation is perpendicular to the magnetic axis ($\beta = \pi/2$). b) Same plasma conditions as in (a) but now the angle of observation is along the magnetic axis ($\beta = 0$).

To demonstrate the magnetic field diagnostic procedures, we consider first the determination of the magnetic field strength. Figure 2a shows the sensitivity of H_α to changes in magnetic field strength for conditions typical of an HDLT plasma. For this line shape, calculated for an angle of observation perpendicular to the magnetic axis, determination of the magnetic field strength is performed by numerically matching the energy difference between the central π-component and a σ-component. This is not the same as inferring the magnetic field from the first-order Zeeman effect alone ($\Delta E_{1/2} = \mu_B B$) since this would be valid only in the absence of plasma broadening. To quantify the error made by neglecting plasma broadening, the magnetic field strength calculated using only the first-order Zeeman formula with $\Delta E_{1/2}$ taken from the $B = 8T$ curve in Fig. 2a would be 8.21T. In this case the result is an overestimate of the magnetic field strength by 0.21T — resulting in a significant difference in the inferred emission region in tokamaks.

Figure. 2b shows the line shape from an angle of observation along the magnetic axis, with the same plasma conditions as in Fig. 2a, and reveals a better candidate for determining magnetic field strength. The absence of the central π-component allows the diagnostic to be applied for lower magnetic field tokamaks. Figure 2b also illustrates the variation in intensity, but not shift, of the line profile components with angle of observation relative to the magnetic axis. It is this feature that can be used to determine β by numerically matching the relative intensity of the π-component to a σ-component.

3.3. Electron density diagnostic

In plasmas without an external magnetic field the spectral line width is commonly used to determine plasma densities [16]. For instance, Griem recommends H_β as an excellent candidate for determining the electron density in plasmas where $n_e > 10^{14}\text{cm}^{-3}$. In this density regime the linear Stark effect dominates over other broadening mechanisms and the FWHM scales as $n_e^{2/3}$. Below an electron density of 10^{14}cm^{-3} fine structure effects begin to enter and the dominant FWHM dependence makes a transition from the linear Stark effect to quadratic ion impact [17].

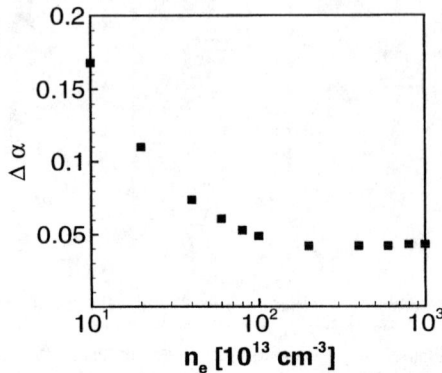

FIGURE 3. A plot of the H_β HWHM including the $n_e^{2/3}$ scaling ($\Delta\alpha = \Delta\lambda_{1/2}/n_e^{2/3}$) as a function of electron density. The curve asymptotes when the linear Stark effect dominates other broadening mechanisms. The plasma properties are: $B = 6\text{T}$, $T_e = 2\text{eV}$ and $\beta = \pi/2$.

In plasmas with an external magnetic field the Stark effect competes with the Zeeman effect. Günter-Könies [18] has investigated the feasibility of H_β as a density diagnostic for low magnetic field tokamaks ($B \sim 2\text{T}$) and found that it is useful provided $n_e > 2 \times 10^{14}\text{cm}^{-3}$. Figure 3, which was calculated for conditions typical of Alcator C-Mod ($B = 6\text{T}$), indicates that H_β is not useful as an electron density diagnostic unless $n_e > 2 \times 10^{15}\text{cm}^{-3}$. Future density diagnostic efforts would have better success considering high-n lines that take advantage of the n^2 Stark scaling, e.g., $H_{8\to2}$.

3.4. Finite plasma slab

HDLT tokamak edge plasmas strongly interact with hydrogen line radiation. Thus, in HDLT plasmas the details of spectral line shapes will determine the spatial and frequency variation of opacity, which in turn influences the spatial variation in level populations and ultimately the transport of energy. Through full integration of TOTALB into a coupled non-local thermodynamic (NLTE) atomic kinetics and radiation transfer code CRETIN [19], the effects of magnetically broadened line shapes can be quantified.

We consider a one-dimensional plasma slab in the x-y plane with a thickness of 1cm and plasma properties: $\mathbf{B} = 6\hat{x}\text{T}$, $T_e = 1.0\text{eV}$, $n_e = 10^{15}\text{cm}^{-3}$ and $n_1 = 10^{14}\text{cm}^{-3}$ (ground state neutral hydrogen density). Figure 4a demonstrates the Ly_α emissivity variation with angle relative to the magnetic axis. Figure 4b plots the spectral radiation intensity escaping the finite plasma slab with (solid line) and without (dotted line) magnetic field

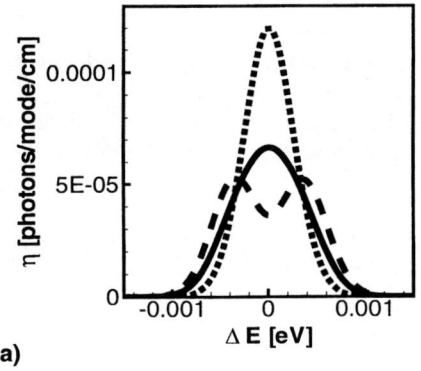
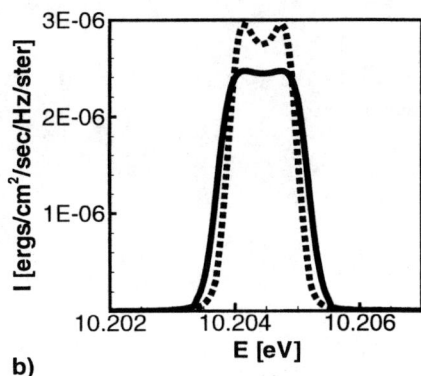

FIGURE 4. a) Ly_α emissivity at the center of the plasma slab. The solid curve ($\beta = \pi/2$) and the dashed curve ($\beta = 0$) are magnetically broadened lines shapes; the dotted line is without magnetic field effects included. b) Ly_α spectral radiation intensity escaping the finite plasma slab with (solid line) and without (dotted line) magnetic field effects.

effects. The integrated intensity increased by approximately 10% — this difference will increase with optical depth. Magnetic field effects will not directly affect the integrated intensity of optically thin lines. All the emitted photons in an optically thin line escape and their distribution relative to line center is irrelevant. However, optically thick lines can affect the excited state populations leading to changes in emission from optically thin lines, as well as ionization, recombination and energy loss rates [20].

4. DISCUSSION

The effects of an external magnetic field have been added to the spectral line broadening model TOTAL [4] and the goal of producing a numerically fast magnetically broadened line shape model for inclusion into large transport codes has been reached. However, while the atomic Hamiltonian modifications are complete, future work in needed to rigorously incorporate magnetic effects into the plasma average.

Since the region of validity of the quasi-static ion microfield is restricted by anisotropic ion dynamics, future research is needed to determine the role of $P(E_{\parallel}, E_{\perp})$, the conditional probability, in regimes where ion dynamics may become important. Since the SLP approximation is out of the limit of validity, research is needed to look at anisotropic electron collision dynamics.

While these areas of research limit the general applicability of TOTALB, there still exist numerous applications to tokamaks edge plasmas, as well as other systems in plasma physics, condensed matter physics and astrophysics. For tokamak applications, simulation of atomic spectral lines, such as carbon and boron, is now possible and can produce numerous direct diagnostic techniques. More interesting to next generation fusion experiments is the potential to quantify radiation effects on edge plasma transport.

ACKNOWLEDGMENTS

MLA would like to thank A.S. Wan (LLNL) and J.P. Freidberg (MIT) for their valuable contributions.

This work was performed under the auspices of the U.S. Department of Energy by the University of California, Lawrence Livermore National Laboratory through contract number W-7405-Eng-48.

REFERENCES

1. Pitcher, C. S., and Stangeby, P. C., *Plasma Physics and Controlled Fusion*, **39**, 779 (1997).
2. ITER Physics Basis, *Nuclear Fusion*, **39**, 2391 (1999).
3. Adams, M. L., Scott, H. A., Lee, R. W., Terry, J. L., Marmar, E. S., Lipschultz, B., Yu. Pigarov, A., and Freidberg, J. P., *JQSRT*, **71**, 115 (2001).
4. Calisti, A., Khelfaoui, F., Stamm, R., Talin, B., and Lee, R. W., *Phys. Rev. A*, **42**, 5433 (1990).
5. Baranger, M., *Phys. Rev.*, **111**, 494 (1958).
6. Kolb, A. C., and Griem, H. R., *Phys. Rev.*, **111**, 514 (1958).
7. Fano, U., *Rev. Mod. Phys.*, **29**, 74 (1957).
8. Fano, U., *Phys. Rev.*, **131**, 259 (1963).
9. Griem, H. R., *Spectral Line Broadening by Plasmas*, Academic Press, New York, 1979.
10. Cowan, R. D., *The Theory of Atomic Structure and Spectra*, University of California Press, Berkeley, 1981.
11. Chen, F. F., *Introduction to Plasma Physics and Controlled Fusion*, Plenum Press, New York, 1984, 2 edn.
12. Cohen-Tannoudji, C., Diu, B., and Laloë, F., *Quantum Mechanics*, John Wiley & Sons, New York, 1977.
13. Bar-Shalom, A., Klapisch, M., and Oreg, J., *JQSRT*, **71**, 169 (2001).
14. Mihalas, D., *Stellar Atmospheres*, W.H. Freeman and Company, New York, 1978, 2 edn.
15. Marmar, E. S., Zweben, S. J., Fulghum, S., and Rostler, P. S., *Rev. Sci. Instrum.*, **70**, 1 (1999).
16. Griem, H. R., *Principles of Plasma Spectroscopy*, Cambridge University Press, New York, 1997.
17. Kelleher, D. E., Wiese, W. L., Helbig, V., Greene, R. L., and Oza, D. H., *Physica. Scripta.*, **T47**, 75 (1993).
18. Günter, S., and Könies, A., *JQSRT*, **62**, 425 (1999).
19. Scott, H. A., *JQSRT*, **71**, 689 (2001).
20. Adams, M. L., and Scott, H. A., *Contrib. Plasma Phys.*, **42**, 395 (2002).

Erosion from Liquid-Metal Plasma-Facing Components in Future Fusion Devices

Jean P. Allain

Plasma-Material Interaction Group, Department of Nuclear, Plasma and Radiological Engineering, University of Illinois at Urbana-Champaign, Urbana, Illinois 61801

Abstract. The use of liquid metals as candidate plasma-facing component (PFC) materials can enhance the performance and heat flux limits of fusion devices. Their applicability however, is strongly dependent on erosion properties including sputtering and evaporation. In this paper, the viability of candidate liquid metals such as lithium, tin and tin-lithium is considered. Recent experimental results in the (Ion-surface Interaction Experiment) IIAX at the University of Illinois and PISCES-B at the University of California, San Diego have demonstrated anomalous erosion properties for liquid lithium and liquid tin-lithium as the sample temperatures are raised. In IIAX the sputtering yield of liquid lithium and liquid tin-lithium has been measured for 200-1000 eV bombardment of H^+, D^+, He^+ and Li^+ at 45-degree incidence. Experiments in IIAX are carried out for sample temperatures between 200-450 °C. These experiments simulate similar conditions in fusion devices such as deuterium surface treatment, oblique incidence and ion fluence. The secondary ion sputtered fraction from liquid lithium and tin-lithium has been found to be near 60-70% for temperatures ranging from 200-450 °C. This measurement is important in that the sputtered flux in the ionic state will immediately return to the lithium surface due to the sheath potential in a magnetic confinement device. Erosion from liquid lithium both from evaporation and sputtering is tolerable for temperatures ranging from 200-450 °C. When all mechanisms are considered: sputtering, secondary ion fraction and evaporation, candidate liquid metals such as lithium, show promise as viable PFCs under fusion device conditions.

INTRODUCTION

The coupling between the plasma and material components, such as limiters and divertors, is necessary to conductively exhaust the non-radiated thermal plasma power. This creates adverse effects on the plasma performance and imposes serious requirements on the thermophysical and thermomechanical properties of plasma facing materials [1]. Therefore, engineering constraints provided by the complex plasma environment near limiter/divertor regions place restrictions on the design of advanced limiter/divertor systems.

CP635, *Atomic Processes in Plasmas: 13th APS Topical Conference*, edited by D. R. Schultz et al.
© 2002 American Institute of Physics 0-7354-0090-3/02/$19.00

The material in question must be able to withstand peak thermal loads of the order of ≥ 10 MW/m^2 for future next-step advanced fusion devices [1]. This condition therefore requires good thermal shock resistance, high heat conductivity for effective cooling and good brazing properties. In addition to high thermal loads, plasma-facing materials must contend with tokamak disruptions and "off-normal" events such as edge localized instability modes-ELM's and runaway electrons. The thermal load from such events can drastically increase locally, delivering energy densities of about 2-10 MJ/m^2 in a short (0.1-3 msec) interval of time over a limited surface area. Due to these stringent edge conditions, requirements for a feasible plasma-facing surface include several thermomechanical properties (high thermoconductivity, high thermal stress resistance, long fatigue lifetime, high fracture toughness and creep strength [2,3]) and thermophysical properties (high specific heat capacity, melting point, evaporation rate, among others [1]).

The radiation of impurity atoms and molecules from the surface is a strong function of Z as discussed earlier. Species other than the products and reactants of a D-T plasma, for example, is inevitable since the plasma will interact with surfaces such as the first wall or divertor plates in a tokamak configuration. Therefore, impurity level limits are imposed for a variety of elements. In fact, the radiation power function varies greatly among the elements. The fractional impurity level, which produces a radiation power equal to half of the alpha-heating power can be about 25% for lithium and down to about 0.03% for tin [4]. Levels of radiative power from impurities of this magnitude would make ignition impossible, ruling out practical fusion reactors [5].

Therefore, utilizing a low Z material or a low erosion rate high Z material is desired, especially low Z materials since for these materials a higher impurity level can be tolerated. The self-sputtering yield of candidate materials must also be minimized since momentum transfer is maximized in such a case, thus leading to impurity yields that may exceed unity, triggering runaway effects. Other limitations may include chemical sputtering and at high surface temperatures, RES. In addition, temperature limitations on the wall have to be imposed if the plasma facing material exhibits a high vapor pressure. Another aspect of the plasma facing material will be its potential to getter incoming particles. For example the ability to strongly absorb incident D-T ions may lead to new low recycling regimes with edge temperatures of 10^2-10^3 eV and edge densities near 10^{19} m^{-3} [6].

In order to address some of these issues a working group was developed under the Fusion Science Technology division of the Department of Energy in the United States, named: ALPS (Advanced Limiter/divertor Plasma-facing Surface). Under the ALPS group, a plasma-material interaction working group is comprised of the IIAX facility at the University of Illinois, Urbana-Champaign, the PISCES-B facility at the University of California, San Diego and the ARIES facility at the Sandia National Laboratory in Livermore, California. The focus of ALPS has remained two-fold [7]:

- To establish the scientific and technological base for innovative plasma-facing systems that can significantly advance fusion science and improve the vision for fusion as an energy source.
- Provide advanced plasma-facing systems and technology to the plasma physics community to enhance the performance and understanding of plasmas existing and near term devices.

Therefore, the ability to achieve low-recycling regimes as described above can provide advanced PFC technology that may enhance plasma performance in current devices. To this end, ALPS has considered and currently studies several liquid surface candidates including, lithium, tin-lithium, tin and molten salt flibe (Li_2BeF_4) [7,8]. Conceptually these potential liquid surface candidates would be implemented as manifested in figure 1.4 where liquid jets are substituted in place of solid divertor collector plates. The jet velocity ~ 10 m/s, and the length of the jet is 30-40 cm before entering the capture manifolds. These advanced liquid surface divertors would face key issues that need attention including: effect of liquid surfaces on plasma edge and core performance, effects of transient/disruption events, ability to achieve high power density and D-T/He trapping and release from surfaces [7].

EXPERIMENTS OF LIQUID-METAL EROSION

Recently a number of experimental results of candidate liquid metals have been studied. The Ion-surface Interaction Experiment (IIAX) facility at the University of Illinois has continued work in the measurement of liquid-metal erosion for the past three years. Erosion measurements including sputtering and evaporation measurements of candidate liquids including lithium and tin-lithium have been measured under a variety of experimental conditions. The experiment is designed to simulate fusion device-relevant conditions such as oblique incidence (45-degrees) and for low incident particle energies ranging from 100-1000 eV [9-11]. In ion-beam experiments the surface is typically treated with a local deuterium plasma to simulate similar fusion device conditions. These treatments have led to a reduction of lithium sputtering from the solid and liquid phase due to preferential sputtering of implanted deuterium atoms [9].

In addition, the secondary sputtered ion fraction has also been measured for lithium and tin-lithium liquids at a variety of temperatures. Figure 1 shows results for lithium sputtering from D^+, He^+ and Li^+ bombardment at 45-degree incidence. These results show a fairly high secondary ion sputtered fraction expected from alkali metals. A high ion fraction is important in that the sputtered flux in the ionic state will immediately return to the lithium surface due to the sheath potential in a magnetic confinement device. Therefore, the flux of lithium-sputtered atoms in the ionic state is "invisible" to the edge and core plasma. This result is significant since from the total sputtering values (ions + neutrals) only the sputtered lithium neutrals would be considered in the erosion balance [9].

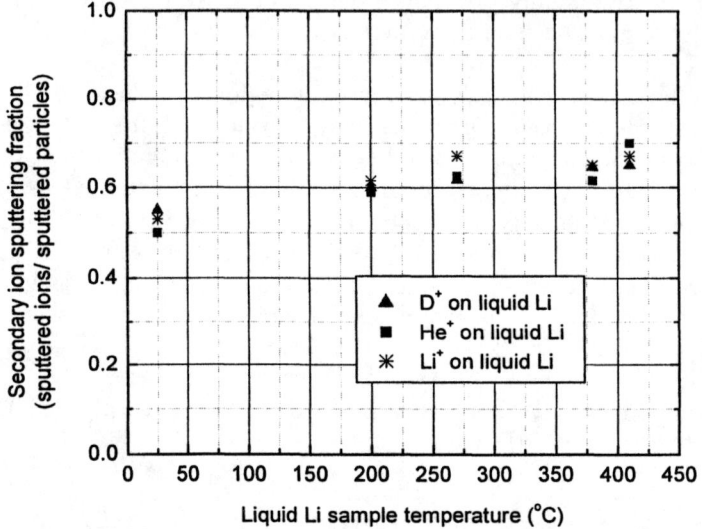

FIGURE 1. Secondary ion sputtered fraction of lithium bombarded by 700 eV D^+, He^+ and Li^+ ions at 45-degree incidence against several target temperatures.

Other experimental facilities that have measured liquid-metal erosion include the linear plasma device PISCES-B. Erosion of liquid lithium and liquid gallium by D^+ and He^+ bombardment have been conducted and show enhanced erosion of lithium at temperatures exceeding 250 °C. Similar results have been obtained in the IIAX facility and these results are presented elsewhere.

LIQUID-METAL EROSION RESULTS AND MODELING

Another aspect of liquid metal erosion is the dependence of lithium sputtering on the incident angle of incidence of the bombarding particle. Typically, this angle averages around 45 degrees with respect to normal [12]. In Figure 2 the sputtering yield of lithium bombardment at oblique incident particle angles of incidence are shown for 100 and 500 eV bombardment. The lithium yield shown is for both ions and neutrals. If the secondary ion sputtered fraction is ignored we note that unity sputtering, which would lead to runaway erosion conditions, is reached with about a 65-degree, 500 eV incident lithium particle at temperatures near the lithium melting point. However, when the ion fraction is taken under account all yields against all incident angles fall below unity sputtering. In addition, if incident angles are closer to normal as would be found in the magnetic configuration of a spherical tokamak machine such as NSTX (National Spherical Tokamak Experiment), the lithium sputtering yields are found to be lower than at incident angles near 45-degrees.

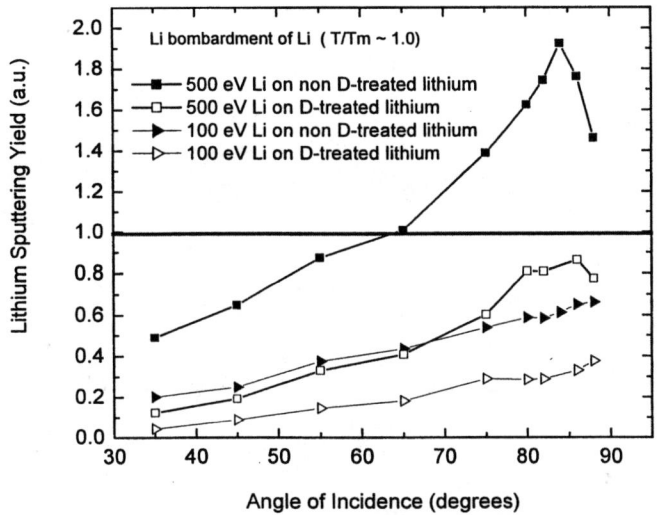

FIGURE 2. Total (ions + neutrals) lithium sputtering yield as a function of angle of incidence (measured with respect to normal) for 100 and 500 Li bombardment at 200 °C simulated by VFTRIM-3D using a model for liquid lithium from the literature [10].

Lithium erosion although tolerable does lead to relatively high evaporation fluxes when surface temperatures near 450-500 °C. Therefore, other candidate liquids are being considered such as liquid tin [13] and liquid tin-lithium [11]. These liquids provide lower evaporation fluxes and still exhibit similar thermophysical properties to liquid lithium. Figure 3 shows VFTRIM-3D simulation results for tin sputtering in the solid phase for bombardment by 45-degree incident Sn, He and D particles. Although, the sputtering yields are relatively low for the energy range shown, liquid tin is expected to be relatively higher and its secondary sputtered ion fraction is yet unknown. Therefore, any reduction to this effect for tin or liquid tin is still not known and awaits investigation [13]. Also, tin is less tolerable in a fusion plasma due its high Z and much lower sputtering yields are desired if it is to compete with liquid lithium. We note that D and He sputtering seem reasonable, however tin self-sputtering could be a problem given that unity sputtering is reached at energies near 1 keV. It should be added that a low-recycling regime is not expected for tin due to its low solubility for hydrogenic species and thus incident particle energies would be of the order of 100 eV, therefore leading to an expected lower self-sputtering yield.

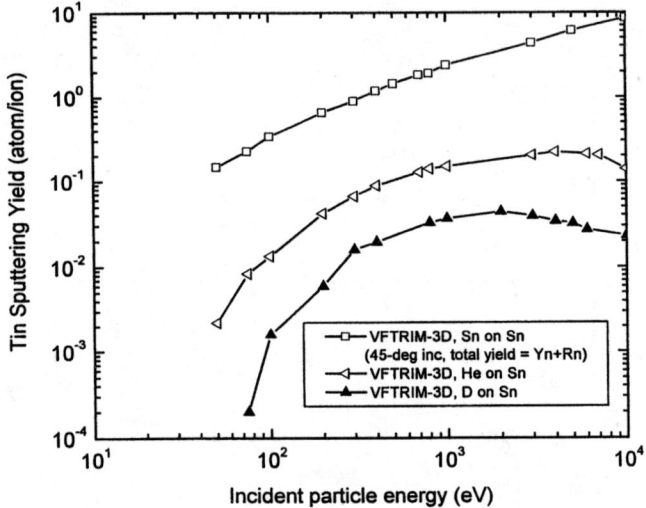

FIGURE 3. Tin sputtering yield from Sn, He and D bombardment at 45-degree incidence with respect to normal against incident particle energy simulated by VFTRIM-3D.

LIQUID-METAL EROSION UNDER FUSION DEVICE CONDITIONS

To consider both the effect of erosion from incident hydrogenic species from the plasma and self-particle bombardment from sputtered species that are ionized and return to the emitting surface an expression for the effective sputtering is used [14]:

$$Y_{eff} = \frac{Y_D}{(1 - Y_S)} \qquad (1)$$

The expression includes the sputtering yield due to incident deuterium particles, Y_D, and the self-sputtering yield, Y_S. In Figure 4 the effective neutral lithium sputtering yield is shown for a variety of deuterium bombardment lithium sputtering yields against lithium self-sputtering. As lithium sputtering due to deuterium bombardment grows, the effective sputtering yield is increased by a large amount of self-sputtering. In order to maintain tolerable impurity limits, these sputtering yields should be carefully controlled.

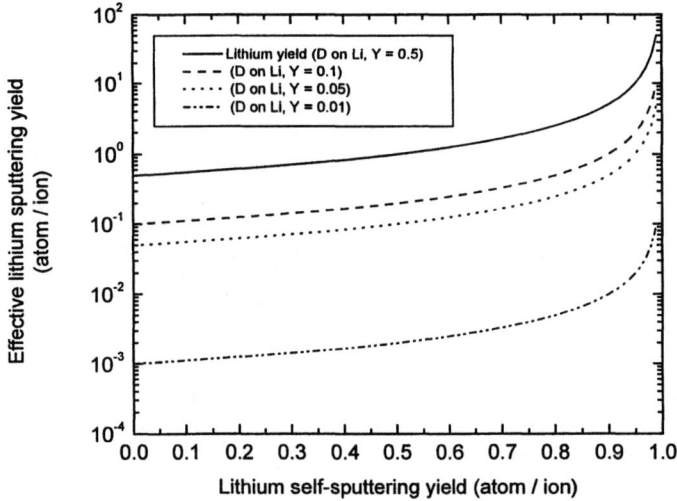

FIGURE 4. Effective lithium sputtering yield for various lithium sputtering yields from deuterium bombardment against lithium self-sputtering yields.

The effective lithium-sputtering yield can also depend on surface temperature. Previous results show strong temperature dependence for both deuterium and lithium bombardment when surface temperatures exceed the melting point [10,14]. Figure 5 shows the temperature dependence of the effective neutral lithium-sputtering yield against surface temperature. The lithium-sputtering yield from deuterium and lithium bombardment at 45-degree incidence is taken from results from IIAX. As shown in Figure 5, the effective yield can reach unity sputtering for surface temperatures near 450 °C. This result is important since it includes the high secondary sputtered ion fraction in the calculation of the effective lithium-sputtering yield. Therefore, surface temperatures near this limit should be avoided. However, if particle angles of incidence are closer to normal (perpendicular to emitting surface), then a relatively higher (~50-100 °C) temperature limit could be tolerated.

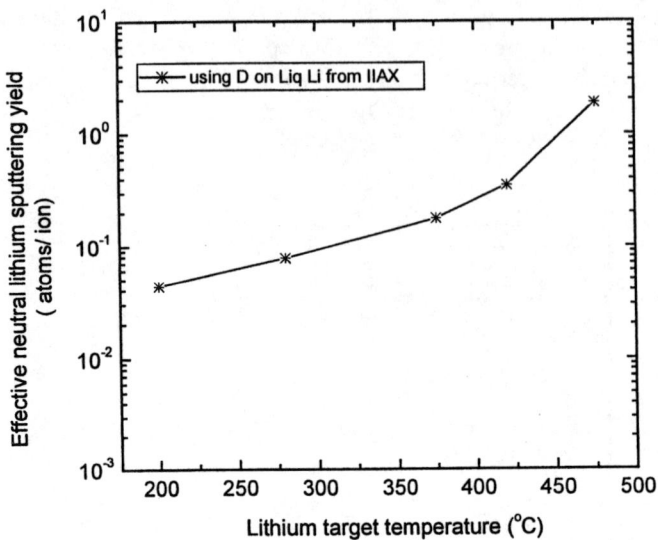

FIGURE 5. Effective neutral lithium sputtering yield as a function of target temperature. The temperature limit is set by self-sputtering runaway conditions where the effective yield exceeds unity leading to a temperature limit near 450 °C.

SUMMARY AND CONCLUSIONS

A brief study has been presented on liquid-metal erosion of candidate future fusion device PFCs. The study focused on results obtained mainly for liquid lithium, which shows promise as a plasma-facing material from an erosion standpoint for temperatures ranging 200-450 °C. This temperature limit can exceed by about 50-100 °C depending on the incident particle angle and its energy. For low-recycling regimes where high edge temperatures can be achieved, particle energies could reach up to 2-5 keV, which would consequently lead to *lower* lithium sputtering yields since these reach a maximum for D^+ and Li^+ bombardment at energies of 200 and 700 eV, respectively [9,10].

In addition, observed enhancement in lithium erosion is still tolerable due to the high secondary ion sputtered fraction measured to be between 55-70% for most incident particle energies and surface temperatures.

Other candidate liquid metals such as liquid tin-lithium and liquid tin have also shown promise for other aspects such as lower evaporation rates. Details of these materials are beyond the scope of the paper and are left to respective references.

ACKNOWLEDGMENTS

The author would like to acknowledge the financial support of the ALPS/DOE contract. In addition, a special acknowledgment is made to Ahmed Hassanein and J.N. Brooks for helpful discussions.

REFERENCES

1. G. Federici, et al., *Nuclear Fusion*, **41**, 12R, 1967-2137 (2001).
2. K.D. Leedy and J.F. Stubbins, *Mat Sci Eng a-Struct*, **297**, 19 (2001).
3. G. Li, B.G. Thomas, and J.F. Stubbins, *Metall Mater Trans A*, **31** (10), 2491 (2000).
4. J. Wesson, *Tokamaks*, 2nd Edition, Oxford University Press, Oxford, 1997.
5. P.C. Stangeby, *The Plasma Boundary of Magnetic Fusion Devices*, edited by P. Scott and H. Wilhelmsson, Plasma Physics Series, Institute of Physics , Bristol, 2000.
6. J.N. Brooks, et al., *J. Nucl. Materials*, **290-293**, 185 (2001).
7. R.F. Mattas, et al., *Fusion Engineering and Design*, **49-50**, 127 (2000).
8. H. Moriyama, et al., *Fusion Engineering and Design*, **39-40**, 627 (1998).
9. J.P. Allain, D.N. Ruzic, *Nuclear Fusion*, **42**, (2) 202 (2002).
10. J.P. Allain, M.R. Hendricks, D.N. Ruzic, *J. Nucl. Materials*, **290-293**, 180 (2001).
11. J.P. Allain, M.R. Hendricks, D.N. Ruzic, *J. Nucl. Materials*, **290-293**, 33 (2001).
12. J.N. Brooks, *Phys Fluids B – Plasmas*, **2** (8) 1858 (1990).
13. M.D. Coventry, et al., *J. Nucl. Materials*, to be submitted (2002).
14. R.P. Doerner, et al., *J. Nucl. Materials*, **290-293**, 166 (2001).

Advanced Spectroscopic Methods
for Diagnosing Magnetic-Fusion-Plasmas

Eugene Oks

Physics Department, Auburn University, Auburn, AL 36849, USA

Abstract. The following topics are presented: 1) diagnostic of the temperature and of the electron density at the edge plasma of tokamaks, including the divertor region; 2) mapping microwave fields in tokamak plasmas; 3) effect of plasma environment on charge exchange.

DIAGNOSTIC OF THE TEMPERATURE AND OF THE ELECTRON DENSITY AT THE EDGE PLASMA OF TOKAMAKS) INCLUDING THE DIVERTOR

The method is based on measuring widths of several highly-excited spectral lines of hydrogen or deuterium. Early diagnostic attempts of this kind [1] relied on Stark broadening theories that were not sensitive enough to the temperature T: they yielded only the electron density Ne and only as a rough estimate. This is because the accuracy of those theories (e.g., the Griem's theory [2]) was poor: they resulted in discrepancies with benchmark (non-fusion) experiments up to a factor of two.
- Now it is possible to employ the same experimental technique to measure simultaneously the temperature T and the electron density Ne with a much higher accuracy - due to the development of the most advanced code for Stark broadening of spectral lines. The code is based on three major theoretical breakthroughs.
- It incorporates a strong indirect coupling between the electron and ion broadenings, the coupling being carried out via the radiating atom acting as an intermediary [3].
- It allows also for a strong direct coupling between the electron and ion broadenings, represented by the acceleration of the perturbing electrons by the ion nearest to the radiator [4].
- It provides a significantly improved treatment of the ion dynamics' based on the analytical solution for this problem [5].

This code eliminated dramatic discrepancies between previous theories and benchmark experiments by various experimental groups at different plasma sources [6-11]. The reliability of the temperature and density measurements based on this code is proven over broad ranges of T and N_e [10-12].

CP635, *Atomic Processes in Plasmas: 13th APS Topical Conference*, edited by D. R. Schultz et al.

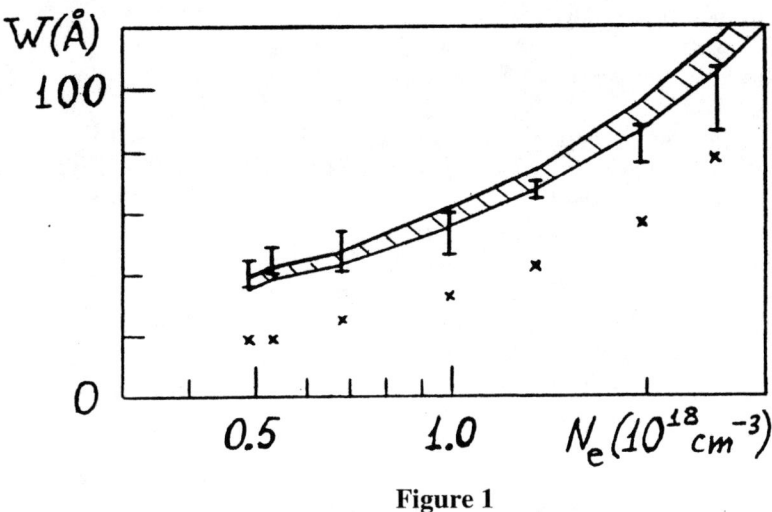

Figure 1

As an example, Fig. I shows the width (FWHM) of the hydrogen Hα measured in the latest benchmark experiment in the gas-liner pinch by the Kunze's group [7]. The experimental width is given by vertical bars. The crosses represent the results of the Griem's theory [2]. The dashed band shows the output of our code (the widths of the band represents an estimated theoretical error). It is seen that our code is in a good agreement with the benchmark experiment [7], while the Griem 's theory yields a discrepancy of a factor of two.

As another example, we compare experimental widths of high-n hydrogen Balmer lines H_n (n=6-12, 15), emitted from a radio-frequency discharge, with theoretical widths yielded by our code [11] (see Table 1).

Table 1. Widths of high-n hydrogen Balmer lines H_n (n=6-12, 15).

n^1	W_{ex} (Å)^2	W_{KG}(Å)^3	D_{KG}^{4}	W_{our}(Å)^5	D_{our}^{6}
6	.115	.137	19%	.122	6.0%
7	.140	.160	14%	.149	6.4%
8	.185	.207	12%	.173	-6.7%
9	.223	.249	12%	.226	1.3%
10	.284	.316	11%	.257	-9.4%
11	.335	.372	11%	.331	-1.2%
12	.397	.451	14%	.379	-4.4%
15	.673	FAILS	100%	.659	-2.1%

^{1}n - upper principal quantum number (the lower one is 2)
^{2}W_{ex} - Experimental HWHM
^{3}W_{KG} - HWHM by the Kepple-Griem (KG) code [13]
^{4}D_{KG} - Discrepancy between the KG code and the experiment: 100 $(W_{KG}-W_{ex})/W_{ex}$
^{5}W_{our} - HWHM by our code
^{6}D_{our} - Discrepancy between our code and the experiment: 100 $(W^D_{our} - W_{ex})/W_{ex}$

It turned out that the root-mean-squared discrepancy (i.e., the discrepancy averaged over all 8 Balmer lines) yielded by the Kepple-Griem (KG) code was quite dramatic: 38%.. In distinction, our code resulted in the root-mean-squared discrepancy of only 5%.

The above examples illustrate that the Griem 's theory becomes more and more inaccurate with the increase of the principal quantum number n and/or of the electron density N_e. The primary reason for this failure of the Griem's theory is that, as n and/or N_e grows, the coupling between the electron and ion microfields (disregarded in the Griem's theory) increases. Our theory is built by going BEYOND THE PERTURBATION THEORY, on which the Griem's approach relied. Therefore, our theory by its design is much more accurate than the Griem's theory and is free from the above shortcomings of the Griem's theory.

We applied our code for interpreting the widths of the high-n Balmer lines measured at the tokamak Alcator C-Mod [1]. In [1], Stark widths (FWHM) of experimental profiles of the deuterium Balmer lines D_8, D_9, D_{10}, and D_{11} were deconvoluted to be 4.9, 5.9, 7.5, and 9.6 Å, respectively. Then Griem used the KG code [13] and found from these Stark widths: N_e = 5.4, 5.6, 5.3, and 5.6×10^{14} cm^{-3}.

The KG widths have a relatively weak temperature dependence (no allowance for the ion dynamics). Therefore, Griem was not able to deduce the temperature from the experimental widths based on that code. Instead, he assumed the temperature to be T=4 eV.

In distinction to this, our code accurately incorporates the ion dynamics and, thus, is much more sensitive to the temperature than the KG code. Therefore, while analyzing the same experimental widths, we were able to derive both the density $N_e = 6.5 \times 10^{14}$ cm^{-3} and the temperature T=7.7 eV. It should be emphasized, that at T=7.7 eV we obtained the same value of $N_e = 6.6 \times 10^{14}$ cm^{-3} from each Balmer line individually.

To further demonstrate the superior quality of our theory, we calculated the ratios of Stark widths $\Delta\lambda_{n+1}/\Delta\lambda_n$ of the adjacent Balmer lines and compared them with both the experimental results and the Griem's theoretical results. The idea behind calculating these ratios is that they should be less sensitive to the electron density than the widths of the individual lines. Therefore, the ratios are conventionally considered as a tool for evaluating the intrinsic self-consistency of a particular theory. It turned out that the root-mean-squared deviation of our ratios from the experimental ratios was almost 4 times smaller than for the Griem's ratios.

Thus, while a relatively small scatter of densities deduced by Griem from different lines was not alarming, the ratios of widths really demonstrate a dramatic difference in the self-consistency of the two theories. The most important is that our code allows also to deduce the electron/ion temperature from the experimental widths.

A possible extension can be a diagnostic of plasma turbulence. This could be achieved by conducting polarization measurements of highly-excited spectral lines of hydrogen or deuterium [14].

MAPPING MICROWAVE FIELDS IN TOKAMAK PLASMAS

The method employs the laser-induced fluorescence (LIF) from a hydrogen or deuterium beam. The first draft of the method was published by us previously [14,15] and implemented in a model experiment [16]. That version required a calibration at the zero microwave field. However, for tokamaks, the regime "microwaves on" and the regime "microwaves off" are characterized by significantly different plasma parameters. Therefore, an attempt to implement this version of the method in tokamaks would yield quite ambiguous results.

In distinction to that, the method we proposed recently [17], does not require a calibration at the zero microwave field and also is more accurate. From the theoretical point of view, the method is based on our analytical solution for an atom interacting with two strong electromagnetic waves characterized by significantly different frequencies. Therefore the solution obtained is valid in a much broader range of fieldstrengths than an analogous solution that could have been found by the usual time-dependent perturbation theory.

The laser frequency ω_L should be tuned to a multiquantum resonance

$$\omega_L \cong \epsilon_{21} + k\omega_M , \qquad\qquad k = 0, \pm 1, \pm 2, ..., \qquad\qquad (1)$$

where ϵ_{21} is the atomic transition frequency and ω_M is the microwave frequency. We showed that the wavelength-integrated intensity of the LIF J_f and the laser field intensity J_L are related as follows

$$J_f^{-1} \propto [1 + g(k, E_M)J_L^{-1}] ,$$

where E_M is the microwave amplitude and the function $g(k, E_M)$ was found by us analytically. By measuring the ratio of slopes $g(k_1, E_M)/ g(k_2, E_M)$ of the experimental dependencies $J_f^{-1}(J_L^{-1})$ at two different values of k one can experimentally determine the microwave amplitude E_M inside the plasma.

To provide an alternative, we also obtained analytically a dynamic Stark shift $S(k, E_M)J_L$ of the multiquantum resonance (1). Therefore, by measuring the ratio of the experimental Stark shifts $S(k_1, E_M)J_L / [S(k_2, E_M)J_L] = S(k_1, E_M)/ S(k_2, E_M)$ of the resonance (1) at two different values of k one can also determine experimentally the microwave amplitude E_M inside the plasma.

The method can be used for microwave powers $P_M = 30$ kW – 10 MW (i.e., $E_M = 3 – 60$ kV/cm). The required laser intensity is $J_L = 3 – 30$ kW.

Effect of Plasma Environment on Charge Exchange

Charge exchange in tokamaks occurs at the presence of relatively large electric fields. Indeed, at $N_e = 10^{15}$ cm^{-3} the most probable value of the ion microfield is $6 – 8$ kV/cm (depending on Z_{eff}). More importantly: for example, 10 keV ions moving across a 10 T magnetic field experience larger electric fields: F=40 kV/cm for carbon ions, F=35 kV/cm for oxygen ions.

Only few theoretical papers analyzed the effect of electric fields F on charge exchange. Olson and MacKellar [18] used the Classical-Trajectory Monte Carlo (CTMC) method to simulate F≠0-collisions of Rydberg H-atoms (n=10, n=20), but only with protons. Salasnich and Sattin [19] used the CTMC method to simulate F≠0-collisions of H-atoms only in the ground state with fully-stripped ions of nuclear charges Z=2, 3, 6, 8. In particular, they showed that a (Z,n)-scaling, true for the F=0 case, breaks down for the F≠0 situation (see Appendix).

Thus, there are no published calculations for F≠0-collisions of H atoms in states of n=2 and n=3 (important for fusion plasmas) with multi-charged ions. The latter cannot be obtained by scaling the results of Salasnich and Sattin [19], who simulated F≠0-collisions of H-atoms only in the ground state. Besides, their simulations for the ground state were performed only for electric field much stronger than those relevant to fusion plasmas. Of course, there are NO ANALYTICAL quantal results on F≠0-collisions.

The paradigm is that charge exchange is an inherently quantal phenomenon [20]. Recently we disproved this paradigm [21, 22]. Based on first principles without any model assumptions, we presented a purely CLASSICAL ANALYTICAL description of anticrossings (avoided crossings) of energy terms (levels) that lead to charge exchange [21, 22].

In [21, 22] we considered a ZeZ'-problem: one electron shared by two fully-stripped ions of charges Z and Z'. Our analytical calculations of classical energy terms yielded astonishing results: 1) there are several (!) energy terms of the same symmetry; 2) two of these classical energy terms undergo an anticrossing, 3) at large internuclear distances, for one of the crossing terms the electron is centered at the Z-ion, for the other crossing term - at the Z'-ion. Thus, this situation classically depicts charge exchange.

In year 2002, within this classical non-model approach, we analyzed the effect of electric fields F on charge exchange. We obtained analytical results valid not only for weak fields F, but also for strong fields $F \geq Z/R^2$, where the perturbation theory fails (R is the internuclear distance at the anticrossing).

Below we present examples of the effect of electric fields F on charge exchange between H-atoms in states of n=2 or n=3 and fully-stripped carbon and oxygen ions (for T = 10 kev, B = 10 T). It should be emphasized that charge exchange from n=2- and n=3-states is just as important as from the ground state. While the populations of n=2- and n=3-states are ~1% of the total, cross-sections of charge exchange from these states are by two orders of magnitude larger than from the ground state [23].

Table 2. Effect of the electric field on charge exchange into states of C VI.

Initial state n of H-atoms	Final state n' of C VI	Reduction (in %) of the cross section
3	17	100% (total suppression)
3	16	31%
2	11	12%

Table 3. Effect of the electric field on charge exchange into states of O VIII.

Initial state n of H-atoms	Final state n' of O VIII	Reduction (in %) of the cross section
3	23	100% (total suppression)
3	22	65%
3	21	29%
2	15	28%

Let us discuss a PHYSICAL MECHANISM of the field-caused PARTIAL suppression of charge exchange. The COMPLETE suppression of charge exchange is due to the field-caused ionization from the final state [18, 19]. However, a PARTIAL suppression is due to the field-caused shift of the anticrossing to a smaller internuclear distance. This mechanism was not revealed by the Classical-Trajectory Monte Carlo method used in [18, 19].

We emphasize that a quasiclassical description for states of a principal quantum number n has a relative error $\delta \cong 1/(\pi n)^2$ (see, e.g., [24]). So, for n=3 we get δ and for n=2 we get $\delta \cong 3\%$.

Thus, we classically obtained analytical results showing how important is the effect of the plasma environment on charge exchange in magnetic fusion plasmas. A practical purpose of these calculations is two-fold. First, they should stimulate more quantal simulations of this effect (though the quantal simulations are computationally expensive). Second, for diagnostics, whose input requires to compute simultaneously hundreds of different cross-sections of charge exchange in a real or reasonable time, our analytical results, which are computationally super-fast, can be directly used with a sufficient accuracy as a part of such large codes. The latter is similar to the situation in computing opacities: thousands of spectral line profiles are needed simultaneously and therefore each profile is calculated by a fast, simplified recipe, rather than by a slow, full-blown formalism.

CONCLUSIONS

- By measuring widths of several highly-excited spectral lines of hydrogen or deuterium, it is now possible to determine simultaneously the temperature T and the electron density Ne (while previously only Ne was determined and only with a poor accuracy). This is feasible due to the development of the most advanced code for Stark broadening of spectral lines, which eliminated dramatic discrepancies between previous theories and benchmark experiments.
- It is now possible to map microwave fields inside tokamak plasmas by employing the laser-induced fluorescence from a hydrogen or deuterium beam. This advanced method does not require a calibration at the zero microwave field and also is more accurate than its first experimentally-implemented version.
- Finally, we showed how important is the effect of the plasma environment on charge exchange in magnetic fusion plasmas. This could lead to a revision of the

results previously obtained from various diagnostics, which employed charge exchange, but did not allow for the effect of the plasma environment.

APPENDIX

For zero-field conditions (F=0), charge exchange cross-sections obey the (Z, n)-scaling as follows. If the energy E is scaled to $E_s = En^2/Z^{1/2}$, then

$$\sigma_s(E_s, Z, n) = \sigma(E)/(Zn^4) \ .$$

For non-zero-field conditions (F≠0), if the field F would be scaled as $F_s = F/n^4$, then it might be expected that

$$\sigma_s(E_s, F_s, Z, n) = \sigma(E,F)/(Zn^4) \ .$$

However, Salasnich and Sattin [19] showed that the (Z,n)-scaling, true for the F=0 case, breaks down for the F≠0-situation.

ACKNOWLEDGMENTS

Thanks are due to M. Pindzola for useful discussions of the charge exchange section of this paper.

REFERENCES

1. Welch, B.L., Griem, H.R., et al, *Phys. Plasmas* **2**, 4246 (1995); AIP Conf. Proc. 381, New York, 1996, p. 159; AIP Conf. Proc. 386, New York, 1997, p. 113.
2. Griem, H.R., *Principles of Plasma Spectroscopy*, Cambridge Univ. Press, Cambridge, 1997; *Spectral Line Broadening by Plasmas*, Academic, New York, 1974.
3. Ispolatov, Ya., and Oks, E., *JQSRT* **51**, 129 (1994); Touma, J.E., Oks, E., et al, *JQSRT* **65**, 543 (2000).
4. Oks, E., *JQSRT* **65**, 405 (2000); *J. Phys. B* **35**, 2251 (2002).
5. Oks, E., *Phys. Rev. E*, Rapid Communications **60**, R2480 (1999).
6. Böddeker, St., Günter, S., Könies, A., Hitzschke, L., and Kunze, H.-J., *Phys. Rev. E* **47**, 2785 (1993).
7. Büscher, S., Wrubel, T., Ferri, S., and Kunze, H.-J., *J. Phys. B* **35** (2002); Ferri, S., Calisti, A., Stamm, R., Talin, B., Büscher, S., Wrubel, T., and Kunze, H.-J., AIP Conf. Proc. 559, New York, 2001, p. 45.
8. Flih, S.A., and Vitel, Y., AIP Conf. Proc. 559, New York, 2001, p. 30.
9. Escarguel, A., Ferhat, B., Lesage, A., and Richou, J., *JQSRT* **64**, 353 (2000).
10. Escarguel, A., Oks, E., Richou, J., and Volodko, D., *Phys. Rev. E* **62**, 2667 (2000).
11. Oks, E., Bengtson, R.D., and Touma, J., *Contrib. Plasma Phys.* **40**, 158 (2000).
12. Oks, E., *J. Phys. B* **35**, 2251 (2002); AIP Conf. Proc. 559, New York, 2001, p. 54.
13. Kepple, P., and Griem, H.R., *Phys. Rev.* **173**, 317 (1968).
14. Oks, E., *Plasma Spectroscopy: The Influence of Microwave and Laser Fields*, Springer Series on Atoms and Plasmas, v. 9, Springer, Berlin, 1995.
15. Gavrilenko, V.P., and Oks, E., *Sov. Phys. Tech. Phys. Lett.* **10**, 609 (1984).

16. Polushkin, I.N., Rjabikin, M.Yu., Shagiev, Yu.M., and Yazenkov, V.V., *Sov. Phys. JETP* **62**, 953 (1985).
17. Gavrilenko, V.P., and Oks, E., *Rev. Sci. Instrum.* **70**, 363 (1999).
18. Olson, R.E., and MacKellar, A.D., *Phys. Rev. Lett.* **46**, 1451 (1981).
19. Salasnich, L., and Sattin, F., *J. Phys. B* **29**, 751 (1996).
20. Hutchinson, I.H., *Principles of Plasma Diagnostics*, Cambridge Univ. Press, Cambridge, 1987, p. 286.
21. Oks, E., *Phys. Rev. Lett.* **85**, 2084 (2000).
22. Oks, E., *J. Phys. B* **33**, 3319 (2000).
23. Isler, R.C., *Plasma Phys. Control. Fusion* **36**, 171 (1994).
24. Migdal, A.B., *Qualitative Methods in Quantum Theory*, Benjamin, Reading, MA, 1977, Chapter 3.

Role of Atomic Physics in Models for Enhanced Edge Transport

J. Hogan

Fusion Energy Division, Oak Ridge National Laboratory, Oak Ridge, TN 37831

Abstract. Since regulation of helium ash accumulation in future magnetic confinement D-T burning plasma experiments requires control of recycling helium, the quest for improved magnetic confinement performance leads to a paradoxical necessity to degrade some confinement properties selectively in the plasma edge. This paper describes the role of atomic and molecular data in current models for several proposed ways to do this: 'Type I ELM-y H-mode', the ergodic divertor, and enhanced radiation operational modes. The same constraint requires better control of intrinsic impurities. The use of atomic and molecular data in the comparison of intrinsic impurity generation rates with observation is also illustrated.

INTRODUCTION

Burn requirements for D-T magnetic confinement plasmas impose a strong need for control of the accumulation of α-particle reaction products. The problem is made acute by the fact that helium ash preferentially recycles back to the core plasma and thus must be actively exhausted. D-T ignition typically requires a low value of the relevant figure-of-merit, $\tau_{He}^* / \tau_E < 10$ (τ_{He}^* is the helium residence time in the plasma, and τ_E is the energy confinement time) [1]. Since fusion device design typically maximizes τ_E, lowering the value of τ_{He}^* is the best available route to achieve this. The helium residence time, $\tau_{He}^* = \tau_{He}^{\alpha} + \tau_{He}^{edge} / (1-R_{He})$, where τ_{He}^{α} is the confinement time in the plasma core of the original fusion-produced αs, τ_{He}^{edge} is the characteristic residence time for helium subsequently re-cycled at the edge, and R_{He} is the helium recycling probability. Given present exhaust capabilities, $R_{He} \sim 1$, so there is a strong incentive to reduce τ_{He}^{edge} as much as possible while maintaining high core confinement. Species-selective degradation of edge confinement for helium alone is not yet practical, so general edge confinement degradation is indicated.

The ELMy H-mode (Edge Localized Mode High confinement mode) operational scenario for future burning plasma experiments does just this. Periodic instability bursts at a fixed edge spatial location expel edge plasma and leave the core plasma relatively unchanged [2]. A complementary scheme, explored on the Tore Supra tokamak, introduces static edge-localized magnetic perturbations with 3-D spatial modulation, to selectively degrade edge particle confinement [3, 4]. However, any degradation of edge confinement strongly increases the heat efflux, so attempts have been made to reduce the heat flux burden by increasing edge radiation through

CP635, *Atomic Processes in Plasmas: 13th APS Topical Conference*, edited by D. R. Schultz et al.

injection of extrinsic impurities, such as neon and argon, to produce Radiatively-Improved (or RI) modes. [5]. However, the ignition requirement for τ_{He}^{*} / τ_E is made much more stringent when these, and also intrinsic impurities (mainly carbon in present devices) are included [6]. Thus, increased attention is being given to identifying the sources of intrinsic impurities and reducing them.

Examples illustrating the application of atomic and molecular data in current studies of edge optimization are described.

ELMY H-MODE IMPURITY DYNAMICS

Helium exhaust is facilitated by increased enrichment of the helium concentration in the divertor relative to that in the core plasma. (enrichment $\eta = c^{core} / c^{div}$, where c is the helium relative concentration). An impurity dynamics simulation using the solps4.0 version of the b2-Eirene code [7] illustrates the enrichment process during an ELM event for neon, chosen for similarity to helium. The code couples a 2-D fluid plasma model (b2) with a 2-D neutrals code (Eirene) to make a time-dependent simulation of ELM behavior for a DIII-D lower single-null divertor case. Periodic ELM events are modeled by enhancing the electron and ion radial diffusivity five-fold during a 100 μsec interval, at the experimentally observed frequency. Figure 1 shows the magnetic geometry of the region modeled, near the DIII-D lower, outer baffle, and also shows a snap-shot of the deuterium neutral density near its maximum. Figure 2 shows the evolution of the divertor electron temperature (T_e^{div}) and of the total neon ion density through one ELM period. For each ELM there is a temperature crash just after ELM onset which produces $T_e^{div} \leq 1$ eV and a detached strike point (low-or zero-current to the divertor plates). T_e^{div} then recovers and rises until the next crash and, during the recovery phase, the plasma is attached. Since can T_e^{div} drop below the ionization potential for NeI at the outset of the ELM cycle, the neon ion density also crashes at first, and then, as T_e^{div} rises, neon is re-ionized after having been dispersed throughout the region. During the ELM recovery phase considerable neon recycling from the front face of the baffle is shown, due to the prior dispersal of neon during the detached phase. The dispersal, and concomitant redirection of recycling from the divertor plate (where pumping is optimal) to other surfaces acts to reduce the enrichment. Because of the large deuterium neutral density at during the detached phase (Fig. 1) this situation raises the possibility that ion dynamics may be strongly affected by deuterium charge exchange with neon ions. As shown in Fig. 3 (time-independent calculations with the b2 divertor code) there is a strong sensitivity of expected enrichment to variation in the charge exchange rate. Present modeling calculations for neon adopt the general scaling $R_{cx} = 2 \ 10^{-15} \ Z^{1.14} \ T_i^{0.35}$ [8] for deuterium charge exchange with ions of charge Z. Better (species-specific) charge exchange rates would improve the simulation. Much previous spectroscopic data has been obtained only after averaging over several ELM events. Since data with better time resolution are becoming available, the model comparison and validation should be improved in the near future, but until the basic processes governing enrichment can be better quantified, predictive capability will remain limited.

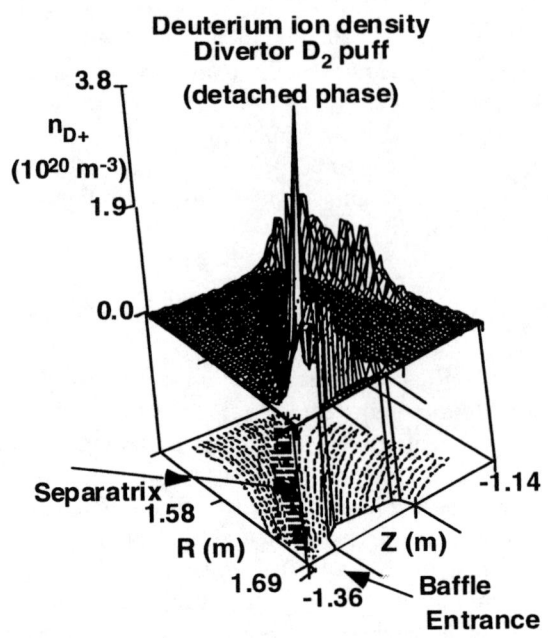

FIGURE 1. Magnetic geometry near DIII-D baffle, deuterium density at its maximum value in time.

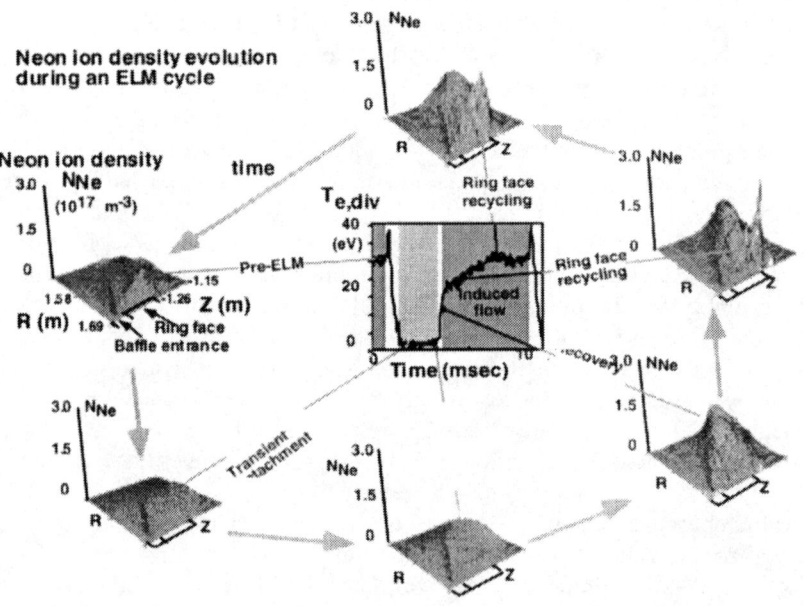

FIGURE 2. Neon ion ELM cycle dynamics simulation with solps4.0 (b2-Eirene code).

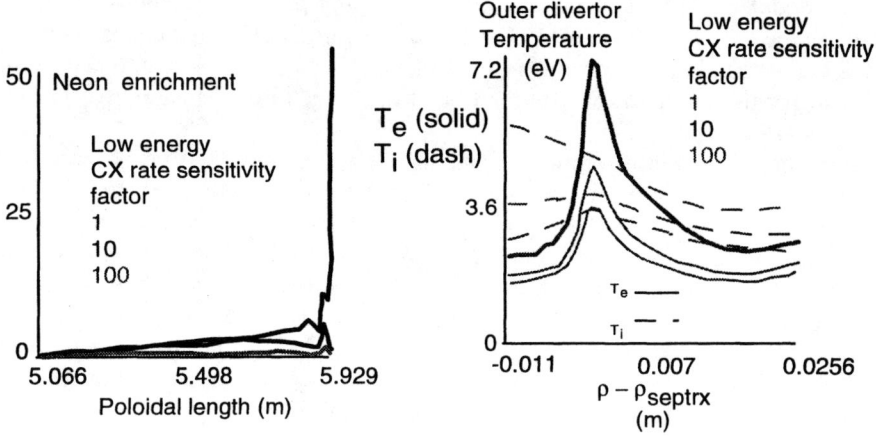

FIGURE 3. Sensitivity of enrichment (function of parallel length) and divertor T_e (vs major radius) to deuterium CX rate.

ERGODIC DIVERTOR IMPURITY TRANSPORT

Ergodic divertor experiments in Tore Supra have also demonstrated an edge impurity regulation effect [4]. The static edge confinement degradation due to ergodic magnetic perturbations has been examined in two cases in which neon spatial profiles were measured. The profiles were obtained with a grazing incidence VUV duochromator with a vibrating mirror in the spectrometer line of sight which measures spatial (poloidal) profiles of the impurity emission lines for the lower half of the Tore Supra plasma [9]. The duochromator finds a complex signal, with the usual (Abel-invertable) peak near the tangency radius where the chords intersect the plasma flux surface with the peak density of the ion in question, but also a second prominent peak near the equator. Figure 4 (curve 'DATA') schematically illustrates the situation. It is found that by augmenting the Abel-invertable symmetric radial shell of emitting ions with an additional component due to localized recycling from the face of the nearby outboard pump limiter, a reasonable quantitative model for the duochromator signal can be constructed [10]. The MIST radial impurity transport code [11] has been used to calculate the poloidally symmetric contribution to the duochromator signal. The asymmetric (limiter) contribution is modeled with the 3-D BBQ scrape-off layer transport code [12]. The model for the profile uses recently developed excitation rates for neon ions [13] allowing discrimination between the behavior of T_e-sensitive ($\Delta n=1$) and T_e-insensitive ($\Delta n =0$) emission. In addition to these rates, used for NeVII and VIII, the empirical $\Delta n=0$ excitation rates from [14] have been used for NeIV and NeV ions. Using measured radial profiles of n_e and T_e, the expected duochromator signal is modeled. Figure 5 (a-d) shows the resulting match between the BBQ/MIST model and the data for Ne IV, V, VII and VIII. The composite model has been able to provide both a qualitative (as shown in Fig. 4.) and, when used with a transport model developed for the ergodic divertor conditions (next section), a quantitative prediction of the duochromator signal

The ergodic divertor also induces systematic poloidal modulations observed by the duochromator when the plasma is removed from contact with the outboard limiter and instead placed in contact with the inner wall. In this case results from the composite model (axisymmetric core and non-symmetric edge contributions) shows that, in addition to the expected direct effect of T_e modulations arising from the laminar field of the ergodic divertor, there is also strong evidence for the effect of impurity charge exchange. The inner-wall limited configuration gives rise to significant neutral density in the observation zone. The level of neutral density is estimated with respect to previous measurements made near the divertor neutralizer [15], using results from a deuterium collisional-radiative model [16]. Using this estimate for the neutral deuterium density and the previously cited charge-exchange reaction scaling laws [8], the modulations in the poloidal profiles are found to agree with observation [17].

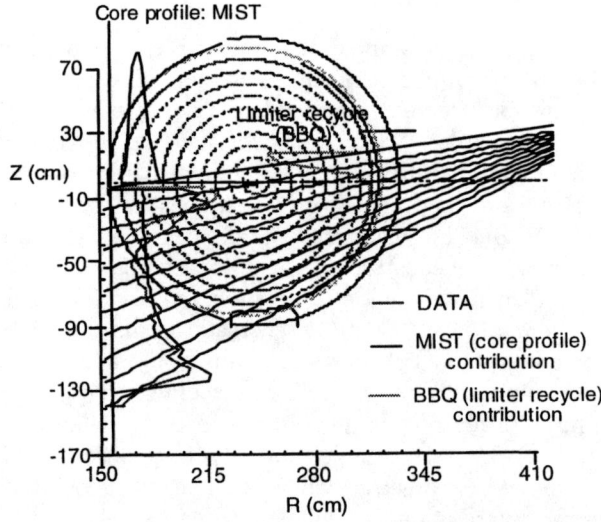

FIGURE 4. Schematic view of Tore Supra VUV duochromator poloidal porofile measurement. Data profile is compared with contributions from symmetric ('MIST') and asymmetric ('BBQ') sources.

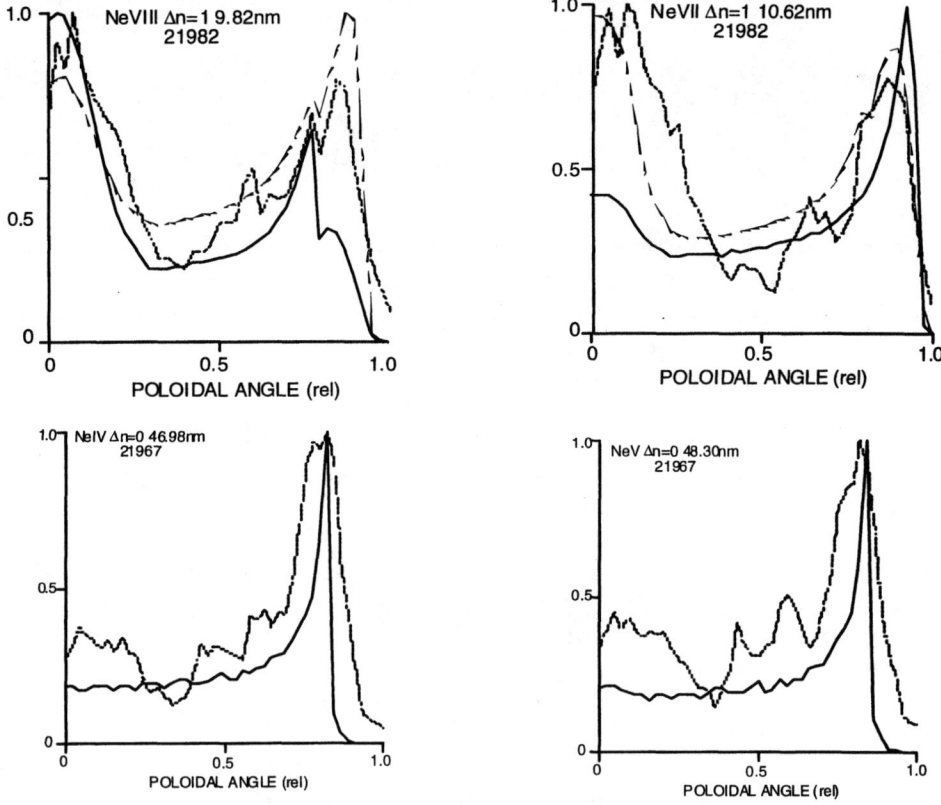

FIGURE 5. Comparison of measured duochromator profiles for NeIV, NV ($\Delta n=0$) , NeVII, NeVIII (Δ n=1) with results from composite model.

ENHANCED EDGE RADIATION MODES

Injection of selected high-recycling impurities (typically neon and argon) has been pursued to reduce the heat flux while maintaining high core confinement. Tore Supra experiments have been carried out to compare the added radiation arising from injection of extrinsic neon and argon impurities with that in the case of nitrogen, which is efficiently pumped. Modeling of this experiment with the combined edge (BBQ) and core (MIST) codes shows that the model gives a reasonable description for the enhanced edge transport under ergodic divertor conditions. In this case treatment of the influx of impurities into the core plasma as ions, rather than as a neutral influx, as is the conventionally assumed, plays a key role in calculation of the radiated power.

The core impurity code models the evolution of the flux-surface averaged impurity density in terms of the flux-surface averaged radial diffusion equation

$$\frac{\partial n_{Zi}}{\partial t} + \frac{1}{\rho}\frac{\partial}{\partial \rho}\left(\rho\Gamma_{Zi}\right) = S_{Zi} \qquad \text{with} \qquad \Gamma_{Zi} = -D_A(\rho)\frac{\partial n_{Zi}}{\partial \rho} + V_A(\rho)n_{Zi}$$

where Γ_{Zi} is the impurity radial flux density for species i, ρ is the normalized radius (in toroidal flux coordinates), S_{Zi} is the local source of species i, and D_A and V_A are the anomalous diffusivity and radial convective velocity. The source term $S_{Zi}(\rho)$ represents the local source of ions with charge Z_i arising through local ionization and recombination processes and also through direct influx *at radius* ρ from transport processes in the ergodic layer.

Core impurity transport rates have been determined from cases for which charge-exchange recombination radial profiles were measured at full ergodic divertor strength [18], and these provide local radial impurity transport coefficients, $D_A(\rho)$, $V_A(\rho)$. Radiative efficiencies are compared for comparable scenarios with 45 kA divertor current and application of 4 MW ICRH power, followed by a short impurity injection pulse. The total radiative power for carbon, oxygen and the relevant extrinsic impurity is calculated from the model. For the case of neon, the rates are taken from the ADAS database [19]. [adf11/plt/plt91_ne.dat with updates by R. Dux (IPP-Garching)]. Nitrogen radiation is calculated using the ADPAK atomic physics data from the MIST code and argon rates are based on the STRAHL model [20] and including modifications from HULLAC [21]. In each case the background carbon and oxygen content is fit prior to the application of ICRH heating, using the Z_{eff} increase to help determine the isotopic composition between carbon and oxygen. Figure 6 shows the comparison of modeled total radiative loss and that measured by the Tore Supra horizontal bolometer for the cases of Ne, N and Ar injection. A comparison of radiative efficiency ($\varepsilon \equiv$ maximum radiated power density / impurity density at the position of maximum radiation density) shows that Ar is the most effective radiator ($\varepsilon = 1.6 - 2.3 \ 10^{-11}$ W / particle, and that N and Ne have smaller and comparable values ($\varepsilon \sim 0.9 \ 10^{-11}$ W/particle).

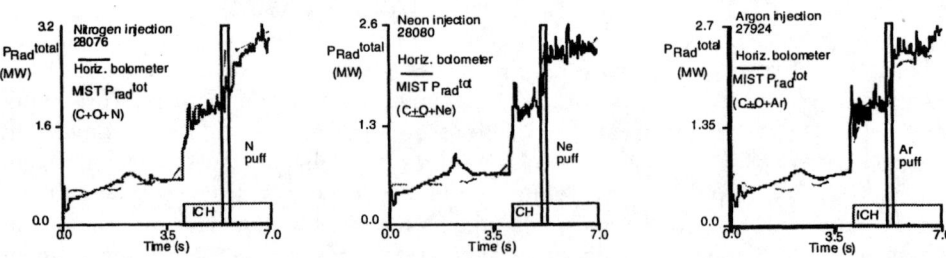

FIGURE 6. Comparison of radiated power calculated from the composite model with Tore Supra horizontal bolometer measurements.

INTRINSIC IMPURITY SOURCES

The identification of intrinsic impurity sources is complicated by transient behavior (ELM activity) in likely scenarios. Further, most current machines rely on non-cooled plasma-facing components, so that an initially small, localized hot spot will grow in magnitude as the discharge continues. Spectroscopic observation, typically integrated over several ELMs, has been used to make a first characterization of sources. The identification of characteristic CD and C_2 band emission is used to distinguish chemical (former) and physical (latter) graphite erosion processes. The measurement of molecular rotational temperature is a further clue. For this problem the recent improvement of rate databases for break-up of complex hydrocarbons [18, 19] has advanced modeling capability. Figure 7 shows a calculation from the BBQ Monte Carlo impurity transport code[12] of the break-up process for acetylene (C_2D_2) originating at a simulated hot spot in the DIII-D device. Molecular spectroscopy finds carbon molecular temperatures to be in the range of 0.1 eV [20]. Simulating the break-up process beginning with a neutral C_2D_2 source, the calculation finds that, while molecular temperatures vary up to 1 eV in the observation volume, the density-weighted temperature is 0.1 - 0.3 eV, which is in the measured range. However, there is still substantial uncertainty in the rates used in modeling the evolution of hydrocarbon erosion products, and this is an area where improved molecular data would be quite helpful.

CONCLUSIONS

The increased fuel particle and impurity efflux resulting from edge confinement optimization seems to lead inexorably to a condition in which impurity charge exchange with deuterium will be important. For modeling of ELMs and the ergodic divertor charge exchange between low energy deuterium neutrals and partially ionized noble gases, such as neon and argon, is observed to be an important aspect. At present modeling codes employ rather crude scaling relations for these rates, so that species-resolved charge exchange rates for recycling impurities such as neon and argon would be of great benefit. Although recent results for emissivities of these ions have proven quite useful, the radiative rates which are used in modeling RI-mode evolution are still qualitative. In the study of intrinsic impurity sources, the availability of improved rate data for breakup of complex hydrocarbons has enabled better understanding of generation processes. However, there are a great many reaction pathways and the opportunities for direct validation of the rates in the plasma environment are limited.

ACKNOWLEDGMENTS

Supported by the U.S. DOE under contract No. DE-AC05-00OR22725 with UT-Battelle, LLC.

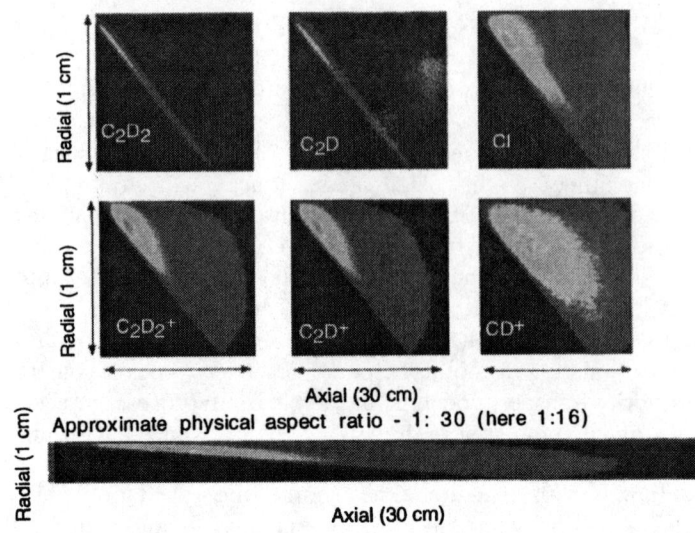

FIGURE 7. Density of C_2D_2 break-up products originating at a simulated hot spot, calculated with the BBQ code.

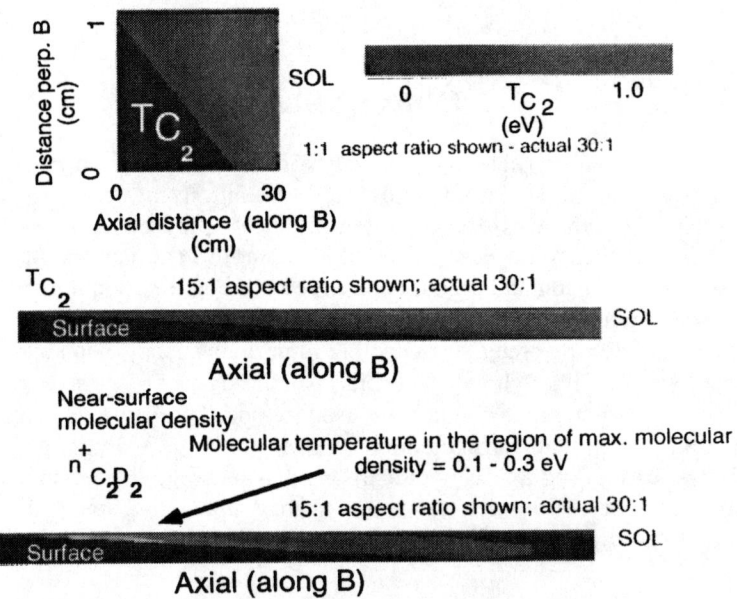

FIGURE 8. BBQ calculation of carbon molecular temperature.

REFERENCES

1. Reiter, D., Wolf, G.H., and Keve, H., *Nuclear Fusion* **30**, 2141 (1990).
2. Wagner, F. , Becker, G., Behringer, K., et al., *Phys.Rev. Lett.* **49**, 1408 (1982).
3. Feneberg, G., and Wolf, G.H., *Nucl. Fusion* **21**, 669 (1981).
4. Samain, A., Grosman, A., Belnski, T., et al., *J. Nucl. Mater.* **128&129**, 395 (1984).
5. Messiaen, A., et al., *Nucl. Fusion* **36**, 39 (1996).
6. Samm, U., Tokar', M.Z., and Unterberg, B., 22nd EPS (EPS, Bournemouth, UK, 1995), vol. 19C II-201-204.
7. Schneider, R., Reiter, D., Zehrfeld, H.P., et al., *J. Nucl. Mater.* **196&198**, 810 (1992).
8. Piuatti, M-E., et al., *Plasma Phys.* **23**, 1075 (1981).
9. Breton, C., De Michelis, C., Finkenthal, M., and Mattioli, M., *J. Phys. E: Sci. Instrum.* **12**, 894 (1979).
10. Hulse, R., *Fusion Technol.* **3**, 259 (1983).
11. Hogan, J., De Michelis, C., Monier-Garbet, P., et al., *J. Nucl. Mater.* **290-293**, 628 (2001).
12. Hogan, J., Klepper, C., and Harris, J., 16th Int. Conf. Plasma Physics Controlled Nuclear Fusion Research (Montreal, 1996), IAEA: Vienna, **2** 625.
13. Fournier, K., Mattioli, M., et al., *Phys. Rev. E* **60**, 4760 (1999).
14. Mewe, R., Schrijver, J., and Sylwester, J., *Astron. Astrophys.* **87**, 55 (1980).
15. Escarguel, A., Pegourie, B., Hogan, J., et al., *Plasma Phys. Controlled Fusion* **43**, 1733 (2001).
16. Fujimoto, T., Sawada, K., and Takahata, K., *J. Appl. Phys.* **66,** 2315 (1989); Sawada, K., Eriguchi, K., and Fujimoto, T., *J. Appl. Phys.* **73,** 8122 (1993); Sawada, K., and Fujimoto, T., *J. Appl. Phys.* **78**, 2913 (1995).
17. Hogan, J., De Michelis, C., and Monier-Garbet, P., *Plasma Phys. Controlled Fusion* (to be published).
18. Hess, W., Farjon, J.L., Guirlet, R., and Druetta, M., *Rev. Sci. Instrum.* (to to be published) .
19. Summers, H.P., *Atomic Data and Analysis Structure (ADAS),* Report JET-IR(94)06 (1994) JET Joint Undertaking, Abingdon (England); *ADAS User Manual,* version 2.2 2000.
20. Behringer, K., et al., *Plasma Phys. Controlled Fusion* **31**, 2059 (1989).
21. Mattioli, M., Fournier, K., Piuatti, M. E., et al., *Plasma Phys. Controlled Fusion* **44**, 33 (2002).
22. Allman, D., Ruzik, D., Brooks, J., et al., *Phys. Plasma* **7**, 1421 (2000).
23. Janev, R.K., et al., "Cross sections and rate coefficients for electron-impact ionization of hydrocarbon molecules", NIFS-DATA-68, Oct. 2001, National Institute for Fusion Science, Japan.
24. Isler, R.C., Colchin, R., Brooks, N., et al., Phys. Plas. (to be published) .

PARTICIPANTS

Joe Abdallah
Los Alamos National Laboratory

Mark L. Adams
Massachusetts Institute of Technology

Nigel G. Adams
University of Georgia

Helmar G. Adler
Osram Sylvania

Kanti M. Aggarwal
The Queen's University of Belfast

Alejandro Aguilar
University of Nevada, Reno

Jean Paul Allain
University of Illinois
  at Urbana-Champaign

Miron Ya Amusia
Racah Institute of Physics

James F. Babb
Harvard-Smithsonian Center
  for Astrophysics

Mark D. Baertschy
JILA, University of Colorado

James E. Bailey
Sandia National Laboratories

Connor Ballance
Rollins College

Mark E. Bannister
Oak Ridge National Laboratory

Avraham Barshalom
NRCN & Artep

Claire Bauche-Arnoult
Laboratoire Aimé Cotton

Jacques Bauche
Laboratoire Aimé Cotton

Peter Beiersdorfer
Lawrence Livermore National
  Laboratory

H. K. Carter
Oak Ridge National Laboratory

Miodrag Cekic
Fusion UV Systems, Inc.

Claude Chenais-Popovics
Laboratoire LULI

Hyun-Kyung Chung
Lawrence Livermore National
  Laboratory

James S. Cohen
Los Alamos National Laboratory

James P. Colgan
Auburn University

Thomas E. Cravens
University of Kansas

Michael Crisp
U.S. Department of Energy

Raymond C. Elton
University of Maryland

Anatoly Faenov
MISDC of VNIIFTRI

Zineb Felfli
Clark Atlanta University

Gary Ferland
University of Kentucky

Kevin B. Fournier
Lawrence Livermore National
 Laboratory

Thomas F. Gallagher
University of Virginia

Jean-Claude J. Gauthier
LULI, École Polytechnique

Dirk Gericke
Los Alamos National Laboratory

Alfred E. Glassgold
University of California

Gianluca Gregori
Lawrence Livermore National
 Laboratory

Hans R. Griem
University of Maryland

Donald C. Griffin
Rollins College

Joyce A. Guzik
Los Alamos National Laboratory

Clifford Harris
University of Nevada, Reno

Charles C. Havener
Oak Ridge National Laboratory

Rainer Hippler
University of Greifswald

John T. Hogan
Oak Ridge National Laboratory

Larry Hudson
National Institute of Standards
 and Technology

Denis Humbert
LPGP, Universite Paris-Sud

Roger Hutton
Lunds University

Verne L. Jacobs
Naval Research Laboratory

Takako Kato
National Institute for Fusion Science

Konstantinos Katsonis
Universite Paris XI

John J. Keady
Los Alamos National Laboratory

John Kielkopf
University of Louisville

Thomas J. Killian
Rice University

Yong-Ki Kim
National Institute of Standards
 and Technology

Marcel Klapisch
ARTEP, Naval Research Laboratory

Marcus D. Knudson
Sandia National Laboratories

Herbert F. Krause
Oak Ridge National Laboratory

Predrag S. Krstic
Oak Ridge National Laboratory

Hirotaka Kubo
Japan Atomic Energy Research
 Institute

Kenneth J. LaGattuta
Los Alamos National Laboratory

Richard W. Lee
Lawrence Livermore National
 Laboratory

Jiri Limpouch
Czech Technical University, FNSPE

Stuart David Loch
Auburn University

Joseph Macek
Oak Ridge National Laboratory

Joseph J. MacFarlane
Prism Computational Sciences

Roberto C. Mancini
University of Nevada, Reno

Steven T. Manson
Georgia State University

Mark May
Lawrence Livermore National Laboratory

Stephane F. Mazevet
Los Alamos National Laboratory

Diana McCrorey
University of Nevada, Reno

Ronald H. McKnight
U.S. Department of Energy

Fred W. Meyer
Oak Ridge National Laboratory

Howard M. Milchberg
University of Maryland

Tatsuya Minami
University of Tennessee

Alfred Z. Msezane
Clark Atlanta University

Alfred Mueller
Universitat Giessen

Izumi Murakami
National Institute for Fusion Science

Eugene Oks
Auburn University

Ron Olson
University of Missouri

Joseph Oreg
NRCN

Sergei Ovchinnikov
Oak Ridge National Laboratory

Frederik B. S. Paerels
Columbia University

Olivier Peyrusse
CEA-DIF

Michael S. Pindzola
Auburn University

Abdur Rahman
Colorado State University

Yuri V. Ralchenko
Weizmann Institute of Science

Carlos Reinhold-Larsson
Oak Ridge National Laboratory

Jorge J. Rocca
Colorado State University

Gregory A. Rochau
Sandia National Laboratories

Frank B. Rosmej
GSI-Darmstadt

William L. Rowan
The University of Texas at Austin

Ulyana I. Safronova
University of Notre Dame

Edward B. Saloman
National Institute of Standards
  and Technology

Akira Sasaki
Advanced Photon Research Center

Daniel Wolf Savin
Columbia Astrophysics Laboratory

David R. Schultz
Oak Ridge National Laboratory

Howard Scott
Lawrence Livermore National
  Laboratory

Ronnie Shepherd
Lawrence Livermore National
  Laboratory

Manolo E. Sherrill
University of Nevada, Reno

Apostolos Siskos
University Paris XI

Augustine Smith
Morehouse College

Timothy J. Sommerer
General Electric Research

Phillip C. Stancil
The University of Georgia

Daren P. Stotler
Princeton Plasma Physics Laboratory

Hiroshi Tanaka
Sophia University

Randy Vane
Oak Ridge National Laboratory

Peter L. G. Ventzek
Motorola

Philippe F. Weck
The University of Georgia

Leslie A. Welser
University of Nevada, Reno

Wolfgang Wiese
National Institute of Standards
and Technology

Honglin Zhang
Los Alamos National Laboratory

Author Index

A

Abdallah, J., 32, 82
Adams, M. L., 261
Adams, N. G., 182
Adler, H. G., 42
Agarwal, A., 125
Alexeev, I., 233
Allain, J. P., 273
Astapenko, V., 3
Audebert, P., 71
Auguste, T., 82

B

Babcock, L. M., 182
Baertschy, M. D., 162
Bailey, J. E., 32
Bannister, M. E., 125
Bar-Shalom, A., 92
Behar, E., 135
Beiersdorfer, P., 135
Bitter, M., 135
Blasco, F., 82
Boswell, C. J., 251
Boyce, K. R., 135
Bradley, P. A., 194
Brown, G. V., 135
Butzbach, R., 61

C

Calisti, A., 101
Cattolica, R., 52
Chen, H., 135
Chung, H.-K., 52, 71, 261
Clark, B., 32
Colgan, J., 145
Cravens, T. E., 173

D

Dalhed, H. E., 61
Delettrez, J. A., 61

Ditmire, T., 52
Dobosz, S., 82
D'Oliveira, P., 82
Dorchies, F., 82

E

Edwards, M. J., 52
Elder, J. D., 251
Eletskii, A., 3

F

Faenov, A. Y., 82, 101
Fakhreddine, K., 155
Fan, J., 233
Filuk, A., 32
Förster, E., 61
Fournier, K. B., 52, 71
Fujita, K., 61

G

Gallagher, T. F., 22
Gauthier, J. C., 71
Geißel, M., 101
Gendreau, K. C., 135
Glassgold, A. E., 221
Golovkin, I. E., 61
Griem, H. R., 113
Griffin, D. C., 145
Gu, M.-F., 135
Guzik, J. A., 194

H

Havener, C. C., 125
Hey, D., 135
Hoffmann, D. H. H., 101
Hogan, J., 290
Hulin, S., 82

305

Stout, P. J., 3
Swenson, D., 125
Szymkowiak, A. E., 135

T

Talin, B., 101
Terry, J. L., 251
Thorn, D. B., 135
Tweed, R. J., 155

U

Uschmann, I., 61

V

Ventzek, P. L. G., 3

W

Welser, L. A., 61
Widmann, K., 71

Y

Yan, D., 125

Z

Zhang, D., 3